PENGUIN EDUCATION

Newer Uses of Mathematics

Logic 308 Dic

Sir James Lighthill, F.R.S., is Lucasian Professor of Applied Mathematics at Cambridge University.

R. W. Hiorns is University Lecturer in Biomathematics at Oxford University.

S. H. Hollingdale is former Head of the Mathematics Department at the Royal Aircraft Establishment.

R. B. Potts is Professor of Applied Mathematics at the University of Adelaide.

R. E. Beard is Special Professor in the Department of Industrial Economics at the University of Nottingham.

B. H. P. Rivett is Professor of Operational Research at the University of Sussex.

D0715214

Newer Uses of Mathematics

Edited by James Lighthill

Penguin Books

Penguin Books Ltd, Harmondsworth,
Middlesex, England
Penguin Books, 625 Madison Avenue,
New York, New York 10022, U.S.A.
Penguin Books Australia Ltd, Ringwood,
Victoria, Australia
Penguin Books Canada Ltd, 2801 John Street,
Markham, Ontario, Canada L3R 1B4
Penguin Books (N.Z.) Ltd, 182–190 Wairau Road,
Auckland 10, New Zealand

First published 1978
Copyright © Institute of Mathematics and its Applications, 1978
All rights reserved

Made and printed in Great Britain by
Richard Clay (The Chaucer Press) Ltd, Bungay, Suffolk
Set in Monotype Times

Except in the United States of America, this book is
sold subject to the condition that it shall not, by
way of trade or otherwise, be lent, re-sold, hired out,
or otherwise circulated without the publisher's prior
consent in any form of binding or cover other than
that in which it is published and without a similar
condition including this condition being imposed on
the subsequent purchaser

Contents

3 Methods of Operational Analysis
Dr S. H. Hollingdale, FIMA

4 Networks
Professor R. B. Potts, FIMA

Foreword

There are several good and highly successful books that explain much mathematics in non-technical terms. The sales that a few of them have attained show that, despite recent educational discussion that suggests that mathematics is not a popular subject, a very large number of the general public find mathematics interesting. This book similarly attempts to cater for the general public who may not have done any mathematics since school days, but it deals not simply with the intellectual satisfaction of mathematics but gives a survey of the most recent applications of the subject. Most people know that mathematics is very useful. Very few will have any idea of the enormous range of society's everyday life in which mathematics is not only useful but is crucially so. To explain the methods that have been devised to bring mathematics into some aspects of the everyday world is a more difficult task than to explain mathematics itself.

The authors who have attempted this daunting task are all authorities in their own fields. To all of them the Institute of Mathematics and its Applications (widely known as the IMA), which has been responsible for the production of the book, records its great appreciation. Special thanks should be expressed to Sir James Lighthill, Past-President of the Institute, from whose original idea the book has sprung, and to Dr Hollingdale, who has undertaken much of the technical editing. Special thanks, too, should be recorded to the Institute's Deputy Secretary, Catherine Richards, who has also contributed to the editing, has read the proofs and has seen the volume through the press.

The Institute of Mathematics and its Applications
Maitland House, Warrior Square
Southend-on-Sea
Essex SS1 2JY

NORMAN CLARKE
Secretary and Registrar

Preface

Recently, mathematics has more and more been used to help with many of the things that really matter to us most. We care about the air we breathe, and its freshness and the climate that it brings us. We care about the water that quenches our thirst and cleans us and irrigates our crops. We care about the earth, as a source of food and of raw materials. We depend upon fuel for warmth and mobility. We want health and the conditions of life that promote it, and we want to see a balanced and healthy wild-life around us. All those things are what people mean when they speak about the *environment*. In the fight to improve it, mathematics is much used.

Again, we care about jobs. That means that we need our industries to be competitive, and to avoid wasteful use of resources. Prosperity for all of us depends upon work being organized economically. We need to communicate with each other, so we want good telephone systems and good transport systems. Indeed, we want *networks* distributing to our homes and factories those essential services and many others, such as power. There is a direct benefit to standards of living from *efficiency* in all of these areas. In the fight to improve it, mathematics is much used.

We depend also upon trade. We all want shops where we can find what we need, and money which shopkeepers will accept for goods we buy and which people will pay us for services we can give. We care about our industries earning enough from the goods they sell to be able to pay their employees and buy their raw materials and pay for re-equipment. So we expect them to concentrate on marketing: making goods available that customers really want, and letting customers know about them. We need money set aside for our pensions. We want to be protected from various possible disasters by insuring against them. Money saved, or put aside for insurance or pensions, must be invested where benefits will be high. In commerce and industry, then, and indeed also on the national scale, where of course (because of sheer size) the problems are hardest, we need sound *finance* and *planning*. In the fight to improve them mathematics is much used.

We outline in this book all these Newer Uses of Mathematics. We begin with its uses to improve the *environment* and man's relation to it – being concerned in Chapter 1 with the physical environment (the air; fresh water and the sea; the earth and its resources) and in Chapter 2 with the living environment (both the wanted organisms, such as livestock, crops and wildlife, and those unwanted organisms that threaten health). A general survey of how mathematics is – or could be – used to promote greater industrial *efficiency* is given in Chapter 3, while Chapter 4 describes the contribution of mathematics to improving distribution networks of all kinds. Chapter 5 outlines some of the mathematics of *finance*, including investment and insurance, and the book ends with Chapter 6 on the use of mathematics in *planning*, both within a company and at the national level.

By calling this book *Newer Uses of Mathematics* we admit that some good practical uses of mathematics which are well established and well known – for example, in all branches of engineering, and in statistics – are left out. Nor do we deal with applications of mathematics to pure science. Rather, we want to outline some of the many ways of using mathematics for significant practical purposes that have only become widespread in the past thirty years.

In writing the book we have been careful to avoid assuming any knowledge of mathematics beyond what might commonly be picked up, in their middle teens, by pupils with some interest in the subject. We have avoided all technicalities. Certainly, we have not tried to write a textbook. Our aim, rather, has been to quicken the interest of many readers in mathematics, by showing the big scope it offers today through so wide a variety of uses important to men and women generally.

JAMES LIGHTHILL

1 Mathematics and the Physical Environment

1. The Air

FORECASTING

'Counting', perhaps, is the first mathematics people learn, and 'adding up' is often the next. In environmental problems, one important use of mathematics is to help observers be precise about 'counting'; that is, to help their treatment of observations to become quantitative in a really systematic way. An even more important use is an extended kind of 'adding up', whereby a vast amount of knowledge obtained by scientists in laboratories can be put together so that information of real value about our environment results.

Mathematics is an outward-looking body of knowledge, then. It continually collaborates with observational science and laboratory science. Through its power to quantify and to organize knowledge, it makes possible the use of that knowledge in problems on a vastly bigger scale than that of the scientists' own laboratories.

In Chapters 1 and 2 we want to show to readers without specialized knowledge of mathematics or science how this works. Obviously, the laboratory science relevant to the physical environment (the subject of Chapter 1) is physics. We assume no detailed knowledge of laboratory physics, however. Readers either familiar with, or prepared to take on trust, a few statements about fundamental properties of matter, established during centuries of patient investigation in laboratories, will find themselves left free to concentrate on the main message of this chapter: how mathematics helps us to 'add up' vast quantities of small-scale knowledge to produce information on the scale of, say, a weather pattern.

Weather patterns, indeed, are a good example with which to start illustrating this process. The weather has a big effect on each of us personally. It also strongly influences many large-scale activities on which we depend: shipping and other forms of transport, the production of food, fuel supply for heating purposes, and so on. We all know from weather maps how changes, important in these connections, come in large patterns extending over hundreds or thousands of kilometres. Here, then, is a physical process of tremendous significance to human beings operating on a vastly bigger-than-human scale.

Notoriously, weather patterns change in a most unreliable way both from day to day and from year to year. Yet that might seem quite surprising from the following simple-minded point of view.

The planet earth, of 12 700 kilometres diameter, rotates on its axis once a day and is covered with enough air (mainly oxygen and nitrogen) to form a relatively thin layer (nine-tenths of it is within 20 km of the earth's surface). For what we call weather (winds, clouds, rain) the only source of energy is the sun, round which the planet moves once a year. The sun heats the earth's surface (which in turn warms the air) unequally: more near the equator than near the poles, and more in the northern hemisphere than the southern around June–July, but *vice versa* around December–January. Air expands when heated, and winds result. At sea, the sun's heat produces also evaporation, so that the air contains a small fraction of water vapour. However, when winds carry the vapour to where it can cool, it may condense into clouds or rainfall. From such a simple-minded point of view, why should all this weather not remain always the same at any particular season (that is, while the sun's input is nearly constant)? Furthermore, why should it not follow, each year, the same seasonal course?

These are reasonable questions; and a great deal of combined work by physicists and mathematicians was needed over the years to provide good answers to them. In science generally, that kind of work, looking for answers to 'reasonable questions' which curiosity may suggest, often brings very useful new insights, and this work was no exception. Furthermore, mathematics played a leading rôle in these investigations on a global scale, showing the wide variety of kinds of *instability* associated with these processes of unequal atmospheric heating. The word 'instability' means that big fluctuations are found * because the system is such that lots of tiny variations automatically turn into big fluctuations. The system can never 'settle down' to a constant course because some variations that occur, far from being 'damped out', are frequently amplified. In certain parts of the world, which include the country where this book is being produced, the variety of different kinds of instability is particularly large.

The book's subject, however, is the practical use of mathematics. Accordingly we omit the mathematical theories of why air movements, and other features of weather patterns, are so unstable and concentrate on the resulting practical problems. Unfortunately, any possibilities of usefully modifying the weather look now just as remote as ever: anything human beings could do would be minutely small compared with the enormous energy of atmospheric motions; admittedly, that small variation might be amplified, but vast numbers of other small variations would be getting amplified at the same time and very soon swamping its effect.

The only practical response, then, to the weather's variability is to set up weather-prediction services. A reliable weather forecast for just 24 hours ahead can be of enormous practical value, not only to the general public but also to whole industries. It helps outdoor industries like farming and construction to plan their day. It helps power-supply industries to estimate demand. It helps ships to plan their operations. It is particularly important for aircraft safety, and for planning the choice of aircraft routes, provided that the winds at the heights where the planes fly are also forecast.

For these and many similar reasons, governments all over the world long since decided to put large resources behind the building-up of weather-forecasting services. For many decades the method used was the 'synoptic' one. This means that data on

* Here, big weather fluctuations from day to day or from week to week, and also big changes in the seasonal course of weather from year to year.

the weather at one particular instant (e.g., 12.00 Greenwich Mean Time) are simultaneously gathered from large numbers of observing stations and recorded on charts in a standard notation; then, a highly experienced forecaster uses the charts to take in all this information and, reasoning from his knowledge and experience, forecasts the charts and the associated weather for, say, 12.00 GMT tomorrow. Results were variable: often impressively good but, nevertheless, highly dependent on the skill and judgement of the individual forecaster.

In 1922 a far-seeing thinker called L. F. Richardson first asked whether all this might not be replaced, in the future, by a mathematical process of the kind described below; one which 'objectively' analyses today's data and then forecasts tomorrow's, essentially by 'adding up' the results of applying appropriate physical laws separately to thousands of different 'slices' of air at very many intermediate instants of time. The necessary calculations were far too extensive to be done by the mechanical calculators of 1922, however, or even on the first generation of electronic computers that began to be introduced in the middle of the century.

It was the Meteorological Office of Great Britain, Richardson's own country, that was the first weather-forecasting service of any country in the world to acquire a computer, the IBM 360/195, big enough to carry out regular weather prediction with a highly detailed and refined process along, essentially, these lines. It was acquired in 1972, with the other equipment needed, and performs over ten million operations a second. By 1974, it was being used continuously to produce regular forecasts for the air of the whole northern hemisphere, including the air at all the heights at which planes fly. Not only were 24-hour forecasts of very high quality produced but also forecasts up to 3 days ahead, of real value to the main 'customer' industries (farming, construction, power supply, transport, etc.).

In describing in detail this very modern and in many respects satisfying logical process, we shall omit the highly intricate but in many ways impressive history of numerical weather forecasting in the 1950s and 1960s. Then, the very limited size and speed of the computers available forced workers in the subject to all kinds of ingenuities and complications to make possible even quite a limited forecast. Members of the British Meteorological Office shone in both periods: especially F. H. Bushby, strongly influenced in his early work by R. C. Sutcliffe and J. S. Sawyer, and working more recently with Margaret S. Timpson and G. R. R. Benwell. In the practical realization of the methods, G. A. Corby and E. Knighting played key rôles, while the outstanding leadership given by the Director General, Dr B. J. Mason, was essential to the project's success. The whole area of research has been highly international, however: at all stages fine work in other countries (for example, that of J. G. Charney and J. Smagorinsky in the USA and of G. I. Marchuk in the USSR) was actively influencing and being influenced by the British work.

It is appropriate to begin this chapter by describing in some detail this important use of mathematics for weather forecasting, partly because the essential ideas of the method are those arising wherever mathematics is applied effectively to problems of the physical environment. This allows us, later in the chapter, to indicate much more briefly how certain other practical problems are tackled: designing chimneys for power-stations or factories so that proper standards of clean air are maintained in their neighbourhoods; working from weather forecasts to forecasts of threats by 'big seas' to coastal defences; and many other such matters.

THE AIR ABOVE US IN TEN EQUAL SLICES

We explained that mathematics can overcome difficulties arising from the sheer size of the weather-prediction problem through an extended kind of 'adding up'. This involves combining vast amounts of knowledge on a small scale to produce large-scale information.

Those remarks apply to times as well as to distances. For example, 24 hours is 'a long time' from the point of view of weather; during 24 hours, a lot of weather change may take place! We can, however, divide that long time-span of 24 hours

into a large number of equal time-intervals of a few minutes.
Each of those is 'a short time', in the sense that only smal
changes to the weather pattern happen in a few minutes. As we
shall see, the fundamental properties of air and water vapour
established in physics laboratories are essentially rules about
their behaviour in *small* changes. Knowledge of those properties
can be used in forecasting, then, but only for the purpose of
'edging' the forecast forwards a few minutes at a time.

To start with, the change that will occur in the first of those
equal time-intervals of a few minutes must be calculated; then
the next small change, occurring in the second; then the next
(in the third), and so on. Gradually the forecast, by 'adding up'
all those small changes successively, will 'edge' forwards until
finally it has reached the desired total time-span of 24 hours
(or perhaps 72 hours for a 3-day forecast). We shall see that such
calculations, in which the time-span of 24 hours is divided into
240 equal parts of 6 minutes each, are completed in only 12
minutes, so there is no chance of the forecast losing value
through having gone out of date!

Just as a long *time* is dealt with by dividing it into many equal
short times, so also with *distances*. We remarked that weather
maps show important patterns extending horizontally over
thousands of kilometres. Such a map, representing a huge area
of the globe, can however be divided into many equal small
areas, represented by a grid of squares on the map* (Fig. 1.1).

The method of forecasting concentrates on predicting conditions (for example, the wind's speed and direction) at the
corners of those squares. This is useful provided the squares are
small enough, so that knowing wind speed and direction at all
the corners of all the squares gives one a fair idea of their distribution over the whole map.

Actually we shall find that, to calculate from the properties of
air how the wind's speed and direction will change at one such

---

*The 'stereographic' projection of the curved earth's surface on to the
flat surface of the map has the property that these equal squares on the
map *do* represent square-shaped areas of the earth's surface, though not
all to exactly the same scale.

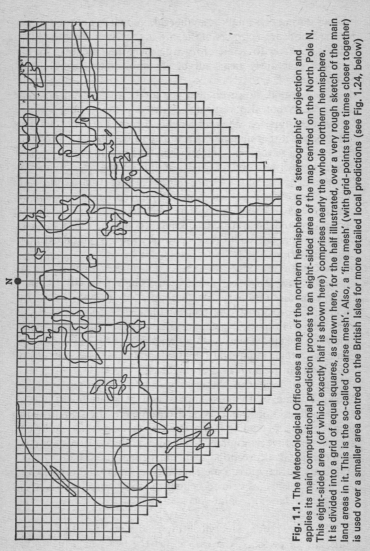

Fig. 1.1. The Meteorological Office uses a map of the northern hemisphere on a 'stereographic' projection and applies its main computational prediction process to an eight-sided area of the map centred on the North Pole N. This eight-sided area (of which exactly half is shown here) comprises nearly the whole northern hemisphere. It is divided into a grid of equal squares, as drawn here, for the half illustrated, over a very rough sketch of the main land areas in it. This is the so-called 'coarse mesh'. Also, a 'fine mesh' (with grid-points three times closer together) is used over a smaller area centred on the British Isles for more detailed local predictions (see Fig. 1.24, below)

corner during one small time-interval, knowledge of their values at the four nearest corners of the grid is necessary. The prediction process is able to 'edge' forward, then, because each of the corners generously helps all of its immediate neighbours at each stage!

It is essential, however, that the air be divided up not only horizontally but also vertically. There is a tremendous amount of air above us! At different heights it moves at quite different speeds, a fact that matters enormously to the planes that may be flying at those different heights. For that reason alone, weather forecasters wish to predict wind at a sufficient number of different heights to give a fair idea of its vertical distribution. A still more compelling reason is that winds at different heights influence each other, as we shall see; this means that an effective prediction process would have to take proper account of winds at many different heights even if the object were merely to predict ground-level conditions.

For these reasons, the prediction process divides the air above us into ten equal slices, each including exactly the same *weight* of air. Ordinary life, perhaps, brings 'weight' to our notice more often in the context of apples than of air! – yet we often feel the wind giving us a good push. Actually, a cubic metre of the air we are breathing weighs about 1·2 kilogrammes: much less than the weight (1000 kilogrammes) of a cubic metre of water, but by no means negligible.

The *total weight* of the air above us is measured very precisely by one of the best-known and most useful of all the meteorologist's instruments: the barometer. The unit in which the barometer's reading is expressed by meteorologists is the bar, which for convenience they divide into tenths, called decibars, or hundredths, called centibars, or most often thousandths, called millibars. Strictly it is a measure of pressure (weight divided by area): it tells us (as defined below) the total weight of air above any particular horizontal surface, divided by the area of that surface.

The one essential fact which the reader needs to know about this unit of pressure, the bar, is just the fact which makes it a good unit for meteorologists. This is that a good average value of the pressure measured at sea-level is 1 bar; that is, 1000 millibars. Barometers in normal use all over the world measure the pressure of air accurately to the nearest tenth of a millibar. Those situated at sea-level are recording values which range, usually, from about 960 millibars to about 1040 millibars. The lower values are found in cyclones and the higher ones in anti-cyclones (see Fig. 1.2). In between is the typical average sea-level pressure of 1000 millibars, or 1 bar. The reader need *not* know either how barometers work or the definition of the bar; but, actually, the pressure is 1 bar when the air above a horizontal surface of area 1 square metre exerts on it a downward force of exactly 100 kilonewtons – in Imperial units, around ten tons.

Weather stations do not just measure pressure at the ground: at least twice a day (at 00.00 and 12.00 Greenwich Mean Time) they send up balloons that carry 'radiosondes' including barometers and a lot of other equipment to a height of about 30 kilometres. Consequently, we know in some detail about how pressure gets less and less as height above sea-level increases. At each particular level this tells us, of course, what is the total weight of air above that level.

At a place where the sea-level pressure is exactly 1 bar, which is 10 decibars, the air above can be divided exactly into ten equal slices (that is, ten slices of equal weight) if we know the levels where the pressure takes the value 9 decibars (that is, where nine-tenths of the air is above the level in question and one-tenth below), 8 decibars (where eight-tenths is above and two-tenths below), 7, 6, 5, 4, 3, 2 and finally 1 decibar. Only one-tenth of the air is above that last level, while nine-tenths is below.

About how high would those levels be? It depends on air *temperature*. At a comfortable warm temperature of 20°C, the bottom tenth of the air's weight fills a slice (where pressure falls from 10 to 9 decibars) of *thickness** about 900 metres (3000 feet). We have already remarked, however, that air expands when

---

* By the *thickness* of a slice, meteorologists mean its total depth from top to bottom.

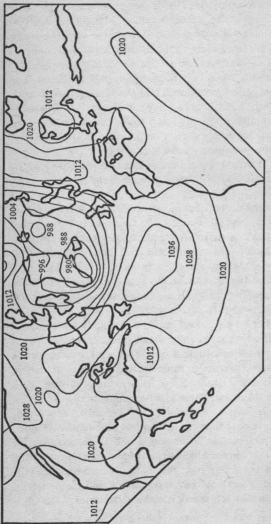

Fig. 1.2. Pressures at sea-level at 00.00 Greenwich Mean Time on 25 November 1975. The lines, drawn over the same very rough sketch map as in Fig. 1.1, are *isobars*; this means that each line joins up points where the sea-level pressure at that time took one and the same value: the value (in millibars) marked on the line. An anticyclone (where the pressure took the high value 1036 millibars) was present between Newfoundland and Spain, while to the north of that, between Greenland and Iceland, there was a cyclone (where the pressure took the low value 980 millibars)

heated. Tropical heat of 40°C raises the thickness, in fact, to about 960m. Conversely, at freezing point (0°C) the thickness falls to 840m: then, the pressure would be 9 decibars at an altitude 840m above sea-level. The thickness shrinks to 780m and 720m, respectively, in temperatures characteristic of cold winters (−20°C) and very severe continental winters (−40°C).

The thickness of the *next* slice up (where pressure goes from 9 to 8 decibars) is a bit different for two reasons, of which the more important tends to make it *greater*. This is the 'elastic' property of the air, represented by Boyle's Law: with *about* one-tenth less air weighing down on it, that slice can take up *about* one-tenth more volume. This can mean that it is about 10% thicker (which at 20°C brings it to 990m thickness).

We do, however, have to combine this idea with the earlier one, that air at a given pressure 'expands when heated'. Occasionally the air hundreds of metres above sea-level may be hotter than at sea-level: for example, because air warmed by blowing over a hot continent in summer finds itself over a sea whose temperature is only moderate. That would thicken still *more* the second slice.

It is commoner, however, for air temperature to decrease with height. The average temperature in the second slice might be 3°C lower, or 6°C lower, or even 9°C lower than in the first. In these cases its thickness would be, respectively, only 9%, 8% or 7% above that of the bottom slice.

There is, however, a quite fundamental reason why the second slice cannot be *more* than 9°C cooler than the bottom slice! This needs to be appreciated if the weather-prediction process is to be understood.

The reader will recall the basic principle that *work* and *heat* are different forms of *energy*. When some air rises to a region of lower pressure it expands, as we have seen, to take up more volume. That means it does *work* on the air around it and its lost energy has to come from its *heat;* in other words, it cools. Many readers will know about this cooling of upward-moving air as a major possible cause of rain.

The actual drop in temperature involved is about 9°C when the pressure falls by one-tenth. That's a lot of cooling! – and normally, when it happens, the air which has risen finds itself colder than the *rest* of the air at that level. Accordingly it takes up less volume in relation to its weight and so is less buoyant and falls back to its original level. This is a *stable* situation: vertical motions are 'damped out' because rising air tends to fall back whence it came.

On the other hand, if the average temperature in the second slice were more than 9°C cooler than in the first, vertical motions would build up to an enormous extent. Rising air, cooling by only 9°C, would still find itself hotter (and therefore more buoyant) than its surroundings and tending therefore to rise still further. Conversely, falling air would continue to fall; and the whole mass of air would be *unstable*: very quickly, all these up and down motions would mix up the air in the two slices till the temperature difference came down to 9°, and only then could they stop. *That* is why 9°C is the greatest temperature drop you ever observe between the two slices! As we saw, it means that the second slice's thickness has to be at least 7% greater than the first slice's.

Similar considerations applied to all ten levels show that the slowest possible increase in slice thickness as we go up is as illustrated in Fig. 1.3(*a*). This corresponds at each level to a proportional volume increase just seven-tenths of the proportional pressure drop: the natural volume change in vertically moving air. The prediction process must take into account the fact that if ever the ratio of a slice's thickness to that of the slice below it became *less* than in Fig. 1.3(*a*), such natural vertical motions would immediately exchange air between them until this minimum ratio was reached.

Actually, it never happens that the thickness increases as slowly as in Fig. 1.3(*a*) *all the way* up to the 1 decibar level. If it did happen, it would correspond to a 150°C total temperature drop! The biggest temperature drop normally found is about 110°C, from a tropical ground temperature of 40°C to a low upper-air temperature of −70°C above the tropics. In temperate zones it is smaller: not only are the ground temperatures less

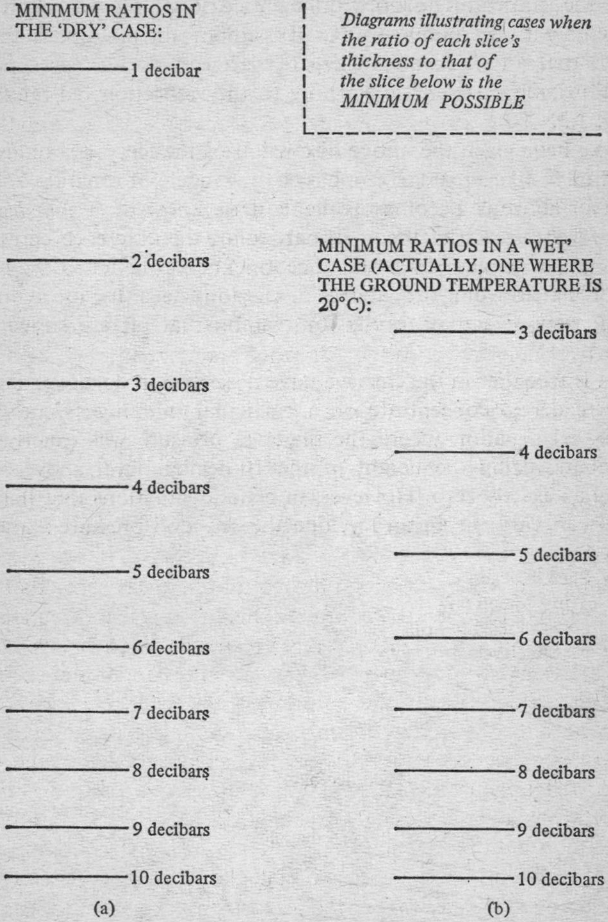

MINIMUM RATIOS IN
THE 'DRY' CASE:

————————— 1 decibar

*Diagrams illustrating cases when
the ratio of each slice's
thickness to that of
the slice below is the
MINIMUM POSSIBLE*

————————— 2 decibars

MINIMUM RATIOS IN A 'WET'
CASE (ACTUALLY, ONE WHERE
THE GROUND TEMPERATURE IS
20°C):

————————— 3 decibars

————————— 3 decibars

————————— 4 decibars

————————— 4 decibars

————————— 5 decibars

————————— 5 decibars

————————— 6 decibars

————————— 6 decibars

————————— 7 decibars

————————— 7 decibars

————————— 8 decibars

————————— 8 decibars

————————— 9 decibars

————————— 9 decibars

————————— 10 decibars

————————— 10 decibars

(a)

(b)

Fig. 1.3. Diagram (*a*) represents slices of air such that the ratio of each
slice's thickness to that of the slice below is the minimum possible. For
reasons to be discussed later (p. 56), the restrictions on this ratio are
still more severe where the air is saturated with water vapour. The greater
minimum ratios applying in such a case are illustrated in diagram (*b*)
(which actually omits the top two slices, where no significant amounts
of water vapour are normally present)

but the temperature drop with height ceases at a lower height, at a temperature around $-50°C$. It is important to emphasize, in fact, that an *enormous* variety of patterns of thickness distribution are possible, subject only to the restriction indicated in Fig. 1.3.

We have given the above discussion of the ten pressure-levels from 1 to 10 decibars to emphasize how much information about the air above us becomes available if we know the *height above sea-level* for each of them. In particular, the difference between two such heights gives the slice thickness which (as may be appreciated from the above discussion) tells us its average temperature; actual graphs for reading that off are shown in Fig. 1.4.

It is true that in the case we have described in detail we asked the reader to concentrate on a particular (although somewhat average) situation where the sea-level pressure was exactly 10 decibars, fixing the height of the 10-decibar level above sea-level as exactly zero. However, in general situations that height also can vary: in an anticyclone the sea-level pressure exceeds

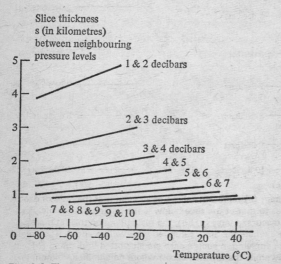

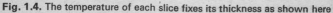

Fig. 1.4. The temperature of each slice fixes its thickness as shown here

10 decibars, so that the 10-decibar level has a *positive* height above sea-level (though hardly ever more than 400 metres). Conversely, in a cyclone the sea-level pressure is less than 10 decibars, which we can represent mathematically by saying that the 10-decibar level has effectively a *negative* height above sea-level. (In other words, if we *extrapolated* the pressure distribution a certain definite distance *below* sea-level, we would reach 10 decibars.)

For all the reasons given in this section, then, the prediction process seeks to calculate the heights above sea-level of all our ten pressure-levels (from 1 to 10 decibars), together with the speed and direction of the wind at all of those levels, at every one of the corners of our square grid covering the map. In the next two sections, we give a first idea of how this is done.

DRY MATHEMATICS

Out of the two leading characters in that drama which is the weather, we have so far introduced only one: the air. The other, of course, is water in its various alternative disguises: either as vapour, which mixes perfectly with the air and which until it condenses we cannot see at all, or as any of its condensed forms such as fog, cloud, rain, hail or snow. Certainly, if our object were to describe weather science, we should need at this stage to outline the physics of evaporation and condensation.

Our actual aim, however, is to explain how mathematics is used in weather prediction. With this aim in view, it is best to proceed in two stages. We first describe the mathematical prediction process as it would be *if all possibilities of condensation were neglected*. This simplified form of the process is the form which it actually does take in all those areas where the weather is 'dry' in the extreme sense that no condensation at all (even as cloud droplets) occurs ...

In this and the following two sections, then, we describe a mathematics which is 'dry' (though, we believe, none the less interesting!). It gives an excellent introduction to the full, 'wet' mathematics which we are able to describe later by means of a limited list of modifications to the 'dry' prediction process. Readers should be warned, however, that where condensation does occur even the *air* movements are greatly affected (in the first place because water condensation causes the surrounding air to expand). Around those areas, then, even the winds are not predicted right until *after* the 'dry' mathematics has been modified to allow for condensation effects.

Since we aim to make predictions at every corner of our grid of equal squares covering the map, we must number each of those corners. We use the simple numbering system shown in Fig. 1.5, based on the bottom left-hand corner which we choose to number (0, 0). Then the corner we reach by going 5 squares to the right from (0, 0) and then 3 squares up the page is numbered (5, 3). The corner numbered (2, 7) is reached by going 2 squares to the right from (0, 0) and then 7 squares up the page.

Any corner of the grid will have some such number, of the form (x, y) where x could be any whole number and y could be any other whole number. This very standard numbering system for a grid will be recognized by many readers as the way of doing geometry introduced by the great Descartes. It is easy to extend it so that map positions not on the *corners* of the grid are described in the same way (x, y) but with x and y not whole numbers; thus, (4·21, 5·09) is reached from (0, 0) by going 4 squares and twenty-one hundredths of a square to the right and then going 5 squares and nine hundredths of a square up the page.

One of our main aims is to predict the speed and direction of the wind at each of our ten pressure-levels (from 1 to 10 decibars) at every corner of the grid. Remember that we also divide the *time* of the forecast into a large number of equal time-intervals of a few minutes, and plan to make predictions, first, of the wind speed and direction at the end of the first of those time-intervals, then at the end of the second, and so on. Thus, we aim to calculate the winds at each of a sequence of equally spaced instants of time.

What exactly do we mean by speed at a particular instant of time? The obvious answer, 'distance gone divided by time taken', is vague unless we say over which period of time the 'distance gone' is measured. The best solution is to choose a period *centred* on the instant we are interested in! Then speed at that instant

NUMBERING SYSTEM FOR A GRID OF SQUARES

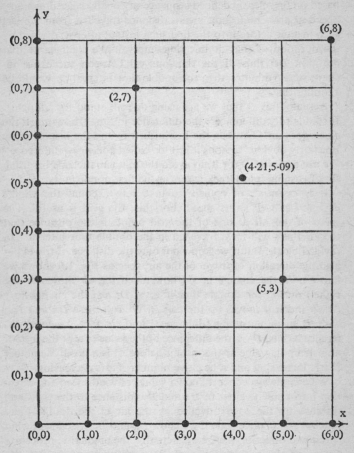

Fig. 1.5. Introducing the map coordinates *x, y*, both when they are whole numbers (and thus represent grid-points), and when they are not whole numbers

depends on distance travelled from just before it to just after it.

This idea becomes very precise when our whole calculation is based on one clearly defined sequence of equally spaced instants. Speed at any one instant means distance travelled from the last such instant before it to the first such instant after it, divided by the duration of that double time-interval. We shall name that duration Dt; thus, if the time-interval between successive instants were 6 minutes, then the double time-interval Dt would be 12 minutes.

Actually this D that we are using doesn't stand for 'double'! It stands for 'difference': the difference in time ('how much the clock goes on') between the instant before and the instant after. For fairly obvious reasons it is often called a '*centred* difference'. We use a capital D for it, to avoid the mathematician's besetting sin of rushing into Greek letters at the first opportunity ...

It is essential, of course, to take into account the wind's *direction* as well as its speed. For this, the grid is useful. The *speed* of the air at one of the grid points is the distance that the air there will have travelled in the double time-interval Dt, divided by Dt. But if we know not only the distance of travel but also the direction of travel of the air, then as Fig. 1.6 shows we know both the change in x ('how much the air moves to the right') during the double time-interval Dt *and* the change in y ('how much it moves up the page'). We call these Dx and Dy; with D again meaning 'difference' and, in fact, '*centred* difference': thus, Dx is the difference of the value of x at the instant after from its value at the instant before. It is a positive number if x is increasing and a negative number if x is decreasing.

We can always convert known values of wind speed and direction into known values of Dx and Dy (distance to the right and distance up the page travelled by the air in the double time-interval Dt). Conversely, if we know Dx and Dy we can get wind speed and direction (Fig. 1.6). Actually, the prediction process is based on the exclusive use of Dx and Dy throughout.* We

---

*We should remark, however, that the actual notation used in the Meteorological Office's publications differs from the deliberately simplified form used here.

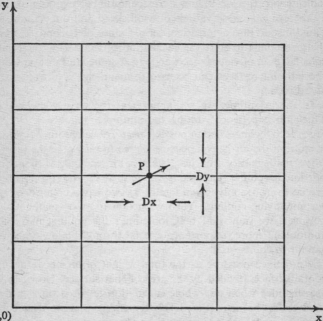

Illustrating Dx (the change in *x* from the instant before to the instant after) *and Dy* (the change in *y* during the same double time-interval, *Dt*) for air which passes through a grid-point *P* at the *centre* of the double time-interval. The arrow points in the *direction* of the wind. The *length* of the arrow gives the distance travelled; that is, the *speed* of the wind multiplied by *Dt*.

Fig. 1.6. Knowing wind speed and direction fixes the length and direction of the arrow in this diagram and, therefore, fixes *Dx* and *Dy*. Alternatively, knowing *Dx* and *Dy* fixes the position of this arrow centred on *P* and, therefore, fixes wind speed and direction.

For example, a wind of 20 metres per second (about 40 knots) would give (with *Dt* = 12 minutes = 720 seconds) an arrow representing a distance of 14·4 km. If the wind direction is 60° east of north, we deduce that the changes *Dx* and *Dy* between the two ends of the arrow must represent distances of 12·5 km and 7·2 km. If the local map scale were 1 grid length in 100 km, then *Dx* = 0·125 and *Dy* = 0·072

indicate later (pp. 62–9) how the calculation begins by converting all the available observations of wind speed and direction at the initial instant into a consistent set of values of Dx and Dy at all the grid-points to start the calculation off; and how the finally calculated set of values of Dx and Dy at some later instant which specially interests us can be used to make charts of wind speed and direction.

In the meantime, we ask about Dp: the change in *pressure* which air experiences during the double time-interval Dt. We are especially interested in possible *negative* values for Dp, which mean that the air is experiencing a pressure *drop*. In the present 'dry' mathematics this just lets the air expand, with a pro-portional volume increase seven-tenths of the proportional pressure drop: only seven-tenths, as we saw, because of the accompanying cooling. When we go on to 'wet' mathematics, however, we shall find that sometimes the cooling may have additional, more important, effects through allowing water vapour to condense.

We defined pressure as the total weight of air above a hori-zontal surface, divided by the area of that surface. How can we use this idea when our whole calculation is based on a grid of squares on the map (Fig. 1.1)? Remember that we decided to estimate 'speed at one instant' as the distance travelled in a double time-interval Dt *centred on that instant*, divided by Dt. Similarly, we estimate 'pressure at one grid-point' as the weight of air above a double square *centred on that point* (Fig. 1.7), divided by the double square's area.

The pressure is dropping during the interval Dt if the weight of air above that double square is getting less because of *net escape of air*. This means that winds take more air out of the square than they put in. The loss of weight above, say, the 10-decibar level is found by adding up the net escape of air from each of the ten slices lying above that level. Similarly, the loss of weight above, say, the 7-decibar level is found by adding up the net escape of air from each of the seven slices lying above *that* level. In each case, we divide the weight loss by the area of the double square to obtain the pressure drop during the interval Dt at the level in question.

It is easy to see (Fig. 1.7) what the net escape from a slice of air above the double square would be if all winds blew to the right. Winds blowing to the right can take out air only from the right-hand side of the square, where they remove air represented on the map by the area $2Dx$ which that air covers. Here, 2 is the length of that side, and we take the value of Dx at the grid-point R (the centre of that side) to represent an average displacement to the right for the removed air. At the opposite side, also of length 2, displacements are represented by the value of Dx at L, the central grid-point on that side. Those displacements are *pushing into* the slice some quite new air represented on the map by an area $2Dx$. The net escape of air is represented by twice the difference between the value of Dx at the right-hand grid-point R and at the left-hand grid-point L.

This is another centred difference – but this time with regard to position on the map instead of time on the clock. We shall write it $X(Dx)$. Thus, we use a capital letter again, but now X instead of D, to represent the difference between values at a grid-point on the right (where x is greater by 1) and at a grid-point on the left (where x is less by 1). The quantity $X(Dx)$ tells us how much winds towards the right are bigger on the right-hand side of the square than on the left. That will cause net escape of an amount of air represented on the map by an area equal to 2 times $X(Dx)$.

Similarly, we use Y for a centred difference with regard to position up the page; that is, the difference between values at a grid-point where y is greater by 1 and at a grid-point where y is less by 1. The quantity $Y(Dy)$ tells us how much winds directed up the page are bigger on the upper side of the square than on the lower, causing net escape of air represented on the map by an area equal to 2 times $Y(Dy)$. When both kinds of winds are blowing, the net escape from the slice is found by adding up the effects of displacements to the right and up the page, to give 2 times $X(Dx) + Y(Dy)$. Dividing that by the total area of the square (2 times 2 equals 4) we get the proportional loss of air from the slice during the interval Dt as

$$\frac{X(Dx) + Y(Dy)}{2} = \text{divergence:} \qquad (1)$$

meteorologists often call this the *divergence*. It is a measure of how much the blowing of the winds is changing the amount of air above us.

For example, when Dx and Dy take values as follows at the grid-points nearest to P:

Dx = 0.118	0.123	0.129
Dy = 0.073	0.069	0.065
Dx = 0.120 P 0.125	0.131	
Dy = 0.076	0.072	0.068
Dx = 0.122	0.127	0.133
Dy = 0.078	0.074	0.070

then we have $X(Dx) = 0{\cdot}131 - 0{\cdot}120 = 0{\cdot}011$ and $Y(Dy) = 0{\cdot}069 - 0{\cdot}074 = -0{\cdot}005$. In that case,

$$\text{divergence} = \frac{X(Dx) + Y(Dy)}{2} = \frac{0{\cdot}006}{2} = 0{\cdot}003.$$

In words, three-thousandths of the air in the slice is escaping during the double time-interval.

We know that the pressure p at any place is a measure of the weight of air above it. We see now how, for the air at any one grid-point, the change Dp in pressure during the interval Dt can be calculated with the help of knowledge about winds at the four nearest-neighbour grid-points. We get it by adding up the change in the pressure exerted by each of the 1-decibar slices of air that lie above the grid-point in question. That change, for each slice, is a *drop*, equal to 1 decibar times the proportional loss of air given by the divergence (1). The value of Dx used in calculating the divergence (1) for each slice is best taken as the average (half the sum) of its values at the pressure-levels at the top and bottom of the slice; and similarly with Dy. Finally, the

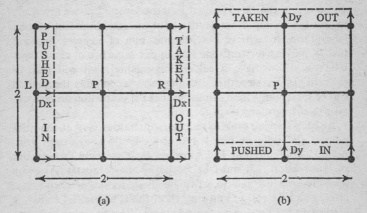

(a)   (b)

Fig. 1.7. Net escape from a slice of air above a double square centred on the grid-point P. (a) Effects of Dx alone (that is, case when all winds blow to the right). (b) Effects of Dy alone (that is, case when all winds blow 'up the page'). In the usual situation, with Dx and Dy both present, we get a combination of both effects.

Case (a): during the interval Dt, the map represents air taken out by an area $2Dx$ with Dx taken at R, and air pushed in by an area $2Dx$ with Dx taken at L. It follows that net escape is twice the difference, $X(Dx)$, between values of Dx at R (where x is greater by 1) and at L (where x is less by 1). Thus,

net escape = $(2Dx$ at $R) - (2Dx$ at $L)$
= $2(Dx$ at R $- Dx$ at $L)$
= $2X(Dx)$.

Case (b): similarly, net escape is twice the difference $Y(Dy)$

loss of pressure at any one level, in decibars, is what we get by adding up the divergence so calculated for all the slices above that level:

$$Dp = -\text{ (added up divergences for all slices above).} \qquad (2)$$

STEPPING FORWARD

We have now set up a sort of mathematical scaffolding, based on the grid and the slices of air above it, and shown how simple properties (which are little more than definitions) of wind speed and pressure can be expressed in terms of it. Now we are ready to use a few important facts about air established in physics laboratories to achieve our essential aim of *stepping forward*: that is, moving forward one step in our sequence of equal time-intervals. If we have a calculation method that will take us forward just one step, then on a big enough computer the method can be used many, many times so that the prediction process can take us many, many steps forward.

At each step we seek to calculate quantities *only* at the grid-points, and only at our ten pressure-levels above each grid-point. This limitation is important as it restricts the amount of information that must be stored by the computer. At every pressure-level we aim to calculate, first, its *height* above sea-level, which we call h; knowing all of these, as Fig. 1.4 shows, is equivalent to knowing air temperatures. We also aim to calculate the amounts Dx and Dy by which air is displaced to the right and up the page on the map in the double time-interval Dt; knowing them, as Fig. 1.6 shows, is equivalent to knowing wind speeds and directions.

Converting any facts found in physics laboratories into such a mathematical step forward on a grid always runs into the same difficulty – actually, a very interesting one. What we know from the laboratory is how a definite mass of air behaves: for example, how it expands when its pressure drops. By contrast, the mathematical process using a grid needs quite different information: how some quantity changes during a time-interval Dt *at a fixed grid-point*. Obviously, if a mass of air was on that grid-point at the beginning of the interval, winds would have blown it right away from the grid-point by the end of the interval! How, then, can we use laboratory results on the behaviour of such a mass of air to predict how the value of some quantity at the grid-point will change?

Fortunately, mathematicians have worked out a standard way of answering this question which is equally effective whatever laboratory results we are trying to use. Suppose that there is some quantity which we shall call q, whose change Dq for a mass of air in the double time-interval Dt is known from physics. (In 'dry mathematics', for example, q could be any of the three quantities we want to calculate at grid-points; or in 'wet mathematics' it could be the concentration of water vapour.) How do we find the change in q at a fixed grid-point during the same interval?

To answer that question, we once more use the X and Y symbols, standing for centred differences between values at grid-points. Thus, Xq is the difference between the value of q at the nearest-neighbour grid-point on the right (where x is greater by 1) and its value at the nearest-neighbour grid-point on the left (where x is less by 1); and Yq depends similarly on the nearest-neighbour grid-points up and down the page, where y is greater or less by 1. We work in the same way with pressure-levels, defining Pq as the difference between the value of q at the nearest level below (where the pressure p is greater by 1 decibar) and its value at the nearest level above (where p is less by 1 decibar); these, of course, are levels above one and the same grid-point on the map. Continuing in this same logical way, we define Tq as the difference between the value of q at the instant after and its value at the instant before; again, all at the same grid-point and pressure level.

This last difference Tq is what we want to find. As we emphasized before, such a change at a fixed grid-point is *not* the same as the change which physics tells us about. That is the change which we called Dq, for a mass of air that 'will not stay put' at the grid-point during the interval in question. In fact, for that mass of air, its position on the grid will change by Dx to the right and by Dy up the page; and we know also how to calculate the change Dp in its pressure. How can we make use of all that information?

A fair estimate of how much the change in q is affected by the change in, say, x is given by our knowledge of Xq. After all, Xq

is the amount by which q changes when x increases by 2 (from the grid-point on the left to the grid-point on the right). Therefore, we can estimate the amount by which q changes when x increases, not by 2 but by the much smaller amount Dx, as *a proportion* $\dfrac{Dx}{2}$ *of that total change* (Xq); which we can write as $(\frac{1}{2}Dx) \times (Xq)$. The best average value to use for (Xq) in this expression will be obtained if we calculate it at the *central* instant of the double time-interval Dt.

For exactly the same reasons, the change (Dy) in y affects the change in q by a proportion $(\frac{1}{2}Dy)$ of the total change (Yq); that is, by $(\frac{1}{2}Dy) \times (Yq)$; and the change in p affects the change in q by $(\frac{1}{2}Dp) \times (Pq)$. Thus, the whole difference between Dq (change in q for the mass of air) and Tq (change at a fixed grid-point) is found by adding up all three of these effects as

$$Dq - Tq = [(\tfrac{1}{2}Dx) \times (Xq)] + [(\tfrac{1}{2}Dy) \times (Yq)] + [(\tfrac{1}{2}Dp) \times (Pq)]. \tag{3}$$

This is the basic formula for 'stepping forward' in time. We can use it for any quantity q when (*i*) we know Dq (how q is changing for the mass of air from its value at the instant before to its value at the instant after) from the physics; and (*ii*) we have the values of q and Dx and Dy at the present instant stored in the computer; so that (*iii*) we can obtain Dp from equation (2). Then from this basic formula (3) we can calculate the difference $Dq - Tq$ and so find the all-important quantity Tq: the change in q at the grid-point from the instant before to the instant after.

In 'dry mathematics', the actual results from physics to which we apply this formula are results about changes in wind speed and direction, and results about what we have called slice thicknesses. It will be sufficient to outline these briefly.

The results about changes in wind speed and direction are concerned with the effects of gravity and of the earth's rotation. For use in our prediction process, they need to be expressed in terms of the changes during the interval Dt in the values of those quantities Dx and Dy which, of course, are being used to represent everything we know about wind speeds and directions. The change in Dx, for example, is written $D(Dx)$. Fortunately, our knowledge from physics about the effects of gravity and earth's rotation on air movements can easily be written in terms of $D(Dx)$ and $D(Dy)$, as

$$\left.\begin{array}{l} D(Dx) = - k(Xh) + f(Dy), \\ D(Dy) = - k(Yh) - f(Dx). \end{array}\right\} \qquad (4)$$

Here, h is the height of the pressure-level above sea-level. When it slopes up as we go to the right (with x increasing), Xh is positive. (The actual slope is $(Xh)/(2a)$, where the scale of the map is 1 in a so that $2a$ is the actual distance on the ground between grid-points $(x - 1, y)$ and $(x + 1, y)$.) That slope up tends to slow down motion to the right (Fig. 1.8), the reduction in velocity in a time Dt being $g(Dt)$ times the slope, where g is the downward acceleration due to gravity (9·8 metres per second in every second). However, Dx is the actual velocity to the right, multiplied by Dt and by the map scale $1/a$. It follows that the terms involving k in the above equations are correct provided that $k = g(Dt)^2/(2a^2)$. (Note that the scale $1/a$ of the map varies gradually across it in a manner depending on the 'projection' used; the Meteorological Office uses a 'stereographic' projection of the northern hemisphere from the South Pole on to a tangent plane at the North Pole, which causes a to depend on latitude, falling at the equator to half its value at the North Pole.)

We go into even less detail about the effect of the earth's rotation, represented by the terms in the above equations involving f. This quantity f is also biggest at the North Pole, where its value is twice the angle (measured in *radians*) through which the earth rotates in the interval Dt. However, f falls as the latitude decreases, becoming zero at the equator. These terms in f represent a general tendency for winds to veer clockwise in the northern hemisphere: thus, the $+f(Dy)$ term makes an air motion 'up the page' generate a motion to the right, and the $-f(Dx)$ term makes an air motion to the right generate a motion 'down the page' (Fig. 1.8).

Often, of course, the effects of gravity and earth's rotation are

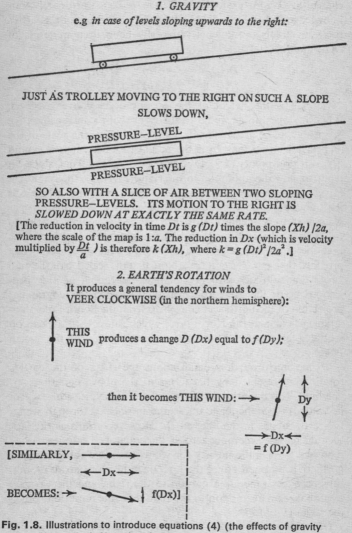

1. GRAVITY

e.g in case of levels sloping upwards to the right:

JUST AS TROLLEY MOVING TO THE RIGHT ON SUCH A SLOPE SLOWS DOWN,

PRESSURE–LEVEL

PRESSURE–LEVEL

SO ALSO WITH A SLICE OF AIR BETWEEN TWO SLOPING PRESSURE–LEVELS. ITS MOTION TO THE RIGHT IS *SLOWED DOWN AT EXACTLY THE SAME RATE.*

[The reduction in velocity in time Dt is $g(Dt)$ times the slope $(Xh)/2a$, where the scale of the map is $1:a$. The reduction in Dx (which is velocity multiplied by $\frac{Dt}{a}$) is therefore $k(Xh)$, where $k = g(Dt)^2/2a^2$.]

2. EARTH'S ROTATION

It produces a general tendency for winds to VEER CLOCKWISE (in the northern hemisphere):

THIS WIND produces a change $D(Dx)$ equal to $f(Dy)$;

then it becomes THIS WIND: →

Dy

→Dx←
= f (Dy)

[SIMILARLY,

←Dx→

BECOMES: → | f(Dx)]

Fig. 1.8. Illustrations to introduce equations (4) (the effects of gravity and earth's rotation). Note that, for the sake of clarity, the slopes of pressure levels have been enormously exaggerated

almost 'in balance'; this means that an *air motion may continue on* 'up the page', say, because *the pressure-level slopes upwards to the right* (Buys Ballot's Law) exactly counteracting the wind's tendency to veer. However, it is departures from that kind of balance that are important in weather prediction, because it turns out that only they can generate the 'divergence' whose importance we saw in the last section.

The equations (4) tell us, then, the change in movement experienced by the air: $D(Dx)$ to the right and $D(Dy)$ up the page. Then, from the basic formula (3) for 'stepping forward' we deduce the quantities $T(Dx)$ and $T(Dy)$ which the prediction process needs to know: how much Dx and Dy change at a grid-point during the interval Dt. It may seem odd that this involves using the formula (3) with the general 'quantity' q replaced throughout either by Dx or by Dy – quantities which themselves appear also somewhere else in the formula – but that is just a good example of the power of mathematics!

We omit details of a slightly modified version of the above procedure used for winds at the 10-decibar level, close to the earth's surface. Actually, the frictional resistance to the wind offered by that surface, in a form slightly different for land and sea, is taken into account at the 10-decibar level.

The remaining result from physics used in the process of 'stepping forward' is one already familiar to the reader. A pressure drop produces an increase in air volume. The *proportional* volume increase is equal to seven-tenths of the proportional pressure drop; that is, to $\left(-0.7 \dfrac{Dp}{p}\right)$ in the case of our known pressure change Dp during the interval Dt. (The minus sign is needed to make volume increase when pressure decreases and *vice versa*.)

In the prediction process, *slice thickness* is the best estimate of the volume occupied by a fixed amount of air. This is because each slice has the same weight of air above a horizontal area of, say, 1 square metre, and its thickness tells us the *volume* of that air. If we write s for slice thickness, the *proportional* volume increase for that air during the interval Dt is $\dfrac{Ds}{s}$. Relating this to

Dp by the rule just stated, we obtain the change Ds as

$$Ds = \left(-0 \cdot 7 \frac{Dp}{p}\right) s. \qquad (5_{dry})$$

(We number that equation (5_{dry}) because it is the only one of our seven numbered equations that will be altered in 'wet mathematics'!) The idea, as usual, is that, given this change Ds, the basic 'stepping forward' formula (3) can be used (with the general quantity q replaced by s) to get Ts.

Note that when we apply this to the slice between, say, the 6-decibar and 7-decibar levels, we should not use the value of $\frac{Dp}{p}$ calculated at either of *those* pressure-levels (the top and bottom of the slice). It is much better to add up those values and divide by 2, to get a logical choice of average value for the air in the slice. Similar averages of values at the top and bottom of a slice are used also for Dx, Dy and Dp when the 'stepping forward' formula (3) is used to obtain Ts.

From each such Ts, which means the change in slice thickness during the interval Dt, we can deduce the value of s at the end of that interval. A check can then be made on whether the new slice thicknesses violate the condition of Fig. 1.3: that is, whether the ratio of any slice's thickness to that of the slice below it is less than the minimum ratio. Often, there will be no such violation. When there is, however, we expect that natural vertical motions will immediately exchange air (and therefore also heat) between the two slices until this minimum ratio is reached. Accordingly, the mathematical prediction process makes just such an adjustment to the values of Ts.

Of course, what we really need is Th; for each of our ten pressure-levels we need to be able to 'step forward' in our knowledge of its height above sea-level, h. Knowledge of h is essential if only because it comes into the equations (4) for changes in Dx and Dy.

Fortunately, the value of h for any of our ten pressure-levels is obtained easily by (*i*) adding up the slice thicknesses s for all slices between that level and the bottom (10-decibar) level; and (*ii*) adding on a small correction for the height h_{10} of the 10-decibar level. Therefore, if we know all the changes Ts and the change Th_{10} we can get any other Th as

$$Th = \text{(added up } Ts \text{ for all slices below)} + Th_{10}. \qquad (6)$$

Meteorologists find Th_{10} by assuming that air movements (that is, winds) near the ground follow the ground's ups and downs. Taking z, at any grid-point, to mean some sort of average *height of the ground* above sea-level in, say the double square centred on that grid-point, they express this assumption in a form like

$$D(h_{10} - z) = 0. \qquad (7)$$

The 'stepping forward' formula (3) then gives $T(h_{10} - z)$, which is the same as Th_{10}; obviously $Tz = 0$ because the ground at any fixed grid-point is *not* heaving up and down!

We omit details of modifications to the procedure for finding Ts applied in the top and bottom slices. We remark, however, that for the bottom slice they include an important extra term in (5) representing expansion of air due to *warming by the earth's surface* (or contraction due to cooling).

To sum up the 'stepping forward' process in dry mathematics, we can say that, knowing the values of h, Dx and Dy at each grid-point (representing the height of each pressure-level, and the wind speed and direction there), we want to be able to compute the *changes* in those quantities in the double time-interval Dt; that is, to compute Th, $T(Dx)$ and $T(Dy)$. To achieve these aims we use our set of seven numbered equations,

$$\frac{X(Dx) + Y(Dy)}{2} = \text{divergence}, \qquad (1)$$

$$Dp = -\text{(added up divergences for all slices above)}, \qquad (2)$$

$$Dq - Tq = (\tfrac{1}{2}Dx) \times (Xq) + (\tfrac{1}{2}Dy) \times (Yq) + (\tfrac{1}{2}Dp) \times (Pq), (3)$$

$$\left. \begin{array}{l} D(Dx) = -k(Xh) + f(Dy), \\ D(Dy) = -k(Yh) - f(Dx), \end{array} \right\} \qquad (4)$$

$$Ds = \left(-0 \cdot 7 \frac{Dp}{p} \right) s, \qquad (5_{\text{dry}})$$

$$Th = \text{(added up } Ts \text{ for all slices below} + Th_{10}), \qquad (6)$$

$$D(h_{10} - z) = 0. \qquad (7)$$

Then, given h, Dx and Dy, equations (1) and (2) allow us to compute Dp. Also, equations (4) give $D(Dx)$ and $D(Dy)$. From these, the 'stepping forward' equation (3) (with $q = Dx$ and Dy) allows us to compute $T(Dx)$ and $T(Dy)$. That's two of our three aims achieved! Similarly, (5_{dry}) gives Ds, and equation (3) with $q = s$ allows us to compute Ts. In the meantime, equation (7) gives Dh_{10} (since the height z of the ground is known), and equation (3) with $q = h_{10}$ allows us to compute Th_{10}. Finally, from Ts and Th_{10}, equation (6) gives us Th.

KEEPING STRAIGHT

We have indicated now the essential features of how the prediction process makes a step forward. When we know, for every grid-point at one particular instant, the height h of each of our ten pressure-levels (equivalent to knowing air temperatures), and the values of Dx and Dy (equivalent to knowing winds at those levels), we can find the centred difference Th: the change in h at any such grid-point and pressure-level between the instant before and the instant after. We can also get $T(Dx)$ and $T(Dy)$. The method is summed up in the preceding paragraph.

As we remarked earlier, a calculation method which will take us forward just one step can on a big enough computer be used many, many times to take us many, many steps forward – an important idea that lies behind an enormous amount of the successful use of mathematics, not just in weather prediction but in lots of other uses mentioned in this book.

There is, however, one problem that has to be faced whenever that idea is used. A toddler may be able to make a step forward and then another and then many more, and yet be quite unable to use the process to get anywhere because he can't keep straight! In other words, stability is necessary too.

Although we claimed that mathematics helps us to 'add up' vast quantities of small-scale knowledge so as to produce large-scale information, we are not saying that the process is just like adding up one's accounts! In the accounts of a household or a firm, the sum paid out for each item is known exactly and adding all of them up must give the total expenditure. Most measurements of our environment, however, are liable to small experimental errors. In any kind of 'adding up' process there is a danger that lots of small errors will add up to give errors which are so big that the results are useless.

Mathematicians call any kind of 'adding up' process *unstable* when it tends to allow large numbers of small errors (which may be positive or negative) to combine together to give very big errors. Mathematicians make some of their most useful contributions when they invent special ways of doing the 'adding up', which are *stable* in the sense that they tend to cause a cancelling out of different errors.

Perhaps the best-known way of achieving this is through various kinds of averaging. Anyone doing an experiment uses that idea if he repeats the experiment several times and takes the average of his readings (assuming that some cancelling between different random errors, positive and negative, is then likely). We shall see that there is scope for intelligent use of averaging in our prediction process.

However, any process involving repeated 'steps forward' is *particularly* difficult to 'keep straight'. A lot of mathematical skill is used to find stable ways of combining successive steps forward. We illustrate this important branch of the mathematician's art by describing its use in the weather-prediction process.

Readers might wish to interrupt here and complain that we have used the word 'instability' in this section and in the introductory part of the chapter to mean quite different things. Admittedly, we emphasized even in our introduction that there are different kinds of instability, and that all of them take the form of an amplification of 'tiny variations' into 'big fluctuations'. Everything we said there, however, was concerned with the real physical processes that make up the weather, and with the different kinds of real physical instability which they show.

Just now, we have described yet another kind of instability:

one which can arise in a mathematical prediction process, although it still involves the amplification of 'tiny variations' (small numerical errors) into 'big fluctuations' (unacceptably large errors). Often, it is called 'numerical instability' to distinguish it from the previously described 'physical instability'.

How are the two kinds related? After all, we are describing methods for predicting the weather, and we admit that the weather is subject to many kinds of physical instabilities. You might imagine that the numerical instabilities are simply a *mathematical form* of the physical instabilities.

That is not true at all, however. Some of the numerical instabilities are of purely mathematical origin, and can occur even when a completely stable physical process is being calculated. Such instabilities can be particularly nasty in a 'stepping forward' process, because they can lead to the build-up of intolerably large errors after only a moderate number of steps. If that happened after 50 steps, for example, then our weather-prediction process would be useless: nonsensical weather patterns would be predicted for an instant only 5 hours later (50 times 6 minutes).

A badly designed 'stepping forward' process, then, could show numerical instabilities that would develop much too fast to make it useful for weather prediction. Developing as fast as that, they could not just be a mathematical representation of the physical instabilities.

There is in fact a good deal of meteorological evidence that any physical instabilities in weather patterns, on the sort of large scale that we are trying to represent with our grid, take at least a week to develop. A really faithful prediction process should without doubt show *that* degree of instability, which should not, however, affect the value of predictions over periods like the 72 hours used nowadays.

We emphasize, on the other hand, that a process making use of a grid should be expected to reproduce *only* those physical instabilities in weather behaviour that are on a scale at least as big as the grid. Because no variations on any smaller scale *can* be represented, there is really no need for any instabilities which operate on a small scale (and, possibly, quicker) to appear in the mathematical prediction process.

Even so, we may sometimes try to represent 'what those small-scale instabilities do to the large-scale weather patterns'. That seemingly ambitious idea has *already* been illustrated. Whenever the ratio of a slice's thickness to that of the slice below it becomes less than the minimum ratio (see Fig. 1.3), an instability on a *very* small scale results. This is because any upward, or downward, movement of air between the slices quickly becomes intensified by the buoyancy changes it produces. Very soon, all those up and down motions mix up the air in the two slices until the minimum thickness ratio is reached. We saw in the last section how the prediction process can represent, not of course those very small-scale vertical movements of air, but this gross effect of them on the scale of whole slices.

What about instabilities on a 'medium' scale, between that 'very small' scale and the large scale of our grid; for example, local short-term veerings in the direction of the wind and increases and decreases in its speed? Meteorologists believe that these produce a certain amount of mixing-up in the *horizontal*. We saw already that one approach to achieving numerical stability might be to allow a bit of 'averaging'; for example, the value of some quantity at a grid-point might sometimes be represented by an average of its values at the four nearest-neighbour points. Now we see that physical considerations suggest that this sort of averaging may be useful also for representing the gross effects of horizontal mixing by medium-scale instabilities.

In the rest of this section we concentrate upon 'keeping straight': adapting our 'stepping forward' process so as to avoid numerical instabilities. We begin by mentioning a certain famous instability inherent in any naïve use of centred differences.

We indicated that a centred difference gives much the most accurate expression of physical properties, like those of air. Speed *at one instant* is best represented by distance gone between the instant before and the instant after. The force of gravity acting at one instant is best related to wind changes between the instant before and instant after. Then the force changing the wind is given its value at the *centre* of the double time-interval: a much more 'typical' value than a value at its beginning or end.

Nevertheless, centred differences do make it difficult, not only to 'keep straight' but also to 'get started'. Of these two related difficulties, we begin by describing the second.

Suppose that, by the methods described later (pp. 62–9), the computer has converted the available observations from all over the northern hemisphere at one initial instant into a consistent set of values of h, Dx and Dy for that instant. The equations of the last section then tell us how to find Th, $T(Dx)$ and $T(Dy)$ at that instant. These are *centred* differences, which tell us the changes in h, Dx and Dy *between the instant before and the instant after* (for every pressure-level above each grid-point).

Then – we 'get stuck'! This is because we are just as lacking in information about the instant *before* as about the instant after. The weather stations have all made their observations at one initial instant. We cannot know what the readings would have been at the instant before. Indeed, physics suggests that we *shouldn't need to know* those! – in the sense that the properties of air that we have used plus observations at *one* initial instant ought to fix completely what happens after it.

Even so, you might imagine that this is just a 'little local difficulty'. If we could once get over it, by some sort of 'intelligent guesswork' about those values at the instant before, then at least the trouble could never happen again. You would be able then to get the values at the first instant after the initial instant. Next, you could calculate the centred differences *at that first instant*, and this time they would give the values at the *second* instant without trouble (Fig. 1.9). Thereafter, the computer would just play leapfrog! Centred differences at each instant would take us over from values at the instant before to values at the instant after.

Mathematical work has shown, however, that such 'leapfrogging' has a very strong tendency to be numerically unstable. You create one 'population' of values at odd-numbered instants and a different, as it were 'foreign', population of values at

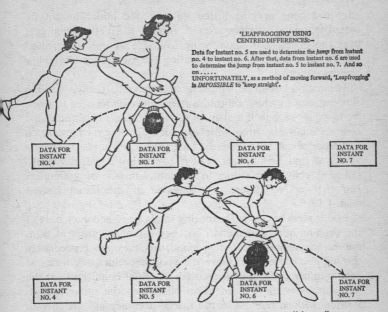

'LEAPFROGGING' USING CENTRED DIFFERENCES:—

Data for instant no. 5 are used to determine the *jump* from instant no. 4 to instant no. 6. After that, data from instant no. 6 are used to determine the jump from instant no. 5 to instant no. 7. And so on . . . . .
UNFORTUNATELY, as a method of moving forward, 'Leapfrogging' is *IMPOSSIBLE* to 'keep straight'.

DATA FOR INSTANT NO. 4

DATA FOR INSTANT NO. 5

DATA FOR INSTANT NO. 6

DATA FOR INSTANT NO. 7

DATA FOR INSTANT NO. 4

DATA FOR INSTANT NO. 5

DATA FOR INSTANT NO. 6

DATA FOR INSTANT NO. 7

Fig. 1.9. Leapfrogging using centred differences is impossible to 'keep straight'

even-numbered instants. You get a value of h at any even-numbered instant by adding on a certain *difference* to its value at the previous even-numbered instant; a value which, in turn, was found from the even-numbered instant before that. The situation is just the same with the odd-numbered instants. Accordingly, it is very easy for the two populations to get completely out of step with each other. Any errors introduced by the initial bit of 'intelligent guesswork' would bring about an initial difference between the two populations which would become amplified until it soon led to intolerably big errors.

All those difficulties are avoided by a method of using centred differences which is purposely designed to work *from values at one initial instant*. Then the prediction process has a much better chance of accurately predicting the *physical* process, because both have the same property; that data at one initial instant fix completely what happens after it. In fact, the method that gets over the problem of data being available at only one initial instant also succeeds in eliminating the numerical instability associated with it.

The method is quite simple to understand. Really, it is the simplest way in which centred differences *could* be used to work from values at one initial instant. It proceeds to values at the *second* instant after that initial instant by adding on the centred difference calculated at the first instant after it from *best-estimate* values at that first instant. Those best-estimate values are obtained directly from values at the initial instant by adding on *half* the centred difference calculated at the initial instant (Fig. 1.10).

In this method one is attempting to get good accuracy only at the even-numbered instants. The best-estimate values at odd-numbered instants are not so good, but they are immediately thrown away after use! Only the accurate values calculated at even-numbered instants are stored, as initial conditions for the next double step forward. This eliminates the numerical instability *without* the loss of the special accuracy associated with centred differences: calculating with best-estimate values gives a good 'typical' value of the difference during the double time-interval, much better than if values calculated at its beginning or end were used.

For example, suppose we applied this technique to get a quick, rough idea of the effect on a capital of £100 of compound interest at 6% per annum, accumulating by ½% monthly for 2 years. From the present value £100 of the capital and the interest rate of 6% we could obtain *without calculation* a crude 'best-estimate' value of the capital after one year (the odd-numbered instant) as £106. This is *not* a good approximation (an accurate value is £106·17). However, if we use it *solely* to calculate the total interest accruing over 2 years as 12% of that best-estimate value £106 at the odd-numbered instant, we get an excellent approximation to the capital after 2 years (the even-numbered instant); namely, £112·72, which is correct to the nearest penny.

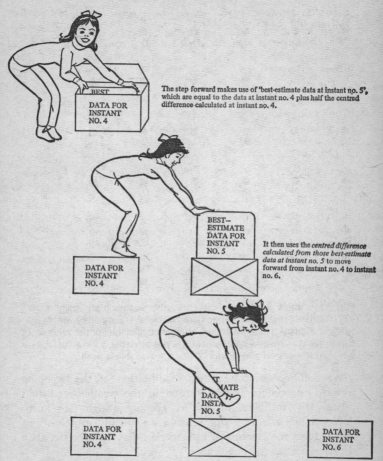

The step forward makes use of 'best-estimate data at instant no. 5', which are equal to the data at instant no. 4 plus half the centred difference calculated at instant no. 4.

It then uses the *centred difference calculated from those best-estimate data at instant no. 5* to move forward from instant no. 4 to instant no. 6.

Fig. 1.10. Double step forward from data known only at a single even-numbered instant, no. 4. This method, which has good accuracy *and* is able to 'keep straight', has been much used by the Meteorological Office. Quite recently, it was superseded by a rather more complicated alternative, not described here, which has the same advantages and produces its results faster, yielding the result quoted on p. 18 (a 24-hour forecast calculated in 12 minutes)

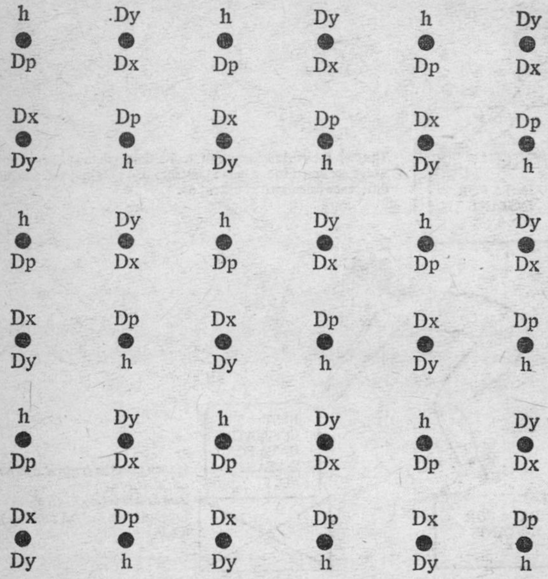

h	.Dy	h	Dy	h	Dy
●	●	●	●	●	●
Dp	Dx	Dp	Dx	Dp	Dx

Dx	Dp	Dx	Dp	Dx	Dp
●	●	●	●	●	●
Dy	h	Dy	h	Dy	h

h	Dy	h	Dy	h	Dy
●	●	●	●	●	●
Dp	Dx	Dp	Dx	Dp	Dx

Dx	Dp	Dx	Dp	Dx	Dp
●	●	●	●	●	●
Dy	h	Dy	h	Dy	h

h	Dy	h	Dy	h	Dy
●	●	●	●	●	●
Dp	Dx	Dp	Dx	Dp	Dx

Dx	Dp	Dx	Dp	Dx	Dp
●	●	●	●	●	●
Dy	h	Dy	h	Dy	h

Fig. 1.11. In this diagram, the quantity written *above* a grid-point is the one calculated there at even-numbered instants. The quantity written *below* it is calculated there at odd-numbered instants. For the value of a quantity at some grid-point where it is not available, an average of its values at the nearest grid-points where it is available is used

We mention just one other modification of the 'stepping forward' procedure which is used in the weather-prediction process. It contributes further to 'keeping straight', by introducing some *averaging* (for reasons we already explained); it also reduces the amount of information that must be stored in the computer at any one time.

Fig. 1.11 shows in fact that at any one instant only one of our four basic quantities h, Dx, Dy and Dp is calculated at each grid-point. The quantity written *above* the grid-point is calculated at even-numbered instants; the quantity written below it is calculated at odd-numbered instants. Whenever our equations need a value of a quantity at some grid-point where it is not available, an average of its values at the nearest grid-points where it is available is used.

Actually, at the points in Fig. 1.11 where Dp has to be calculated, from (2), the required divergences can be obtained immediately, from (1); for example, $X(Dx)$ is available at those points because Dx is known to the right and to the left. This Dp is then used in (5_{dry}) to obtain the change Ds in slice thickness. On the other hand, when the basic 'stepping forward' formula (3) is used to deduce Ts, we need to know quantities like Dx and Xs. That is where we need to use the average of Dx at the two next grid-points to right and left; also, we use the average of Xs at the two next grid-points up and down the page. Similar averaging occurs at many other places in the prediction process. It gives a further very useful contribution to 'keeping straight' (although in some versions of the process a little more averaging than we have described here is also used).

'Keeping straight' does have just one more aspect: it requires that effects due to ignorance of conditions *beyond* the edge of the grid be kept to a minimum. At present the grid covers only the northern hemisphere, largely because there are not nearly enough weather stations in the southern hemisphere to provide the initial data. At the grid's edge, the basic 'stepping forward' procedure cannot be used, for lack of information. Instead, we use an artificial assumption that h, Dx and Dy on the edge of the grid do not change. (Fortunately, that assumption is better in the tropics than in some other places!) Also, an extra amount of averaging is used very close to the edge. Errors so introduced are not believed to extend outside the tropical regions during the 72 hours of the forecast.

WET MATHEMATICS

Readers who have followed the account of 'dry mathematics' given in the last three sections know now how mathematics is used to 'add up' vast quantities of small-scale knowledge to produce large-scale information. At this stage, some of them may prefer to leave weather problems, and look at many other 'newer uses of mathematics' described in this book: if so, they should omit the next two sections. Those pressing on will learn in the present section how the 'wetness' is put into weather prediction; and in the next section how the problems (some of them mathematical) of processing the data from weather stations, and of usefully presenting the results of a forecast, are tackled.

Most air contains water vapour – which, as we remarked earlier, we cannot see at all until it condenses. In our prediction process, which began by dividing the air into ten slices of equal weight, we use r (standing for 'ratio') to mean the ratio of the weight of water vapour to the weight of the air in any slice.

Condensation – that is, conversion of water vapour into some visible form such as fog, cloud, rain, hail or snow – tends to occur whenever r reaches a certain maximum value. We call this r_{sat} because the air is then described as 'saturated'. From physics the dependence of r_{sat} on pressure and temperature is known. That gives us for each slice the dependence of r_{sat} on the slice thickness s shown in Fig. 1.12. These ratios of weight of water vapour to weight of air may seem *small*, but can be very important to us: water vapour making up just one-hundredth of the weight of air in just one slice (that is, where $r = 0.01$) could produce a whole centimetre of rainfall if it were all caused to condense!

Is r_{sat} the *exact* value of r at which vapour condenses in drops (or other forms) big enough to fall out of the slice? The prediction process *does* work on that assumption, partly because it is extremely hard to take any proper account of the two main sources of error in it. On the one hand, r is only an average ratio for a slice, so that a value less than r_{sat} does not prevent some parts of the slice being saturated and producing condensation; on the other hand, physics tells us of a variety of circumstances when air can become 'supersaturated' (with r a bit greater than r_{sat}) without significant condensation occurring. It is fortunate that those two sources of error tend (at least, partly) to cancel out!

Physics is much more precise about the fact that water vapour carries 'latent heat', which it gives up on condensation. That

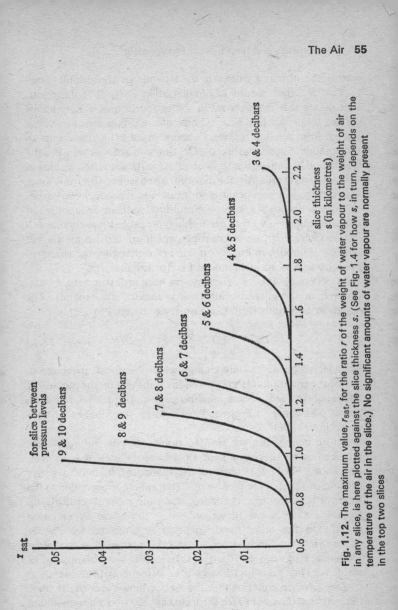

Fig. 1.12. The maximum value, r_{sat}, for the ratio r of the weight of water vapour to the weight of air in any slice, is here plotted against the slice thickness s. (See Fig. 1.4 for how s, in turn, depends on the temperature of the air in the slice.) No significant amounts of water vapour are normally present in the top two slices

warms the air, and causes it to expand. In the double time-interval Dt, the amount of condensation is $(-Dr)$: the minus sign makes this the *decrease* in the proportion of water vapour carried by the air in a slice. The resulting expansion is precisely known from physics: it gives a change in slice thickness equal to the amount of condensation $(-Dr)$ times an amount we will call l (to remind ourselves that it is an effect of latent heat). Values of l, reading upwards for the bottom seven slices (the only ones where there is a significant amount of water vapour), vary from about 8 to about 21 *kilometres*; multiplying them by amounts of condensation $(-Dr)$ that are unlikely to exceed one-thousandth in one interval Dt, we get changes which are only in metres but may have important effects on winds by changing the slopes of pressure-levels (Xh in equations (4), for instance).

In 'wet mathematics', then, we do need an extra term, $(-Dr)$ times l, in the equation for how a slice of air expands. The equation's previous form (5_{dry}) must be changed to

$$Ds = \left(-0 \cdot 7\frac{Dp}{p}\right)s + (-Dr)l. \qquad (5_{wet})$$

That, actually, is the only change which 'wetness' produces in any of the equations (1) to (7) used for summing up the method of 'stepping forward'. These must now be supplemented, however, by an additional rule, needed for finding the amount of condensation $(-Dr)$.

Before giving this, we stop to point out one quite different implication of the extra slice thickening $(-Dr)l$ due to condensation. Fig. 1.3(*a*) showed the minimum ratio of a slice's thickness to that of the slice below. That ratio corresponded to the 'natural' change in s, occurring when air moved vertically up from the slice below (described in a 'dry' case by equation (5_{dry})). On the other hand, air that is saturated with water vapour ($r = r_{sat}$) and moves up into the next slice finds there a lower value of r_{sat} (see Fig. 1.12). Its excess water vapour must condense, the condensation $(-Dr)$ being equal to the drop $(-Dr_{sat})$. When equation (5_{wet}) is used with this idea to calculate the 'natural' change in s for saturated air, the $(-Dr)l$ term makes it come out *bigger* than in Fig. 1.3.

Actually, Fig. 1.3(b) shows the results of this calculation for saturated air. Considerably greater minimum ratios of a slice's thickness to that of the slice below are found. The prediction process uses these increased minimum ratios wherever $r = r_{sat}$, together with ratios adjusting gradually from those to the 'dry' case when (r/r_{sat}) is a bit less than 1. (This gradual adjustment allows for the fact that a grid square with an average value of r a little below r_{sat} must include some parts with $r = r_{sat}$.) Wherever the ratio of thicknesses falls below this minimum, natural vertical motions are expected to exchange both air and the water it carries between the two slices until the minimum ratio is reached.* The mathematical prediction process makes just such an adjustment to the values of Ts *after* the general 'wet mathematics' rules described below have been applied.

These rules take the simplest possible form ($Dr = 0$) at those grid-points where no condensation is occurring; namely, the points where in every slice r continues to be less than r_{sat}. Then there is no change to equation (5_{dry}) and the whole calculation is just as in the last three sections.

Actually, the prediction process, at each grid-point where the heights h of pressure-levels are being calculated at even-numbered instants (Fig. 1.11), makes those calculations first assuming $Dr = 0$. The process then checks whether or not, for each slice below 3 decibars, the new value of r so calculated is less than the 'saturated' value r_{sat} associated with the new slice thickness s. If that is so, then the process accepts the new values.

If not, then the process picks out the highest slice where the new value of r is larger than the new value of r_{sat}. In *that* slice, it calculates *what amount of condensation* ($-Dr$) is needed to make those two new values become equal – using equation (5_{wet}) to get the modified value of s on which the new value of r_{sat} depends.

The process looks next at the slice immediately below. It may be that there also the new r is larger than the new r_{sat}. If *that is so*, then the necessary condensation ($-Dr$) is calculated exactly as before.

If not, then you might suppose that $Dr = 0$ remains the only possibility. Now, however, with water in condensed form coming

*This process is commonly *observed* as a thunderstorm!

down from the slice above, a positive change ($+Dr$), representing partial evaporation of that water, becomes possible. It depends on how *much* less than 1 is the 'relative humidity' (r/r_{sat}). From various collections of observations, meteorologists have worked out rules for calculating the evaporation ($+Dr$) from that relative humidity and the amount of water in condensed form coming down from above.

Moving down the slices one after another, then, we use the rules just mentioned in any unsaturated slices to calculate an evaporation ($+Dr$). That is calculated from the relative humidity (r/r_{sat}) and from the amount of water coming down in condensed form (namely, ($-Dr$) added up for all slices above). In slices where the new r exceeds the new r_{sat}, however, a *condensation* ($-Dr$) is chosen giving modified values of both r and r_{sat} which are equal.

As before, we omit details of modifications used in the bottom slice. We remark, however, that much of the input of water vapour to the atmosphere occurs there. In fact, over the sea the value of r is fixed by the sea temperature, which changes much more slowly than air temperatures. The prediction process uses, accordingly, an unchanging value of the sea temperature at each grid-point, where r takes the saturation value corresponding to that temperature. The 'wet mathematics' for the bottom slice is considerably more complicated over land: for example, it uses an estimate of *fractional cloud cover*, obtained from the values of the relative humidity in all the slices above, to control the assumed warming of air by the surface and, also, the evaporation from the surface.

The pay-off from all the complications of a proper 'wet mathematics' is partly that the predictions of air temperatures and of winds are made far more accurate. Even more important is the fact that the 'total precipitation', that is, all the water in condensed form reaching the ground, is calculated. In fact, the values of ($-Dr$) added up for all the slices give a forecast of the local average precipitation 'in metres', in the sense that if the added-up values of ($-Dr$) come to 0·002 that means 2 millimetres of rainfall!

STARTING AND FINISHING

We have now described all the main mathematical ideas involved in converting knowledge of the physics relevant to weather into a prediction process. We have omitted many details (mainly details of the physics) but none of the essential mathematical tools used in the conversion.

Again, we have not described how a computer works, but we have indicated the main types of mathematical operations which it must be programmed to perform. Besides arithmetical operations, they include the ability to test whether or not one quantity (say, r) is greater or less than another (say, r_{sat}) and to initiate in either case a *different* set of calculations.

The computer's capacity to *store* information is also most important. Each double 'step forward' in the prediction process starts from air-and-water data at one of the even-numbered instants. The air data consist of the values of h and Dx and Dy at all ten pressure-levels above the grid-points indicated in Fig. 1.11; the water data are values of r for slices below 3 decibars at grid-points where h is known. During the whole double step forward, all those air-and-water data remain stored in the computer for use as required.

In addition, many other quantities need to be calculated and stored during the stepping forward process. The first is Dp, obtained from (1) and (2). We do not list all the others, but note that, at an intermediate stage of the step forward, the *best-estimate* air-and-water data at the centre of the time interval must be stored (Fig. 1.10). The end-product of the whole extremely intricate process of stepping forward is a new set of air-and-water data for the end of the double time-interval.

At that stage all the previous data are thrown away! This includes both the previous air-and-water data and practically all the intermediate quantities used to find the new air-and-water data. (The only intermediate quantity retained is the total precipitation predicted during the double time-interval in each double grid square.)

The fact that almost all of the large amount of stored information available at that stage is abandoned is important because it limits the total amount of space needed for information storage. Indeed, each quantity calculated in one double time-interval is stored in the space used for the same quantity in the previous interval (and it is in fact when this is done that the information about its previous value gets lost).

At regular intervals (when the prediction process reaches, for example, 06.00, 12.00, 18.00 or 00.00 Greenwich Mean Time) the predicted air-and-water data are copied on to a separate data store before they are used to continue the prediction process. This separate data store is then used to generate forecasts for a wide variety of users.

That process might be called 'finishing'; preparing the actual end-products wanted by customers. The same data store is used to prepare large numbers of end-products, because different customers have different needs.

The data in that store are already in a form close to the needs of one extensive body of users: those planning heights at which aircraft will fly and other features of their routes. In fact, the data can be used readily to give them temperatures (Fig. 1.4), winds (Fig. 1.6) and relative humidities (Fig. 1.12) at many different heights all over the northern hemisphere. This is just the type of information they need.

For other users, a considerable amount of interpretation of the forecast is required. The personal experience of skilled meteorologists remains important for inferring from the predicted *large-scale* pattern 1, 2 or 3 days ahead detailed weather features that may be experienced in particular localities (see, for example, Fig. 1.13) or may be important to particular industries. Going over to a mathematical process for predicting development of the large-scale weather pattern has left such meteorologists free to concentrate upon interpreting its local meaning: for example, as regards where and when ground fog will develop; and local users have benefited greatly from this. In addition, of course, the mathematical process has proved able to give a useful prediction for considerably farther ahead than could be attempted before (roughly speaking, 3 days instead of 1).

FORECAST SURFACE PRESSURE AND PRECIPITATION

24-HOUR FORECAST. DATA TIME=12Z 1/11/75. VERIFICATION TIME=12Z 2/11/75. S-1 METHOD

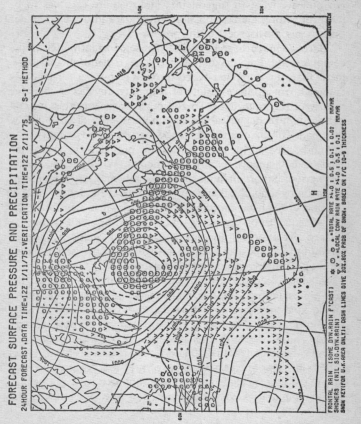

Fig. 1.13. One of the many charts automatically printed by the Meteorological Office's computer as part of its 24-hour forecast for 12.00 Greenwich Mean Time on 2 November 1975. On the basis of all of those many charts, the forecast for South-East England issued by the London Weather Centre at 16.45 GMT on 1 November for use on BBC radio was as follows: 'Here is the weather forecast for the next 24 hours. This evening will be rather cloudy with some more showers for a time, but most places will be dry by midnight with clear periods. A few mist or fog patches will form around dawn, clearing quickly to leave a bright, dry morning. However, it will become mostly cloudy by afternoon with occasional rain spreading to all districts later. Temperatures tonight will fall to about 5° C, rising tomorrow to near 13° C. The wind will become light to moderate, west–north-west, backing to south-westerly and increasing to fresh during tomorrow'

The prediction process has now been developed to the stage when the best chance for improving the quality of the end-product lies in improving the quality of the initial data. We have left until now a description of how predictions *start*: where the weather stations are; how they make measurements; how those are transmitted and received over a special communications network; how they are converted into a suitable initial set of air-and-water data. Partly, we put off such a description until it should be clear what initial data the prediction process needs; but now we want to suggest that almost more care has to be taken in getting the best possible initial data into the computer than in any other aspect of the process.

'Garbage in, garbage out!' is a phrase constantly used in the computer world to make sure everyone remembers one basic fact: that no computer-based process, however ingeniously designed, can produce results of a quality any better than the quality of the input data. To add our emphasis to this, we end our account of numerical weather prediction by discussing how the quality of the input data can be made as good as possible.

Another reason for dealing with this problem *last* is that the kind of mathematics required is different. We suggested earlier that various important kinds of useful mathematics are extensions of 'counting' and of 'adding up'. Now we have seen how ideas of 'adding up' can be enormously extended when a properly designed sequence of steps forward combines the small changes in each to give a forecast for, say, 3 days ahead. Similarly, ideas of 'counting' can be extended to give much more sophisticated ways of getting figures out of observations.

These are needed, for example, to deal with the fact that the grid-points (corners of our mathematically convenient square grid on the map) are not where the observing stations actually are. For example, Fig. 1.14 shows the position of those weather stations within the northern hemisphere which release radiosonde balloons twice daily (at 00.00 and 12.00 Greenwich Mean Time) to obtain upper-air measurements of pressure, temperature and humidity. These include land stations and ocean weather ships. But although it is these stations on whose observations the estimation of all initial data at our upper pressure-levels depends, those initial data are *wanted at the grid-points*!

The whole process of using data at weather stations to get data at other places (such as grid-points) was once treated, essentially, as an 'art'. At any rate, it included a lot of inspired drawing! Contour lines were drawn on the map, by a meteorologist taking into account as best he could all the measurements available to him. From these 'subjectively' drawn contours, values at a particular point would be got by *interpolation* between values on the contours nearest to it.

In this way, contours for pressure at sea-level (as in Fig. 1.2) can be derived rather accurately because of the large amount of data available: about four thousand ground stations make regular observations of ground-level pressure and temperature (from which an equivalent sea-level pressure is easily obtained); in addition, hundreds of merchant ships all over the oceans make pressure measurements at 00.00 and 12.00 GMT daily and feed the results (together with the ship's position) into the main network for exchange of meteorological data. All this information can be used in drawing contours. On the other hand, with upper-air data available at far fewer stations (Fig. 1.14), subjective analysis may involve greater errors, and depend too much on the skill, judgement and dedication of individuals.

Not only for these reasons but also to save precious time in generating the initial data for a forecast, methods of 'objective analysis' have been developed, over two decades, into a fully automatic system for handling the incoming observations. This is now used exclusively to generate the initial air-and-water data. To give readers an idea of the kind of mathematics needed to design such a system, we concentrate on explaining how it deals with the problem illustrated in Fig. 1.14.

This is a problem of horizontal distribution. Thus, we omit any description of how the measurements of pressure and temperature transmitted by each radiosonde balloon as it ascends are converted into useful information on vertical distribution; that would include the application of results like those of Fig. 1.4 to deduce the height h of each of our ten pressure-levels.

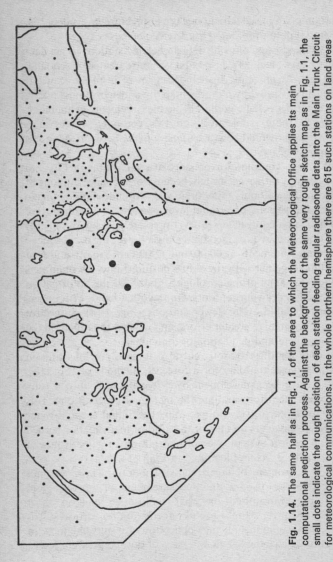

Fig. 1.14. The same half as in Fig. 1.1 of the area to which the Meteorological Office applies its main computational prediction process. Against the background of the same very rough sketch map as in Fig. 1.1, the small dots indicate the rough position of each station feeding regular radiosonde data into the Main Trunk Circuit for meteorological communications. In the whole northern hemisphere there are 615 such stations on land areas (including small islands). The large blobs represent weather ships, of which there are seven in the whole northern hemisphere

We omit any account of how the ground-station 'tracks' the balloon so that winds at those levels can be derived. We concentrate, rather, on describing how the values of h obtained for, say, the 5-decibar level at each station can be 'objectively analysed' to give values at grid-points. (Actually, this separation of the problem into first vertical and then horizontal aspects is not essential: the original method of horizontal analysis, now to be described, has since been extended so as to give a simultaneous objective analysis of the observations in both the horizontal and the vertical.)

To obtain the height h of the 5-decibar level at grid-points, the method uses primarily the values of h obtained at all the weather stations, but supplements them with information of two other kinds. Such additional information is used with the aim of improving accuracy; this must be particularly important in areas where weather stations are scarce.

First, the method uses the values calculated for Dx and Dy (Fig. 1.6) from the speed and direction of the wind observed at the 5-decibar level above each station. Why are those of any use in estimating values of h at grid-points? We gave the answer during our discussion of equations (4): meteorologists know that winds tend to adjust themselves so that the terms on the right-hand side of either of those equations are *nearly* in balance; for example, so that $k(Xh)$ and $f(Dy)$ are *nearly* equal. A known value of Dy at a weather station can accordingly be used to check whether assumed values of h meet this criterion.

Secondly, the method makes use of the values of h obtained at the grid-points by the latest 12-hour forecast (the one starting from initial data fed into the prediction process 12 hours before). Those *predicted* values, as we shall see, are 'given much less weight' than measured values; and rightly so! Nevertheless, in areas where measurements are scarce, they can help to improve accuracy, particularly when the weather has been moving into those areas from others where more observations are available.

We have more than once mentioned the important idea of 'taking an average'. For example, if something has been measured several times and there is no reason to regard any of those measurements as 'better' than the others, what value for the quantity should be used? Statisticians answer: their average.

Among the reasons they give for that answer is one very interesting fact. Any *discrepancy* between a measurement and the value to be chosen may be positive or negative, but its *square* is always positive. *Adding up* the squares of the discrepancies gives us an idea of just how 'badly chosen' the value is, with bad discrepancies properly penalized through being squared. Now, the average value is the 'least bad choice' in that sense: it is the value for which added up squares of discrepancies come out *least*.

On the other hand, when *some* of the measurements are (shall we say) thought likely to be twice as good as the others, they can be given twice the weight. This means that squares of discrepancies from those specially good measurements are doubled before we add up. The chosen value that comes out least bad on that basis of scoring is called the *weighted average*. Fig. 1.15 shows why the word 'weight' is used: the weighted average is represented by that point where you could balance a light rod, carrying weights (that describe how good the measurements were) in positions corresponding to the measured values.

Now we have given the essential idea of the method by which the value of the height *h* at any particular grid-point *P* is chosen. Among a wide range of possibilities (which we list below) we make the least bad choice in the above sense; estimating the badness of the choice by finding the *squares* of discrepancies from it, multiplying them by *weights*, and adding up.

Of course, for choosing the value of *h* at a grid-point *P*, we should attach much higher weight to observations made very close to *P* than to observations made further away. Fig. 1.16 is a graph showing how the weight attached to results from a weather station is allowed to fall off as its distance from *P* increases; Fig. 1.16 shows also the far lower weight attached to values of *h* that are *forecast* at grid-points.

The reader might expect that this method would just give a *weighted average* of the measured values of *h* and of the values forecast. That would be so if the only possibilities tested (to find which was least bad) were possibilities for perfectly *horizontal*

The horizontal positions of the weights represent measured values of a quantity q. When all the measurements are given 'equal weight', the best value to choose is the average (3.305), here represented by the arrow in the upper diagram.
THE LIGHT ROD CARRYING THE EQUAL WEIGHTS.
IN THE POSITIONS SHOWN COULD BE
BALANCED AT THAT POINT.

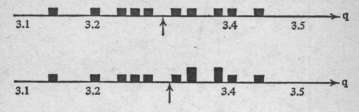

The lower diagram shows the situation when the two measurements that were made under the best experimental conditions are given twice the weight.
THE LIGHT ROD THEN BALANCES AT A DIFFERENT POINT (3.315), REPRESENTING

THE WEIGHTED AVERAGE OF THE MEASUREMENTS.

Actually, this is the *least bad* choice' of value for q if 'badness of choice' is estimated by finding the *squares* of discrepancies from it, multiplying them by the *weights*, and adding up.

Fig. 1.15. Weighted averages

pressure-levels: among such cases of constant h that for which the squares of discrepancies multiplied by weights and added up will be least is the weighted average (Fig. 1.15).

Sloping pressure-levels are far more interesting, however; especially, because of their effect on winds. We described the $k(Xh)$ term in (4) as arising from a slope in pressure-level, and we have just mentioned that it is expected to be nearly in balance with the $f(Dy)$ term. We can use known values of Dy at weather stations to test assumed possibilities regarding the 5-decibar pressure-level, however, *only* if those possibilities allow it to be sloping. For example, the equation

$$h = h_1 + h_2 (x - x_P) + h_3 (y - y_P)$$

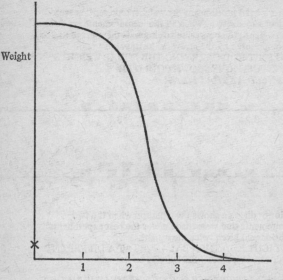

Fig. 1.16. In the process for using observations at weather stations to determine the height *h* of a particular pressure-level at a certain grid-point *P*, the *weight* attached to results from any weather station falls off steeply with an increase in its *distance from P* (that is, its distance on the *map*, where all grid squares have side 1 as in Fig. 1.5). At the same time, 'mere' forecast values of *h* are given an appropriately low weight, represented by the cross

might describe a pressure-level that was *sloping, but flat* (within the region close enough to *P* for the weights in Fig. 1.16 to be significant). Here, h_1 is the quantity we are looking for (value of *h* at *P*), and Xh is easily calculated as $2h_2$. Discrepancies between $k(Xh)$ and $f(Dy)$ at weather stations then depend on the value of h_2 chosen. They can be squared and multiplied by the right weights and added in with the others. So can discrepancies between $-f(Dx)$ and $k(Yh)$, where Yh is $2h_3$.

The result of all the additions is a test of *how bad is the choice* of h_1, h_2 and h_3. Mathematically, it is an easy task to find the values of h_1, h_2 and h_3 which (on that test) make the choice least bad. When that has been accomplished, only the value of h_1 needs to be stored (the required value at P); the process should then be repeated for every other grid-point.

The actual method used is only twice as complicated as that: three more terms

$$h_4 (x - x_P)^2 + h_5 (x - x_P)(y - y_P) + h_6 (y - y_P)^2$$

are included in the equation for h; but the mathematics of finding the least bad choice of h_1, h_2, h_3, h_4, h_5 and h_6 is still easy to carry out on the computer. Once more, it is only the value of h_1 so determined (the value of h at P) which needs to be stored.

As a sample of how initial upper-air data are obtained, we have now outlined the method for h. Actually, the values of Dx and Dy are derived mathematically from h, though the details are complicated. This is because a rather refined matching between initial values of Dx and Dy and initial values of h turns out to be needed. (Without that, equations (4) are found to imply a quite unrealistically large initial growth in the divergence (1), with drastic consequences.) Finally, the initial values of relative humidity are obtained by a method taking into account not only radiosonde measurements but also cloud cover as recorded by weather satellites. By being so careful in all these respects, we avoid putting 'garbage in' – and may, therefore, hope to avoid getting 'garbage out'!

2. Fresh Water and the Sea

RAINFALL AND RUNOFF

We went into substantial detail in Part 1 of this chapter, with the aim of conveying many of the essential ideas of how mathematics is used on large-scale problems like weather prediction. Against that background of substantial detail regarding one such problem, we survey much more briefly in the rest of this chapter the uses of mathematics in many different large-scale problems of the physical environment, showing what a wide range of such problems can be usefully treated by broadly similar mathematical approaches.

We begin with questions that follow on naturally from Part 1. The weather's main 'end-product', which indeed the prediction process is especially geared to calculate, is total precipitation: the amount of water in condensed form falling on the earth's surface (for example, as rain). Precipitation over the sea is admittedly of little practical significance. Over most land areas, however, precipitation is of crucial importance as the only means of replenishing water resources for agriculture, home use, and industry.

This statement is not merely true of visibly stored water resources (as in a reservoir) which may be replenished after heavy rain. Where a well has been drilled, the lower parts of the well *up to a certain level* are full of water which has seeped into it from the 'ground-water' in the surrounding soil or porous rock. This level, below which ground-water fills all available pores in the ground, is called the water-table. Extraction of water from the well, and from other wells in the same area, leads to a progressive lowering of the water-table. Replenishment of the ground-water is ultimately possible only through precipitation on to the earth's surface, followed by *infiltration* of water and percolation down to the water-table.

During a rainless period when no such infiltration is occurring, there are two layers of ground above the water-table which may nevertheless contain significant amounts of water. The force called capillary attraction draws water upwards (through a distance of from a few centimetres in coarse gravels to at most a few metres in certain silts) to fill some of the pores above the water-table. This layer may, or may not, overlap another important layer: a region of fine soil near the surface, found often in association with the root systems of plants, and able to trap water in pore spaces surrounding soil particles.

It is, of course, of great significance to agriculture and forestry that some rainfall can be stored around the root systems of plants in this way; and that the control systems within plants regulate *transpiration* (that is, the seepage of water, carrying some dissolved nutrients from the soil, up the stem, through the leaves and out into the air) in accordance with both the plant's needs and

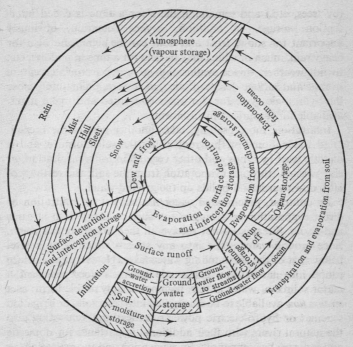

Fig. 1.17. The hydrologic cycle. Read diagram anticlockwise. (Reprinted with kind permission from *Hydrology*, 2nd edition, C. O. Wisler and E. F. Brater, John Wiley & Sons Inc., 1959)

the water's availability. Nevertheless, replenishment of that soil moisture, either by direct rainfall or through irrigation systems bringing in the products of precipitation elsewhere, is essential for healthy plant growth.

The processes by which water resources are depleted, and then replenished as a result of precipitation over land, form part of the *hydrologic cycle*. Fig. 1.17 shows its appropriate representation in circular form, given in Wisler and Brater's excellent book *Hydrology*.

Precipitation from the atmosphere is represented in the upper left-hand part of the circle. The part intercepted above ground (by trees, etc.) and remaining there for a time is called interception storage. In addition, *surface* detention, obviously important for snow, frost and dew, can be significant also for heavy rain, much of which may be held for a time in puddles, etc. In hot weather, direct evaporation of water from interception storage and surface detention may, as it were, prematurely close the hydrologic cycle for the water concerned and make it unavailable for resource replenishment.

Infiltration, more usefully, takes another part of the precipitated water. Some of that replenishes the soil moisture, and is directly used by crops and other vegetation for transpiration, or else 'wasted' through evaporation from the soil; the rest percolates downwards and builds up the ground-water.

By contrast, precipitated water *not* subject to infiltration or direct evaporation is available for surface *runoff* into streams and rivers. Streams do not receive all their water by this route, since ground-water seeps into any stream which is at a level below that of the surrounding water-table. Nevertheless, surface runoff into natural and man-made channels does represent a major source of water in this, its most flexibly usable form: as a *stream flow* available to be channelled into reservoirs or irrigation schemes or hydro-electric power schemes; and generating also the natural rivers with their additional importance for domestic and industrial water supply, for navigation and for recreation.

Over the earth as a whole, the rivers in each year give back to the ocean about 30 000 cubic kilometres of water. Yet the average precipitation over land is about 0·75 metre (30 inches) in each year, representing a total volume of water of about 110 000 cubic kilometres. The difference is 80 000 cubic kilometres, most of which is returned directly to the air through evaporation and transpiration. Thus, the proportion so returned *averages* around three-quarters, although the proportion varies greatly over the earth's surface. In temperate zones, it shows also a marked seasonal variation, since summer *temperatures* promote evaporation, while summer levels of vegetative growth (in response to *sunlight*) promote transpiration. Knowledge on such trends is used by the weather-prediction process for estimating Dr in the bottom slice of air (see p. 58); but it is equally important for the estimation of water-resource replenishment: in Britain, for example, winter rain does far more for water supplies than summer rain can do.

After that brief explanation of the hydrologic cycle (Fig. 1.17), we shall now go on to describe the use of mathematics for relating rainfall and runoff. The rest of this section, then, is concerned with how the flow in some river or stream varies in response to varying amounts of rainfall (in cases where the total precipitation does not contain a significant amount of *snow*).

One important aim of such an analysis is to make possible a system for giving advance warning of flooding dangers. In fact, we concentrate upon this *high-rainfall extreme* of the problem, which is particularly susceptible to analysis in terms of knowledge from physics. By contrast, analysis of the equally important low-rainfall extreme, aimed at giving advance warning of water shortages, depends greatly on the biological properties of transpiration rates, and is omitted from this chapter.

Various means exist for measuring the flow, in cubic metres of water per second, past some instrument placed in a river or stream. (The job is easiest at any *weir*, artificial or natural, where a simple measurement of the height of the water surface above some fixed mark, just upstream of the weir, tells a hydraulic engineer to good accuracy the rate of flow of water over the weir.) If there has been no rain for weeks, then *all* the flow in the stream has seeped into it from the ground-water. Gradually, the stream flow *depletes the ground-water*: each day, the water-table is a little bit lower, and that reduces the rate of seepage into the stream. Instruments recording the stream flow show, therefore, a gradual reduction, known as a *ground-water depletion curve* (Fig. 1.18).

Paradoxically enough, those who want to analyse how a stream flow responds to high-rainfall conditions must begin by thus observing how it behaves in no-rainfall conditions. This gives essential information about the ground-water contribution to the stream flow which, in turn, can be used to estimate surface runoff by subtracting such a ground-water contribution from

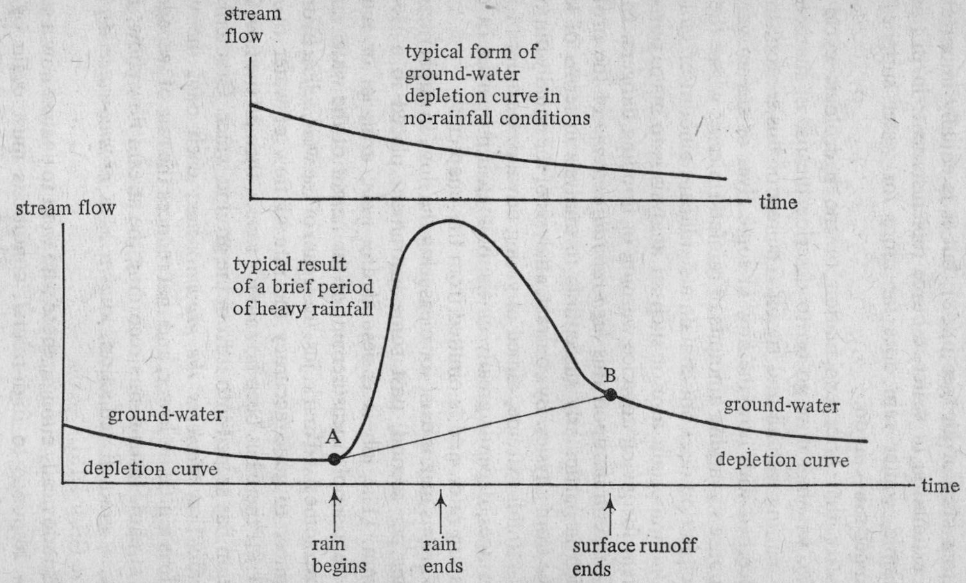
stream
flow
typical form of
ground-water
depletion curve in
no-rainfall conditions
time
stream flow
typical result
of a brief period
of heavy rainfall
ground-water
depletion curve
A
B
ground-water
depletion curve
time
rain
begins
rain
ends
surface runoff
ends

Fig. 1.18. Ground-water depletion curves and their use in analysing surface runoff

the total observed stream flow. For example, the response of the stream flow to a period of heavy rain between two long rainless periods can be seen to consist of a very prominent 'hump' (lying between A and B on Fig. 1.18) joining two ground-water depletion curves. The crudest estimate of surface runoff is the part of the hump that lies above a straight line AB, representing an interpolated ground-water contribution during the intervening period. Mathematics is used, however, to do this sort of thing more accurately, and to solve many other important problems concerned with ground-water movements, by 'adding up' the knowledge available from the physics of seepage through porous soils and rocks.

Surface runoff, then, can be found by *subtraction* of an estimated ground-water contribution from an observed stream flow. Similarly, the part of the rain (called 'effective' rain) which gives the surface runoff can be found by subtraction of an estimated *infiltration rate* from the total rate of rainfall (in millimetres per hour) determined by rain gauges. Here, infiltration is emphasized as the main cause of loss of rainwater from the surface, because evaporation and transpiration are usually small in conditions of heavy rain; indeed, as a consequence of this, hydrologists usually lump together all the mechanisms withholding rainwater from the surface (including interception storage) into a single *so-called* infiltration curve.

The best way to determine an average infiltration rate (in this sense) during a period of heavy rain is by 'trial and error'. We guess a value and assume that any excess rate of rainfall over and above that value, in millimetres per hour falling on whatever area contributes to the stream flow, must become surface runoff. Adding it up for all the hours of rain gives a total. If this is less than the total surface runoff determined from the stream flow, then a lower average infiltration rate must be tried; if greater, then higher; and so on until the two methods of determining total surface runoff agree. Applying this kind of procedure, incidentally, leads to information on the seasonal dependence of infiltration rate: in summer the water being warmer (and so less *viscous*) seeps more readily through the ground, and the denser vegetation promotes soil porosity, so that infiltration is considerably increased.

For large river basins, such a seasonal average infiltration rate may be all that can usably be found out. For smaller basins, it is possible by various means (some of which use mathematical analysis) to determine an *infiltration curve* (Fig. 1.19); this shows how infiltration rate falls from a maximum rate after a period of no rain to a steady minimum rate after a period of 1 to 3 hours (when all soil moisture has become replenished); note that some of the change may result from changing porosity, as when a soil contains clay particles that become swollen after wetting. The behaviour illustrated in Fig. 1.19 has an important consequence for flood prediction: exceptionally intense rain over a short period is more likely to create flood dangers if it happens when the soil is already wet.

At last we can state the basic mathematical problem of rainfall and runoff (the title of this section). Given the *effective rainfall* (rainfall minus infiltration), we want to predict the *surface runoff* (stream flow minus ground-water contribution). By definition, the *total* amount of water in both is the same; the important question, however, is how the passage of that water is spread out in time. In fact, observations (or predictions) of the rainfall in

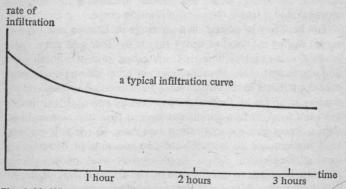

Fig. 1.19. Where the soil is dry, the rate of infiltration is usually greater during, say, the first hour of rainfall than it is later

each of a *sequence of equal time-intervals* of a few minutes should be used to determine, first, the corresponding *effective* rainfall; next, the resulting surface runoff in each of the sequence of equal time-intervals; and, then, the corresponding stream flow. This last quantity is what determines whether flooding may occur.

The question of how best to make the middle step, from effective rainfall in each time-interval to surface runoff in each time-interval, is a mathematically puzzling one. In this section we mention the best-known method, which gives fairly good but not outstandingly good accuracy.

This method has the virtue of being able to take very full account of previous observations, but the defect of ignoring one quite important physical fact: that rainwater makes its way down to the stream (more strictly, to the point where stream flow is being measured) rather faster when there is already a lot of water on the surface. This method, in other words, assumes that each and every *unit* of effective rain (say, a millimetre, which however must be multiplied by the area of the basin to give the total volume of water involved) falling in each and every time-interval produces the *same surface runoff response* (division of that total volume of water among subsequent time-intervals). Fig. 1.20 illustrates this; if the fraction of that unit of effective rain joining the stream flow in the actual time-interval in which it fell is u_0 (usually a *very* small fraction) and the fractions joining in the first, second, third and fourth time-intervals thereafter are u_1, u_2, u_3 and u_4, and so on, this distribution of that unit of effective rain is called the *unit hydrograph* (Fig. 1.20). Adding up all the u's, of course, gives 1.

Suppose that the number of units of effective rainfall in the first time-interval after it started to rain was e_1 (not usually a *whole* number, of course). Suppose also that the numbers in the second time-interval, the third and fourth and so on were e_2, e_3, e_4 and so on. Similarly, suppose that the numbers of units of surface runoff in these intervals were j_1, j_2, j_3, j_4 and so on. Then, j_4 for example (surface runoff in the fourth interval) would consist of four parts: a fraction $u_0 e_4$ of the e_4 units falling in the very same (fourth) interval; a fraction u_1e_3 of the e_3 units falling in the previous interval; a fraction u_2e_2 of the e_2 units falling in the interval before that, and a fraction u_3e_1 of the e_1 units falling 3 intervals back (in the first interval). In symbols, then,

$$j_4 = u_0e_4 + u_1e_3 + u_2e_2 + u_3e_1. \qquad (8)$$

In some particular rainstorm, for example, 0·4, 0·8, 1·0 and 1·1 millimetres of rain might fall in the first four time-intervals (each of 10 minutes). The data of Fig. 1.20 then suggest a surface runoff in the fourth time-interval equal to 1% of 1·1 mm plus 5% of 1·0mm plus 11% of 0·8mm plus 15% of 0·4mm, adding up to 0·21mm. In a basin of area 1000 square kilometres this would mean a flow of 210 000 cubic metres (that is, tonnes) of water in 10 minutes, or 350 cubic metres per second.

It is easy to show similarly that, if n stands for any whole number,

$$j_n = u_0e_n + u_1e_{n-1} + u_2e_{n-2} + u_3e_{n-3} + \ldots + u_{n-1}e_1. \qquad (9)$$

In each term on the right-hand side of (9), the two little numbers

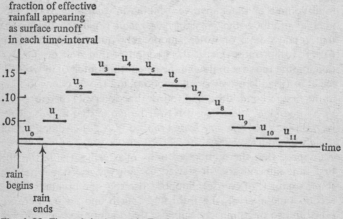

fraction of effective
rainfall appearing
as surface runoff
in each time-interval

Fig. 1.20. The unit hydrograph. Twelve successive time-intervals are shown; rain fell during only the first of these. The fractions $u_0, u_1, u_2, \ldots u_{10}, u_{11}$ add up to 1; they tell us what fraction of the *effective* rainfall appeared as surface runoff in each of the twelve time-intervals

below the line add up to n; this is because, for example, a proportion u_3 of the effective rainfall e_{n-3} in interval $n - 3$ is found 3 intervals later as surface runoff in interval n.

A large set of equations like (9) which ought to be satisfied for each n by the fractions u_0, u_1, u_2, u_3 and so on lends itself admirably to the determination of those fractions from observations. Just as in the last section, we find the least bad choice for the u's, with badness of choice estimated by adding up the squares of the discrepancies between the two sides of equation (9) for the different available values of n. Once that has been done for a particular basin, the u's can be regarded as known for that basin and equation (9) is available for prediction purposes.

In particular, extreme flooding possibilities may be estimated. For this purpose, past data on extremely heavy rainstorms are often extrapolated, by means of the statistical theory of 'extreme values', to suggest a 'once-per-century worst case', from which the 'worst' (that is, maximum) possible stream flows are then derived.

There is scope, however, for mathematicians to develop an improved way of predicting the surface runoff from the effective rainfall. The need is for a method retaining the good features of that described above, yet using all available data on how the time taken for water to reach the point where the stream flow is being measured may show a tendency to decrease as the volume of water making that journey increases.

CLEAN WATER

In the last section we sketched processes by which rain finds its way into rivers, and briefly indicated how mathematics can be used to predict, on the basis of past observations, how the river flow may respond to any future pattern of rainfall. This information is important to those responsible for *river management*, aimed at minimizing flooding dangers, at utilizing the stream flow to the full, and at apportioning its use between many competing interests: domestic, industrial, agricultural, recreational, navigational, etc.

River management also uses mathematics for analysing several important processes that take place as water moves down the river to its estuary, and then out into the open sea. In this section, we describe *one* of these additional uses of mathematics for purposes of river management.

We do not, however, select the one that might appear to follow our last subject most naturally. That would be the widespread and successful use of mathematics for 'flood routing': the prediction of just how a peak stream flow will pass *down* the river, both under natural conditions and under various alternative conditions involving human intervention (through provision of reservoirs, operation of sluice-gates, etc.). We omit flood routing because it uses mathematics very much as we used it in the first part of this chapter: it divides the river into many equal short stretches and divides time into many equal short time-intervals; then it edges the forecast forwards by applying to the water in each short stretch of river knowledge from physics similar to that used earlier (acceleration due to gravity, retardation due to friction, etc.). Broadly speaking, the world's longest rivers need the most refined analysis; much of the progess in improvement of techniques was achieved in connection with the Mississippi and its tributaries.

Yet another important aim of river management, in many cases, is to reduce pollution. We choose in this section to describe how mathematics is used for this end (that is, in the struggle for *clean water*), partly because there is a specially remarkable success story to tell here, and partly because the processes being calculated are different in kind from those discussed in earlier sections.

Actually, there is a vast number of uses of mathematics in pollution control besides the one we shall describe. Not only clean water but also clean air is of great importance to human well-being; furthermore, the ways in which smoke from tall chimneys is carried away from centres of population by winds can be calculated mathematically by techniques generally similar to those described earlier in the chapter; the results of such calculations are used with statistical data on meteorological conditions to design chimneys for factories and power stations.

Here, however, we are concerned only with water pollution, and we concentrate, indeed, on the problems of pollution in an *estuary*; that is, in the part of a river that responds to the rise and fall of the tide. Around Great Britain, for example, the sea exhibits large tidal movements (to be discussed in more detail in the next section): on most exposed coasts the average tidal range (vertical distance between the high-water and low-water levels) is already 2 to 5 metres, but in the numerous estuaries (which generally give an exaggerated response to tidal rise and fall) the average tidal ranges are almost twice as much. The sea flows fast into such an estuary as the tide rises, and ebbs out of it fast as the tide falls.

Actually, much of the population, and of the industry, of Britain is situated near estuaries; the tidal movement in and out of them is therefore a great national asset, which may allow waste

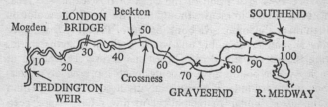

Fig. 1.21. The Thames estuary. Figures denote distance in kilometres seaward from Teddington Weir. There are sewage works at Beckton, Crossness and Mogden. (Reprinted with kind permission from *Advances in Water Pollution Research*, ed. S. H. Jenkins, Pergamon, 1973)

products from homes and factories to be carried away and dispersed effectively among the vast volume (over a thousand million cubic *kilometres*) of the oceans. The more important an asset may be, however, the more necessary it is to *quantify* it: for example, to calculate the amounts of waste products which an estuary can absorb without the water becoming unacceptably polluted.

The Thames estuary (Fig. 1.21) exhibits these problems to a quite exceptional extent: most of Greater London's enormous population (of the order of ten millions) lives close to the narrow upstream half of the estuary's considerable (100-kilometre) length. Domestic sewage is by far the most serious pollutant. The tidal ebb and flow moves water on the average only 14 kilometres down the estuary and then about the same distance back again in one tidal period ($12\frac{1}{2}$ hours).

The remarkable success story we shall tell was achieved by an outstandingly able and well coordinated team of water-pollution research workers during the period 1948–63 before pollution had begun to receive widespread public attention. They showed how a mathematical method of predicting the effect of pollutant discharges into the Thames estuary was an essential tool for estuary management; that is, for determining the measures needed to restore clean water (in a sense which we shall make precise). Their predictions were borne out to rather good accuracy.

'The Thames was dead, but is alive once more' has been true ever since 1964 as a result of the work of this team. The local benefits have been immense. In addition, the work, meticulously written up as the impressive book *Effects of Polluting Discharges on the Thames Estuary*, became widely used all over the world, during the subsequent decade when major exercises aimed at reducing water pollution were being much more widely undertaken, as an example of how to go about such work. (Within Britain, also, similar mathematical methods have since been applied to the management problems in such important estuaries as the Severn, Mersey, Tay, Tees, Humber and Solent.)

Actually, the Thames exercise represented the second time London's river had been cleaned up. For centuries, London's sewage ran directly into the Thames, and, as the volume increased, the resulting stink became more and more offensive. Then the Metropolis Act of 1855 instituted a main sewerage system, discharging the city's sewage into the estuary at two positions (Fig. 1.21) around its mid-point (20 kilometres to the *seaward* side of London Bridge). By 1865 the work was completed, and for some sixty years the estuary gave no cause for complaint. Then, as population growth brought about further massive increases in the volume of sewage discharged, there was more and more concern about the offensively smelly conditions in the estuary. The Thames Survey Committee was set up in 1948, and supervised a fifteen-year exercise centred on the Water Pollution Research Laboratory, where A. L. H. Gameson and W. S. Preddy were among the leading workers in a large and highly effective team admirably coordinated by the then Director, Dr B. A. Southgate.

Oxygen, the element essential to life in the air we breathe, is equally essential to the life of a body of water. The sea, for example, besides containing dissolved salt, contains also dissolved oxygen. A fish in the sea absorbs from the water through its gills the dissolved oxygen that it needs. Water without dissolved oxygen is 'dead' in the sense that it can support no fish or other animal life, *and also* in the sense that it begins to smell offensively. The famous 'rotten egg' smell of hydrogen sulphide, for example, is produced by those abundant bacteria which when oxygen is not directly available obtain it by reducing sulphates to sulphides. It is under such *anaerobic* conditions (without dissolved oxygen) that this class of micro-organisms the bacteria, which in general helps to purify the sewage, changes rôles and begins to putrefy it.

Throughout the 1950s anaerobic conditions were extremely prevalent in the Thames estuary, filling a stretch of 20 to 30 kilometres in the summer months (for example). No fish could be found in the upper 70 kilometres of the estuary, the only exception being an *occasional* eel. After 1964, however, anaerobic conditions were eliminated, and by 1970 fish of more than fifty species were living in all parts of the estuary.

To achieve these results, the essential need was to obtain a reliable mathematical method of calculating the dissolved-oxygen content all along the estuary under any conditions that might occur. Then and only then could the conditions required to bring the Thames to life again (that is, to a permanently aerobic condition) be planned for.

Mathematics (as we emphasized earlier) is an outward-looking body of knowledge: it continually collaborates with observational science and laboratory science. Fortunately, the observational data on the Thames make it one of the best-known and best-documented estuaries in the world; in addition, laboratory research of high quality on the effects acting to increase or decrease dissolved-oxygen content formed an essential part of the programme. The mathematical prediction process was built up taking the fullest account of the conclusions of all this work.

The upper Thames discharges its fresh water over Teddington Weir into the tidal estuary (Fig. 1.21). The volume flow rate (in cubic metres per second) is exceedingly variable: the 50-year average for 1921–70 was 71, but averages over individual years ranged from 20 to 123 and averages over individual *months* from 2 to 360. Indeed, this rate of fresh-water inflow from the upper Thames is among the most important *variable inputs* which the prediction process must take into account. The time spent by pollutants in the estuary is increased if rates of inflow are low; then, for example, water now at London Bridge, although ebbing and flowing an average distance of 14 kilometres every $12\frac{1}{2}$ hours as already noted, may not get clear of the estuary for 3 months, as against 6 weeks with a high rate of inflow. Temperature is *another* important variable input, increasing the rates at which pollutants bring about consumption of dissolved oxygen. Nevertheless, the biggest single reason why the threat of anaerobic conditions is greatest in summer is because the inflow from the upper Thames is then least.

A third variable input (the most significant of all, of course) is the pollutant discharge. Its geographical distribution is important; this comprises inflows from several small tributaries of the Thames that introduce sewage into it, as well as the main inputs from the sewage outfall works at Beckton and Crossness and a smaller pollutant input from works at Mogden; but, above all, its *composition* is important. The mathematical prediction process was unable to obtain good results until it made the vital distinction between the short-term and the long-term oxygen demands of sewage.

Briefly, this means distinguishing between:

(*i*) the amount of *oxidizable organic matter* (that is, matter containing carbon atoms), as measured by the oxygen consumed during its oxidation (typically in a few days) by bacterial activity;

(*ii*) the pollutant's *oxidizable-nitrogen content*; this determines how much ammonia is generated in the above oxidation, after which certain *much* slower processes bring about the oxidation of that ammonia.

Evidently, dissolved oxygen becomes depleted in two stages (one fast, one slow) by the activity under headings (*i*) and (*ii*).

The prediction process uses also one constant piece of information: the way in which the *cross-sectional area of water* varies with position along the estuary in average tidal conditions; that is, at half-tide. This cross-sectional area increases with distance from Teddington Weir more and more steeply, rising to about 10 000 square metres at a distance of 80 kilometres.

The prediction process divides the estuary into many short stretches, each of equal length, and of known volume (length times average cross-sectional area) at half-tide. It then considers what *changes in the composition* of the water in each stretch will occur in a whole tidal period (from one half-tide to the next half-tide but one). Here, the word 'composition' comprises not only dissolved-oxygen content but also dissolved-salt content, and temperature, as well as the characteristics of the pollutants defined under (*i*) and (*ii*) above *and* the amounts of certain important products of their decomposition (ammonia and the oxides of nitrogen).

Four physical effects and four chemical effects are taken into account in estimating changes in the composition of the water in a stretch during a tidal period. The physical effects are as follows.

In each stretch the 'push' during one period from the total volume of all inflows to the estuary situated upstream of it displaces an identical volume of water (and its contents) into the next stretch down river. Secondly (and simultaneously) the tidal movements during one period bring about a certain *mixing between the contents* of any two adjacent stretches. It proves easy to check that this has been well estimated; particularly, because the distribution of dissolved *salt* along the estuary is subject only to the two physical effects just mentioned, and to none of the chemical effects, and so is readily calculated and compared with observations under various inflow conditions. (To readers familiar with, say, Norwegian fjords we should explain here that in the relatively shallow Thames estuary the salt content shows no significant variation with *depth*.)

A third physical effect is the input of each constituent at each of the inflow points. A fourth, highly important one, for the accurate estimation of which much work needed to be done, is the input of dissolved oxygen from the air. This takes place at a rate proportional to the *difference* by which dissolved-oxygen content falls short of its maximum value; that is, its value 'in equilibrium with air', also known as a *saturation* value, and dependent on dissolved-salt content as well as on temperature. Finally, knowledge of the constant of proportionality was obtained by direct experiment.

Two of the chemical effects, (the ones noted in (*i*) and (*ii*) above) are represented mathematically as follows:

(*i*) a definite fraction of the oxidizable organic matter in each stretch is removed during one period (this fraction depends only on temperature) and a corresponding reduction in dissolved oxygen occurs;

(*ii*) a similar fraction of the water's oxidizable-nitrogen content is removed, with a corresponding increase in ammonia content; simultaneously, a decrease in ammonia equal to a much smaller fraction of the ammonia content occurs (this represents its *slow* oxidation, and goes again with a corresponding reduction in dissolved oxygen).

We do not give details of the remaining chemical effects, relating to sulphates and to the oxides of nitrogen, but we want to mention that one of them turned out to have interesting practical implications. Certain abundant bacteria make a transition, from absorbing dissolved oxygen directly to breaking up the nitrate ion (NO_3^-) into oxygen and nitrogen, when the level of dissolved-oxygen content drops to around 5% of saturation! The effects of *their* work on the mathematically predicted distributions of dissolved oxygen were found to be very clearly identifiable, and in rather good agreement with observations. They mean, effectively, that any management procedures derived from the calculations are 'robust'; that is, insensitive to small errors in input assumptions, because the river is 'cushioned' against a bump down to anaerobic conditions by the activities of these nitrate-reducing bacteria.

We have given brief verbal indications of the equations by which the composition of the water in each stretch after one tidal period can be written down in terms of the inflows and the previous composition. Those equations might have been used in various ways, but it was found that the most valuable was also the most simple. It assumed that the variable inputs may be represented by constant, 'average' values during a long enough period for the distribution of dissolved-oxygen content along the river to settle into a steady state. Evidently, this steady state is determined by *solving a large set of simultaneous equations* (that equate the composition in each stretch after one tidal period to the previous composition); a job which computers are particularly good at!

The most significant end-product of such calculations is an 'oxygen sag curve'; that is, the distribution of dissolved oxygen along the estuary, which is given that name because it 'sags' to a minimum value somewhere in the area where pollutants are making their biggest demands on oxygen (Fig. 1.22). The prediction process, when it had been refined, gave remarkable agreement with observed sag curves under a wide variety of average input conditions.

Above all, that process made it possible to look into the future and forecast the changes in pollutant inflow that would be required to make the estuary aerobic. It predicted, correctly, that the additional treatment of the polluting effluent from Beckton, which was already planned for 1955, would not reduce its oxygen demand enough to improve the estuary significantly, but that another change, actually carried out in 1960, *would* make a big improvement; and that a further provision of additional treatment at Crossness, actually carried out in 1963, would remove anaerobic conditions altogether. In crude terms, the 'effective oxygen demand' (including the short-term and long-term demands distinguished under (*i*) and (*ii*) above) was halved in

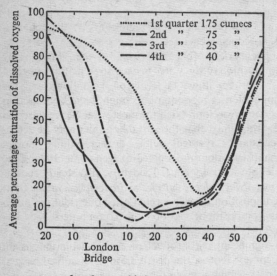

Fig. 1.22. Oxygen sag curves around 1970 (here, 'cumecs', standing for cubic metres per second, are the units used for the average flow of fresh water over Teddington Weir). (Reprinted with kind permission from 'Water Pollution Research', Technical Paper No. 13, HMSO, 1972)

this period; and, as predicted, the tidal estuary proved able to cope with what was left, while remaining aerobic. That is the story of how 'the Thames was dead, and is alive once more'.

SAFE COASTS

Fresh water and the sea, the subjects of this part of the chapter, are (as we have seen) mixed together within an estuary. For a balanced account of 'Mathematics and the Physical Environment', however, the sea proper and a few of the uses of mathematics to deal with problems posed by it must be mentioned.

The oceans and seas, which cover 71% of the surface of our planet, are important to us in many different ways. They are a fount of natural resources: both mineral resources (including oil and natural gas) and living resources (fish and shellfish). Mathematics is extensively used in assessing the resources that can be won from the sea (see pp. 94 ff. for mineral resources, and Chapter 2 for living resources).

In the meantime, world trade has been steadily growing, and is mainly oceanborne. Although both passengers and those cargoes which are of high value in relation to their weight have transferred to the air, the oceans continue (for sound economic reasons) to be the best general medium for transport of goods. To this end, several new types of vessel have been introduced (fast cargo ships, giant bulk carriers, supertankers, etc.) and, although in this book we leave out engineering questions involved in their design, we shall mention some of their environmental problems.

Next, the sea is important to us as a neighbour; sometimes friendly, sometimes threatening. We want to preserve coastal amenities for recreational use. We need also to defend our coasts against threats from the sea's possible encroachment on the land. For both these purposes, mathematics is widely used. Here, we omit any further accounts of its use in pollution studies, although they are important on coasts as well as in estuaries for safeguarding amenities and inshore stocks of fish and shellfish. We concentrate rather on the problems of *keeping our coasts safe* against the threats posed by 'big seas' in the form of exceptional waves and exceptional tides.

Exceptional waves, of course, threaten not only coasts but also shipping and, above all, those fixed structures that are used to extract oil and gas from the seabed. The mathematics of ocean waves is of long standing; it has been much used for making precise the environmental threat in each of these cases. The engineering design of oil production platforms, for instance, is centred around such a mathematical assessment of the maximum wave loads on the structure.

Ships, equally, are designed to withstand exceptionally heavy seas. However, the economics of shipping requires that time lost through encountering storms should, wherever possible, be avoided by careful routing. Fortunately, some well-developed aspects of the mathematics of ocean waves have allowed relationships between regions of heavy seas and regions of strong winds to be made rather precise, enabling the type of weather-forecasting services described earlier to lead to clear ship-routing recommendations, which are much valued by shipping companies.

Coastal structures, again, need careful design to resist the heavy loads imposed on them by exceptional waves. Nevertheless, the biggest threat of all to many coastlines is from the general rise and fall of water-level comprised in the word 'tide'.

To be sure, the world's coasts include some with very regular tides: each year, the highest tide is at about the same level, and no special problem of predicting the maximum threat to coasts arises. By contrast, the maximum tidal levels in a shallow, partly enclosed sea like the North Sea are exceedingly sensitive (in a way we shall discuss) to meteorological conditions and pose flooding threats of a not easily predictable character to any surrounding low-lying country, such as the Netherlands and eastern England. In India, similarly, the strong winds of the south-west Monsoon blowing across the shallow northern area of the Bay of Bengal pose very severe flooding threats around the Ganges delta.

The rest of this section is concerned with illustrating tidal mathematics by describing prediction processes for the North Sea developed in the Netherlands and in England. The title of this section, 'Safe coasts', is chosen to represent the primary aim of this work, with its emphasis on predicting any *exceptionally high* water levels. We may briefly note, however, a significant secondary aim: the method is able also to predict exceptionally low water levels; for example, in shipping lanes where modern deep-draught vessels like supertankers and giant bulk carriers may be threatened with grounding by abnormally low-water conditions.

First, we describe the main tidal movements of the *deep oceans*. These, being practically unaffected by winds, produce on closely adjacent coasts the very regular tides already referred to. Appreciating the main tidal movements in the great mass of the deep Atlantic Ocean is essential if the more complicated rises and falls in the North Sea, driven primarily by influences from the Atlantic moving around the British Isles, and only secondarily by wind action, are to be understood.

Tides in the ocean are *bulk movements* relative to the solid earth, in which all the water from the surface to the bottom takes part almost equally. They are recognized as resulting, principally, from an *excess of attraction* by the moon on water nearer to it than the earth's centre, and a *defect of attraction* on water farther than the earth's centre. Both of these may be described as *tide-raising forces*. The locations of the maximum tide-raising force move, as the earth rotates and the moon pursues its orbit, passing each meridian of longitude every $12\frac{1}{2}$ hours.

The *sun* exerts similar smaller forces, which act nearly in line with them once a fortnight, at full moon or new moon, soon after which we observe a maximum motion called the *spring tide*. Around September and March the forces of the sun and moon come perfectly into line, and accordingly the equinoctial spring tides are the highest of all.

Isaac Newton in the seventeenth century had recognized these facts about tide-raising forces, but it was the great eighteenth-century mathematician P. S. Laplace who first obtained equations (described below) governing the resulting oscillations of water-level. One very significant implication of his equations was that shallow seas (like those bordering the British Isles) respond with a certain delay to the tidal rise and fall in an adjoining deep ocean, the influence of which can move into such a shallow sea only at quite a limited speed of travel (under 100 kilometres per hour).

Evidently, such influences from the Atlantic travel into the North Sea from two directions. The main influence, however, moves into it southward from around the North of Scotland. The other, travelling eastward up the English Channel, affects the North Sea less because of the narrowness of the Straits of Dover.

A prediction process for the tidal rise and fall in the North Sea has to take into account wind forces as well as those other effects (of gravity and earth's rotation, as in Fig. 1.8) allowed for already by Laplace. As we said earlier, tidal movements involve all the water from the surface to the bottom. This, in a deep ocean, comprises *far too much water* for its tidal movement to be modified significantly by the wind forces at the surface. Those forces, however, can modify significantly the movements of the far smaller amount of water in a shallow sea.

Any such prediction process, like that involved in weather forecasting, divides time into a large number of equal time-intervals. It also divides a map of the North Sea into a grid of equal squares which approximately fits the area concerned. The northern boundary of the grid is in water deep enough for the tidal movement across the boundary to be practically uninfluenced by wind and therefore to take the regular, predictable form imposed by the Atlantic tides. Until certain very recent developments (see Fig. 1.23 below) it was further assumed that the flow across all other boundaries is insignificant, especially because trials of alternative assumptions allowing for realistic variations of flow rate through the Straits of Dover showed that their effects were only rather local.

The principal difference from weather forecasting is that each grid square consists of one 'slice' only (instead of ten). This is because, to close approximation, all the water from the surface to the bottom takes part equally in tidal movements. The 'thickness' of the slice can be written

$$s = b + h. \tag{10}$$

Here, h represents the height of the tidal rise above an average sea-level (so that at low tide h would be negative) and b is the *depth of the bottom* below that average sea-level; in practice, some sort of locally averaged depth is used (averaged over a double grid square, for example).

Just as in weather forecasting, the *divergence* (as defined in equation (1)) plays an important rôle: for the sea, it signifies the proportional reduction in slice thickness s for the water filling the double grid square during the time-interval Dt. Thus the equation

$$\frac{Ds}{s} = - \text{(divergence)} \tag{11}$$

replaces equation (2). Next, equation (3) contains one less term than before (in a one-slice system there is no term in Pq representing differences between slices above and below). On the other hand, equations (4) are each supplemented by additional terms representing both the lunar and solar tide-raising forces and the wind forces. Allowance is also made, as before (see p. 41), for the frictional resistance to movement (which is offered, in the tidal case, by the sea bottom).

All the information needed for 'stepping forward' is then present in equations (1), (10) and (11) and the modified forms of equations (3) and (4). At the boundary of the grid, however, a different procedure for stepping forward is needed; in fact, the assumptions made there concerning tidal movement across the boundary are as stated earlier. Beyond the problem of 'stepping forward', there is once more the problem of 'keeping straight'; and it again proved essential to use 'numerically stable' procedures for 'adding up' changes in successive small time-intervals.

Results of applying such prediction processes have made very clear the conditions under which water levels substantially higher than normal spring-tide levels appear. When, as on 31 January 1953 or 2 January 1976, a deep cyclonic depression producing gale-force winds moves southward across the North Sea at a time coincident with the normal high spring tide, serious threats to coastal defences result.

In 1953 the consequent floods both in the Netherlands and eastern England were of appalling severity. Later, the defences were greatly strengthened, leading to only a moderate amount of flooding from the similar storm in 1976.

The Thames estuary is itself very vulnerable to such exceptionally high water levels. Now, a Thames flood barrier is being built at Silvertown, a short distance up river from Beckton

(Fig. 1.21), so as to protect London. The barrier includes, besides some smaller gates, four huge 'rising sector' gates of 61 metres span. Each gate forms a segment of a circle which, when the gate is open, lies horizontally on the bottom in a recessed concrete sill, with its flat face uppermost. In order to close the waterway, the gate is rotated through 90°, until its flat face is vertical and faces seaward. Mathematical prediction processes such as were sketched in this section are to play an important part in determining when the barrier must be used.

At the same time, Britain's Institute of Oceanographic Sciences has been developing an extended computational prediction process for the tides in the whole shallow-water area around the British Isles (see Fig. 1.23). Besides increasing the accuracy of predictions in the North Sea itself, by not neglecting the flow through the Straits of Dover, this will help greatly in the defence of other threatened coasts, especially in the Severn estuary and in the English Channel.

3. The Earth and its Resources

EXPLORING

For centuries, people have been exploring the earth, looking for places to live and places to grow food; looking for sources of raw materials (such as metals) and of fuels; looking for good transport routes and terminals (such as harbours and, later, airports); looking for places to enjoy the beauties of nature. The exploring continues today – and is complicated by one widely recognized need: to plan for *reducing conflict* between all the different uses of the earth and its resources. We end this chapter with a brief outline of the contribution of mathematics to these activities.

For all of them, a prime requirement is accurate mapping: first of the earth's surface, and then of what lies under the surface. Modern land-use planners want maps with *errors under* 0·02%; meaning that points and lines on a map one metre square should be placed to within an accuracy of one-fifth of a millimetre (corresponding, in maps of the largest scale 1:10 000, to 2 metres on the ground).

To get that sort of accuracy in all your local maps, you need a basic *framework* of points, distributed across the country or

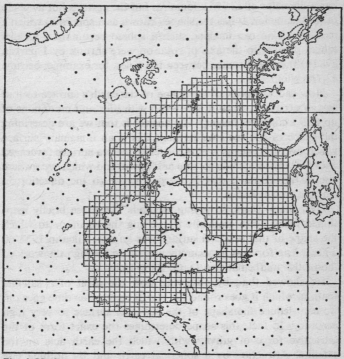

Fig. 1.23. Square grid for the shallow-water area around the British Isles, used for the tidal prediction process described by R. A. Flather and A. M. Davies in the Institute of Oceanographic Sciences Report No. 16 (1975). The broken line encloses sea areas with depths below 100 fathoms (183 metres). Because this map uses a different projection of the earth's surface from that of Fig. 1.1, the dots, representing grid-points on the Meteorological Office's 'fine mesh', do not here form an exactly square grid. This does not, however, prevent them from being used as the locations where inputs of wind force for the tidal prediction process are specified

continent being mapped, whose relative positions are known to the even greater accuracy of 0·002%. That involves getting a very accurate picture of the external shape of the earth's surface. This has long been known to be *flattened at the poles*: the equator has a diameter of 12 756 kilometres but the earth is not a sphere of that diameter. It has the shape (known as a spheroid) which a mathematician can imagine such a sphere becoming if all distances parallel to the axis of rotation were shrunk by 1 part in 298 (so that the distance between the poles, for example, became 12 714 kilometres).

How can we speak of the shape of the earth's surface to that sort of accuracy, when it is covered with hills and mountains of up to 9 kilometres height? The answer is that we are describing the shape of the *sea-level surface*: this is an imaginary surface which over the oceans is at the level of the sea surface (averaged over the year as the tide goes up and down) and which everywhere else is 'horizontal'; that is, perpendicular to the direction in which a weight hangs.

Over many years, surveying methods like those briefly mentioned below were gradually improving the accuracy to which the shape of that sea-level surface was known. Then in 1957–60 a great leap forward in accuracy was made through the work of the British mathematician D. G. King-Hele.

He showed that, by observing how the orbit of a satellite (path in which it goes round the earth) changes as time goes on, we can deduce the shape of the sea-level surface to very high accuracy. In fact, this shape determines the exact form of the attractive force of gravity with which the earth acts on the satellite. Attraction by a spherical earth and by nothing else would give an unchanging orbit (an ellipse with the earth's centre as focus), as Isaac Newton first showed. King-Hele proved that observed changes in the orbit can be completely disentangled into the changes due to departures of the sea-level surface from an exact sphere, and those due to other causes (like the attractions of the sun and moon). The observed changes could therefore be used to find very precisely the shape of the sea-level surface.

The long-term result of using those methods is that the spheroid of shape closest to that of the sea-level surface is now known to have equatorial diameter 12 756·28 kilometres and a flattening by 1 part in 298·25, so that the polar diameter is 12 713·51 kilometres. Fig. 1.24 shows the extent to which the sea-level surface departs from that spheroid (nowhere by as much as 120 metres; that is, 0·12 kilometres).

The accurate framework used by map-makers is called a 'geodetic' (earth-dividing) framework. The aim is that it should specify to the nearest 0·00001° the latitude and longitude of a network of well-defined points, spaced not more than about 100 kilometres apart.

A good mathematical method has been invented for improving the accuracy of geodetic frameworks whenever new observations are made. Such observations consist of (*i*) measurements of *distances* between points of the framework (in terms of the time of travel of a *laser pulse* or some other form of electromagnetic radiation between them, proper account being taken of the refractive index of the air); and (*ii*) measurements of *angles* between an arm of the framework and *either* another *arm* or the north as accurately known from astronomy.

The method is like that described on pp. 65 ff. Given any choice of latitudes and longitudes for points of the framework, every distance (*i*) or angle (*ii*) can be computed from the mathematics of spheroids. Then the squares of their discrepancies from the observed values can be multiplied by weights and added up to estimate the badness of the choice of latitudes and longitudes. (For the *weight to be attached to each measurement*, statisticians recommend a value proportional to the inverse square of its estimated 'standard error'.) Fortunately, the mathematics of going from a not-so-good set of latitudes and longitudes to the set which is 'least bad' (as so estimated) only involves solving a large set of simultaneous (linear) equations: a job that computers (as we remarked earlier) are particularly good at.

It is essential that some points of a geodetic framework are stations facing the *open sea* (not in estuaries or narrow straits) where average readings of tide gauges over, say, a year will give a good determination of sea-level. Then the heights of other points in the framework above sea-level can be found as follows.

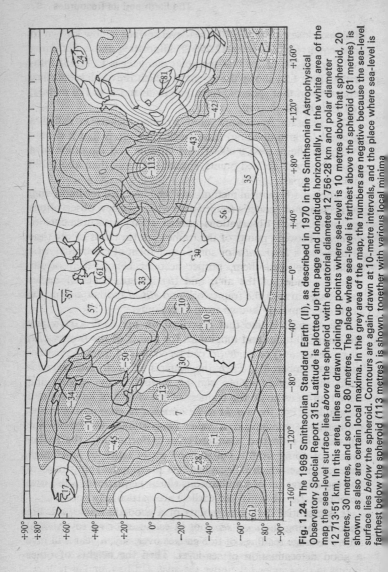

Fig. 1.24. The 1969 Smithsonian Standard Earth (II), as described in 1970 in the Smithsonian Astrophysical Observatory Special Report 315. Latitude is plotted up the page and longitude horizontally. In the white area of the map the sea-level surface lies *above* the spheroid with equatorial diameter 12 756·28 km and polar diameter 12 713·51 km. In this area, lines are drawn joining up points where sea-level is 10 metres above that spheroid, 20 metres, 30 metres, and so on to 80 metres. The place where sea-level is farthest above the spheroid (81 metres) is shown, as also are certain local maxima. In the grey area of the map, the numbers are negative because the sea-level surface lies *below* the spheroid. Contours are again drawn at 10-metre intervals, and the place where sea-level is farthest below the spheroid (113 metres) is shown, together with various local minima

Angles between the line of sight joining any pair of points and the *horizontal* (as determined at each by methods which are extensions of the idea of the spirit-level) are compared with those calculated by assuming particular heights above sea-level. (Only the *difference* in height at those points affects the angles.) Finally, the heights of all points in the framework are chosen 'least badly' as estimated by multiplying the squares of the discrepancies by weights and adding up.

Another technique much used in exploring is to measure the accurate magnitude of g, the downward acceleration due to gravity. Timing a simple pendulum is one of the good ways of doing that. For weather prediction it is sufficient to take g as an acceleration of 9.8 metres per second in every second, but the *next three decimal places* (at least) are what matter in exploring!

Fig. 1.25 shows how g *measured at sea-level* would vary (from 9.7803 at the equator to 9.8322 at the poles) if the sea-level surface was exactly a spheroid and there was no ground above it. Measured values of g tend to differ from those in Fig. 1.26 by up to about 0.003 either way. The exact discrepancies are of great interest after they have been corrected to take into account (*i*) the *height* of the measuring station (which is easy) and (*ii*) the local distribution and density of material above sea-level (which is a laborious calculation of somewhat variable accuracy). What remains is called the *gravity anomaly*.

There are two reasons why people have long been going around the world with various instruments measuring the gravity anomaly. The great nineteenth-century mathematician G. G. Stokes showed that if you knew its value everywhere you could calculate how much the sea-level surface departs from the spheroidal shape. That reason was partly superseded when the satellite observations became the most accurate method of obtaining a picture of the large-scale departures illustrated in Fig. 1.24. Now a modified form of Stokes's equation is used which expresses local small-scale departures from that large-scale pattern in terms of the departure of the gravity anomaly from the one corresponding to the pattern . . .

The rest of this chapter is about exploring what lies beneath the earth's surface; and it begins with the second reason for measuring gravity anomalies. Some very local variations in the gravity anomaly may be important clues to the subterranean position of upward-pointing folds in rock layers, or of *salt domes* (Fig. 1.26); features which are often 'traps' for oil or natural gas (or other minerals such as sulphur) that otherwise would have seeped away through the earth's surface.

The idea is to find some place near the surface where the density of rock is above average for its depth by looking for the excess gravitational attraction which it must exercise. Mathematics plays a part in the search when various possible distributions of excess density are studied to find what may give a good fit with the observed gravity anomaly at the surface; or, in some instances, with the output of another instrument, used by explorers to measure the *difference* in the value of *g* (and in the direction of the vertical) at positions of the order of a metre apart (Fig. 1.27).

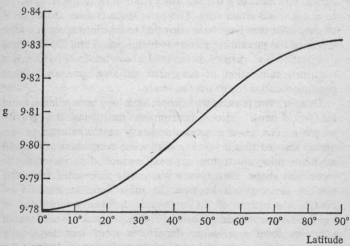

Fig. 1.25. How *g*, the downward acceleration (metres per second in every second) due to gravity, would vary if the sea-level surface was exactly a spheroid and there was no ground above it

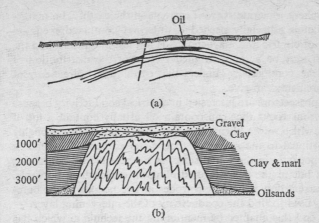

Fig. 1.26. The relevance to exploration for oil of (*a*) rock folds, which may allow oil to accumulate either in the top of a fold of porous rock overlain by impermeable rock, or where the sloping upper surface of some porous rock encounters a *fault* (broken line) ; and (*b*) salt domes, which again may act to confine oil, that may have seeped up from oilsands under the pressure exerted on them by overlying layers. (Reprinted with kind permission from *Applied Geophysics*, 4th edition, A. S. Eve and D. A. Keys, C U P, 1954)

Mathematics is similarly used when minerals are tracked down through their magnetic or electric properties. Any body of ore with high magnetic *susceptibility* (such as certain iron ores) tends to get magnetized by the earth's own magnetic field. The resulting modifications to the magnetic field measured at the surface can be compared with mathematical calculations of the effect of differently shaped ore-bodies of different susceptibility. Other ores (especially sulphides) have high electrical conductivities. Direct current flowing between two electrodes set into the earth a few hundred metres apart has its flow pattern distorted by an intervening ore-body of high conductivity. Measurements of the pattern of electrical potential at the earth's surface can then be compared with various calculations of such distortions.

Magnetic and electric effects are combined when either alternating currents, or unsteady electrical disturbances in the atmosphere, generate *induced* currents in the earth. The mathematician is especially needed to interpret combined measurements of electric and magnetic fields in these situations, and the pay-off may be a more detailed picture of the distribution of electrical conductivity under the earth than direct-current measurements can give.

Oil prospectors are interested in electrical conductivity because the porous rocks that can contain oil usually contain a lot of water which gives them a substantial conductivity (roughly proportional to the *square* of their water content). This fact is used, not only in the initial searches for possible oil-containing layers, but in the next stage: drilling an exploratory borehole. The current flow between electrodes at two heights in the borehole is used to find the conductivity of the intervening layer.

We end this chapter by mentioning the technique which has been most widely used in exploring the North Sea and other

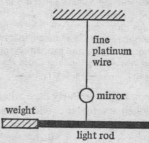

fine
platinum
wire

mirror

weight

light rod

weight

A dense rock formation below the earth's surface on the reader's side of the diagram attracts the lower weight more than the upper weight. This causes the rod, suspended by a fine wire, to twist by an amount detected very sensitively through the deflection of a light beam reflected from the mirror.

Fig. 1.27. The principle of the Eötvös Torsion Balance. At each station, measurements are made with the horizontal rod in several different directions. This leads, after interpretation by an interesting mathematical theory, to a comprehensive picture of how the magnitude of g and the direction of the vertical are varying locally

extensive offshore areas to find deposits of oil and natural gas. This method is to generate an explosion near the surface of the sea at a precisely known time and then to note the signals received by a large number of instruments called *hydrophones* which automatically record sound vibrations in the sea. The hydrophones are arranged in a pattern, at a variety of distances from the explosion, and the total travel time of the sound signal received at each is accurately measured.

The interpretation of the data is based on one simple physical principle. The sound signal follows the path which gives its time of travel (the quantity measured) the least possible value – taking into account the speed of sound in water and the various greater speeds of sound known for the different rocks which may lie under the sea.* However, the mathematics of using that principle to identify rock layers beneath the sea (and above all to find folds or salt domes) can be quite fascinating.

To suggest just a very little of its flavour by describing an extremely simple case, we will suppose that the time of arrival of a sound signal at a hydrophone, plotted against its distance from the explosion, was found to fall closely along three straight lines as in Fig. 1.28. The nearer hydrophones lie on the lowest line, for which distance travelled divided by time taken is u, the speed of sound in water. For the middle group of hydrophones a path diagonally to the bottom, along the bottom, and diagonally up takes least time. Any extra distance † divided by extra time taken for that group is v, the considerably higher speed of sound in the bottom material. However, the fact that a third group of points is present suggests the existence of another layer, say, at a depth d below the bottom, with a still greater sound speed w, given by the extra distance divided by extra time which is observed for that group.

Try your hand at a bit of 'Mathematics and the Physical Environment' by calculating that on those assumptions the

*Yet again, at the later stage when an exploratory borehole has been drilled, the measured times of travel of sound between points at different depths in the borehole are used to get the actual speeds. These help greatly in distinguishing between different kinds of rock.

†By 'extra' distance we mean the distance of some farther hydrophone from the explosion, minus the distance of some nearer one.

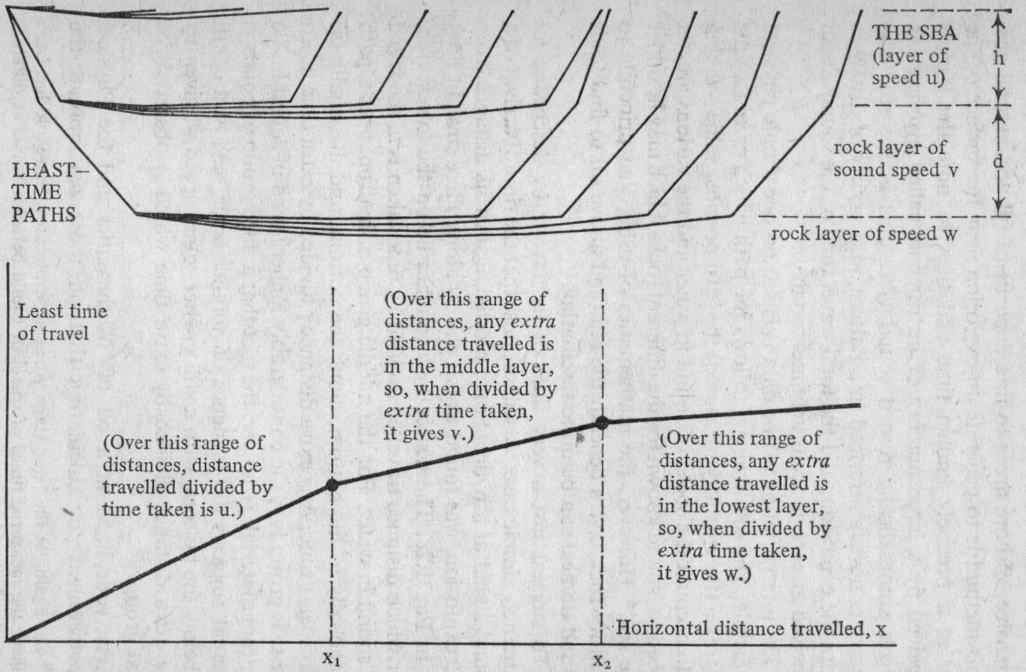
THE SEA
(layer of
speed u)
h
rock layer of
sound speed v
d
rock layer of speed w
LEAST–
TIME
PATHS
Least time
of travel
(Over this range of
distances, distance
travelled divided by
time taken is u.)
(Over this range of
distances, any *extra*
distance travelled is
in the middle layer,
so, when divided by
extra time taken,
it gives v.)
(Over this range of
distances, any *extra*
distance travelled is
in the lowest layer,
so, when divided by
extra time taken,
it gives w.)
Horizontal distance travelled, x
x_1
x_2

Fig. 1.28. Some properties of least-time paths used in the geophysical exploration technique known as *seismic refraction*

hydrophones in the middle group would be those placed at horizontal distances from the explosion between x_1 and x_2, where

$$x_1 = 2h \sqrt{\frac{v + u}{v - u}}$$

and $$x_2 = 2h \frac{v\sqrt{w^2 - u^2} - w\sqrt{v^2 - u^2}}{u(w - v)} + 2d \sqrt{\frac{w + v}{w - v}}.$$

Note that from the *measured* value of x_1, and this calculated value, you could deduce the depth of water h; then, if that checked against knowledge of h from some other source, you could reasonably use the measured value of x_2 to infer the depth d of the lower rock layer beneath the bottom!

BIBLIOGRAPHY AND REFERENCES

1. The Air

Elementary accounts of meteorology are given in:
LONGLEY, R. W., *Elements of Meteorology*, Wiley, New York, 1970.
RIEHL, H., *Introduction to the Atmosphere*, McGraw-Hill, New York, 2nd edn, 1972.

Intermediate-level treatments of some of the applications of mathematics in meteorology are given in:
HOLTON, J. R., *An Introduction to Dynamical Meteorology*, Academic Press, New York, 1972.
HESS, S. L., *Introduction to Theoretical Meteorology*, Henry Holt, New York, 1959.

Advanced accounts of the field as it was before the development of computational forecasting are:
GODSKE, C. L., BERGERON, T., BJERKNES, J., and BUNDGAARD, R. C., *Dynamical Meteorology and Weather Forecasting*, American Meteorological Society, Boston, 1957.
PETTERSSEN, S., *Weather Analysis and Forecasting*, McGraw-Hill, New York, 2nd edn, 1956.

The first book on computational forecasting was:
RICHARDSON, L. F., *Weather Prediction by Numerical Process* (first published in 1922 by CUP), Dover Publications, New York, 1965.

Later books on this subject are:
THOMPSON, P. D., *Numerical Weather Analysis and Prediction*, Macmillan, New York, 1961.
HOLTINER, G. J., *Numerical Weather Prediction*, Wiley, New York, 1971.
BENWELL, G. R. R., GADD, A. J., KEERS, J. K., TIMPSON, M. S., and WHITE, P. W., *The Bushby-Timpson 10-level Model on a Fine Mesh*, HMSO: Meteorological Office Scientific Paper No. 32, 1971.

2. Fresh Water and the Sea

Elementary accounts of hydrology are given in:
WISLER, C. O., and BRATER, E. F., *Hydrology*, Wiley, New York, 1959.
WILSON, E. M., *Engineering Hydrology*, Macmillan, 1969.
VIESSMAN, W., HARBAUGH, T. E., and KNAPP, J. W., *Introduction to Hydrology*, Intext, New York, 1972.

A comprehensive intermediate-level account is in:
CHOW, V. T. (editor), *Handbook of Applied Hydrology*, McGraw-Hill, New York, 1964.

For an up-to-date account of water resources see the Proceedings of the IMA Conference on The Mathematics of Hydrology and Water Resources, to be published by Academic Press, New York.

A broadly based account of modern work aimed at combating water pollution is given in:
JENKINS, S. H., *Advances in Water Pollution Research*, Pergamon Press, Oxford, 1972.

The classic book describing the mathematical prediction process for the Thames is:
Effects of Polluting Discharges on the Thames Estuary, HMSO: Water Pollution Research Technical Paper No. 11, 1964.

A survey of recent mathematical and allied work on estuary pollution problems is:
Mathematical and Hydraulic Modelling of Estuarine Pollution, HMSO: Water Pollution Research Technical Paper No. 13, 1972.

An elementary account of oceanography is:
LIGHTHILL, SIR JAMES, 'Ocean Science', *Bulletin IMA*, Vol. 9, p. 30, 1973.

For the practical applications of the mathematics of water waves, see:
MCCORMICK, M. E., *Ocean Engineering Wave Mechanics*, Wiley, New York, 1973.

An elementary account of tides is given in:
MACMILLAN, D. H., *Tides*, CR Books, London, 1966.

Relatively advanced accounts of waves, tides, currents and the associated mathematics are given in:
NEUMANN, C., and PIERSON, W. J., *Principles of Physical Oceanography*, Prentice-Hall, Englewood Cliffs, NJ, 1966.
DEFANT, A., *Physical Oceanography*, OUP (1961).

3. The Earth and its Resources

Elementary accounts of D. G. King-Hele's work deriving and applying knowledge from satellite observations were given in:
KING-HELE, D. G., *Satellites and Scientific Research*, Routledge and Kegan Paul, 1960.
KING-HELE, D. G., *Observing Earth Satellites*, Macmillan, 1966.

Intermediate-level accounts of the earth's gravity field are given in:
COOK, A. H., *Gravity and the Earth*, Wykeham Publications, London, 1969.
COOK, A. H., *Physics of the Earth and Planets*, Macmillan, 1973.

Advanced accounts of modern geodesy and of the input to it from satellite observations are:
BOMFORD, G., *Geodesy*, OUP, 3rd edn, 1971.
KING-HELE, D. G. (organized by), 'A Discussion on Orbital Analysis', *Philosophical Transactions of the Royal Society*, Series A, Vol. 262, pp. 1–202, 1967.

An elementary account of prospecting methods using various physical effects is given in:
EVE, A. S., and KEYS, D. A., *Applied Geophysics*, CUP, 4th edn, 1954.

An intermediate-level account is in:
PARASNIS, D. S., *Principles of Applied Geophysics*, Chapman and Hall, 2nd edn, 1972.

An advanced account of methods using electrical and magnetic effects is in:
KELLER, G. V., and FRISCHKNECHT, F. C., *Electrical Methods in Geophysical Prospecting*, Pergamon Press, Oxford, 1966.

2 Mathematics and the Biological Environment

1. Introduction

What use is mathematics to the biologist? What are the problems facing the modern biologist? In this chapter we shall explore some of the livelier areas of the biological environment and hope that answers to these questions will appear at every turn. Here what we have to say concerns many of the plant and animal species that survive together on earth. But the central figure on this stage is man, and most of our attention here will be directed to him.

To begin with we shall look at the building blocks which are passed on from one generation to the next by members of any biological species. These are the *genes*, the fundamental units of inheritance. They distinguish men from mice and other species from one another. What is more, they distinguish members of the human race from each other. Each one of us is unique in that no two people possess the same set of building blocks. Of course, we may challenge this statement as one which cannot be proved as a mathematical proposition. However, the random shuffling of a large number of genes which takes place to produce each new individual makes it unlikely that the reader has a living double anywhere on earth, unless he or she is a monozygotic twin. We can go further than this and with some primitive arithmetic show that, except for twins, such a double has never lived and will never appear in the proverbial million years! This issue is not of great concern to biologists because of the highly improbable event referred to here. Should the conditions of our society change dramatically and affect the way we reproduce ourselves then finite rather than infinitesimal chances will make the question of real concern to us all. There are, meanwhile, some questions of current importance to do with differences between human groups. What happens to a population in which marriage brings together partners with similar characteristics? Is natural selection a force to be reckoned with today, and what subtle forms does it take? Do our genes tell the full story of past intermixture between different races? Can we predict the future consequences of population mixture? How closely related are village populations today? Why does a mathematician draw maps using the results of blood tests? These are some of the questions which are tackled in the next section.

We may be concerned, not with the characteristics of individuals in a population, but solely with their number. The growth of population size is of importance to all of us. Not only do we want to speculate about the future size of human populations, there are many species of plants and animals which interact with each other and may hold clues to our own destiny. Projection-like forecasting in any sphere holds a special fascination. We are hopeful that methods of forecasting are improving all the time. Certainly, the involvement of mathematicians is considerable and they are addressing themselves to some major biological problems. How do simple populations grow? Why do some populations follow a cyclic pattern? Why do some species oscillate in phase with one another while others do so out of phase? Why are some human populations very unstable? Why does it take a century to change the structure and pattern of growth of the population of a typical country? How do social classes with different rates for births and deaths from each other affect total population growth? How can social mobility affect population growth? Are there natural laws for the numbers of species found in an area? What is a 'biotic equilibrium'? How does a new species grow and develop and react with the carrying capacity of an environment? How can we manage our food resources by manipulation of the fishing stocks? These questions and some others will be pursued in the third section of this chapter.

Having looked at the characteristics of individuals in human populations as well as the size of these populations, we next turn our attention to our well-being. The field of medicine is vast and has a long history. The rôle of mathematics within it is a new one and the prospects, although having obvious limitations, are exciting. We shall follow a path through the doctor's consulting room, to his treatment table and so to recovery. In the course of the fourth section we shall embrace the theories for infectious diseases, a natural law which applies to cancer and the use of mathematics in clinical trials for new drugs.

This leads us to the final section on pharmacology: we shall consider what happens to us when we take drugs. The ways in which drugs take their effect and how they disperse through the body are described mathematically. The effectiveness and safety of drugs are becoming subjects of great concern to us all. The administration of drugs has to be carefully calculated so that we maintain the effect of a drug without allowing it to reach highly toxic levels in the body. Kidney machines and their use by people with reduced kidney function bring new problems in drug administration. How can mathematics help here? Finally, there are special problems which arise when giving drugs which attempt to eradicate cancer cells. How can we ensure that enough unaffected cells remain active so that the patient may survive the treatment? We would all be ready to acclaim further contributions from mathematics towards the relief of this increasingly serious disease. Let these be soon.

On this forward-looking note, we are almost ready to go on our tour of the biological environment. First, though, a word of warning. We shall see that not all of the important areas of biology can be fairly dealt with in these few pages. The reader is invited to watch for the spirit of the approach in these newer uses of mathematics. This is a common spirit to all applications: it excites mathematicians and biologists in this common cause!

2. Genetics

Genetics is concerned with the inheritance of specific characteristics in plants and animals. Our genes are the basic units in our biological make-up and they largely determine the characteristics we possess. Some of our genes can be identified by chemical tests applied to samples of body substances such as blood and saliva. In diploid organisms such as man, genes occur in pairs and an individual receives one gene from each parent. To begin with, we consider a simple genetic system with just two genes, to be called A and a. With regard to this system, men are of three types corresponding to the three different pairings of the two genes:

$$AA \quad Aa \quad aa$$

(since the parental origin of each gene is not retained Aa is the same as aA, both will be called Aa). Often these 'pairings' are called 'genotypes'.

We next introduce p, the *gene frequency* for the gene A, defined as the proportion of all individuals who possess that gene. Correspondingly, q $(=1-p)$ will be the gene frequency for the gene a. Furthermore, *genotype frequencies* for the three genotypes may be defined. The frequency of the genotype AA is the proportion of those who having received the gene A from one parent also receive the gene A from the other parent, i.e., a proportion $p \times p = p^2$. Similarly, the genotype frequencies for Aa and aa are $2pq$ and q^2, respectively. These frequencies are displayed in Fig. 2.1.

In arriving at these frequencies, we have used a simple argument which would apply only if parents were randomly chosen from the entire population. Clearly this is not always true, for some genes are responsible for characteristics that are sought after when choosing a marriage partner. This *preferential* or *assortative* mating can be positive, as is the case for height (tall people do tend to marry tall people), or negative, as in the case of hair colour, where it has been found that people with red hair marginally prefer not to marry people with red hair. In the event of positive assortative mating, the frequencies of the homozygotes, AA and aa, will be greater than p^2 and q^2 respectively, while the frequency of the heterozygotes, Aa, will be correspondingly less than $2pq$. Negative assortative mating produces more heterozygotes and fewer homozygotes than we would predict from knowledge of the gene frequencies.

If some of the newly born individuals with a particular

genotype fail to reach physical maturity and to produce children then the frequencies of the genes in the population will be affected and a gene may even disappear. This process is called *natural selection* where the causes of the failure to reproduce

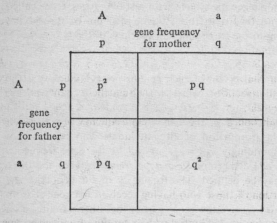

in the boxes are genotype frequencies

for the genotypes   AA   Aa   aa

p^2 $2pq$ q^2

HARDY — WEINBERG LAW says that if these are disturbed in any way (perhaps some *aa* removed)

then p and q may change to p' and q' but the new frequencies maintain the ratios

$$p'^2 : 2p'q' : q'^2$$

in the next generation after RANDOM MATING

THIS LAW IS USED TO DETECT NATURAL SELECTION OR NON-RANDOM MATING IN BIOLOGICAL POPULATIONS

Fig. 2.1. Genotype frequencies from gene frequencies

arise in the natural environment and *artificial selection* if these causes are man-made. The former mechanism has played a major part, as Darwin recognized, in man's evolution, whereas the latter mechanism is used by man in developing plants and animals with desirable characteristics in agricultural breeding programmes.

We may describe mathematically the effects of selection with time upon the frequency of a gene. If some fraction s of the homozygotes aa fail to survive to reproduce in each generation, then the genotype frequencies for AA, Aa and aa, respectively, will change from p^2, $2pq$ and q^2 at birth, with q^2 becoming $q^2 - sq^2$ and the total becoming $1 - sq^2$ instead of 1. Dividing by this new total leads to $p^2/(1 - sq^2)$, $2pq/(1 - sq^2)$ and $(1 - s)q^2/(1 - sq^2)$ for the frequencies at the time of reproduction, after the action of selection. From these frequencies, counting the genes, we see that the AA individuals have two A genes and the Aa individuals one A gene, so that the frequency of the A gene is

$$p' = \frac{p^2 + pq}{1 - sq^2} = \frac{p(p + q)}{1 - sq^2} = \frac{p}{1 - sq^2}$$

and of the a gene is $q' = \dfrac{q(1 - sq)}{1 - sq^2}$.

After reproduction, offspring will be produced with genotype frequencies p'^2, $2p'q'$ and q'^2, respectively. Since $1 - sq^2$ is less than 1, p' will be greater than p, and if this selection continues at a constant rate we see that the frequency of the gene A increases at the expense of the gene a each generation, as depicted in Fig. 2.2. The reader may care to verify that if $p = \frac{2}{3}$, $q = \frac{1}{3}$ and $s = \frac{3}{4}$ then $p' = \frac{8}{11}$.

Ultimately, the position will be reached when the gene a will have disappeared and the gene A is then said to be *fixed* in the population. For the values suggested for verification, the gene frequency p reaches the value 0·99 after 133 generations, and fixation can take a very long time.

In small populations, however, the effects of random sampling can produce dramatic results and genes can be lost by chance. We see that if two Aa heterozygote parents produce only AA children in a small family then the gene a will not be represented in the next generation. If this happens to all families in a small population, then gene A would become fixed in the population at an earlier time than we would predict using the above mathematical model. This process, which leads to deviations of gene frequencies in small populations from those predicted by our analysis, is called *genetic drift*. Although other factors are sometimes present, genetic drift will explain the variations found in

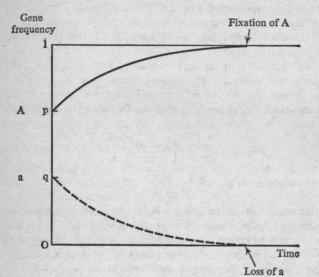

Fig. 2.2. The effect of selection for gene *A*. Where a gene *A* has a simple alternative (or allele) *a*, the frequency of the gene *A* will increase with time from an initial value *p* to 1 if selection acts against the gene *a*

gene frequencies in neighbouring local populations in many biological situations. Where drift is not the major cause of variation, natural selection may be considered as a possible cause, particularly where environments differ markedly and different genes may confer different advantages on individuals possessing them.

We have considered selection against one homozygote in our simple genetic system, but a much more interesting situation has selection acting against *both* homozygotes *AA* and *aa*. In this case the heterozygote is favoured and the ability of heterozygotes to produce new homozygotes means that depleted homozygote groups can be replenished and a stable equilibrium state for the genotype frequencies in this system can be achieved.

We may use symbols S and s to denote the proportion of AA and aa, respectively, selected against in each generation, and this results in genotype frequencies after selection:

$$\frac{p^2(1-S)}{1-Sp^2-sq^2}, \frac{2pq}{1-Sp^2-sq^2} \text{ and } \frac{q^2(1-s)}{1-Sp^2-sq^2}$$

with gene frequencies for A and a of

$$\frac{p(1-Sp)}{1-Sp^2-sq^2} \text{ and } \frac{q(1-sq)}{1-Sp^2-sq^2}.$$

If we ignore the trivial cases where p or $q = 0$ or 1, these frequencies will be equal to p and q, respectively, the initial frequencies, if $p = \dfrac{s}{S+s}$ and $q = \dfrac{S}{S+s}$, and in this event the frequency of the homozygotes AA and aa will be p^2 and q^2 in each generation. Those homozygotes removed by selection in any generation are replenished for the following generation, and the name given to a system in this state is a *balanced polymorphism* (Fig. 2.3).

The stability of the equilibrium in which p and q are as given above may be investigated by introducing a small perturbation e in the frequencies. We suppose that

$$p = \frac{s}{S+s} + e \text{ and } p' = \frac{s}{S+s} + e'$$

are the frequencies of gene A before and after one complete generation of selection and reproduction. Substitution into the expression for the frequency for A above and ignoring small terms of the order of e^2 or less provides the relation

$$e' = \frac{S+s-2Ss}{S+s-Ss} \cdot e$$

from which we can see that $e' < e$ for all values of S and s which are positive proportions ($<$ means 'less than'). Hence, the equilibrium state is a *stable* one, since any perturbation from it will lead to a return to the equilibrium.

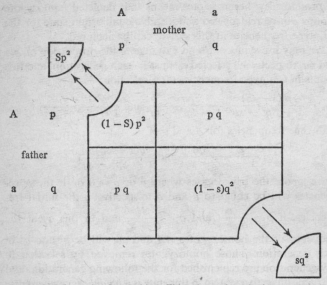

Fig. 2.3. Balanced polymorphism. This arises when the heterozygote Aa is favoured and fractions S and s of the homozygotes AA and aa, respectively, are removed each generation. If S and s are positive proportions between 0 and 1 and also if the frequencies of the genes A and a are p and q satisfying:

$$p = \frac{s}{S+s} \quad \text{and} \quad q = \frac{S}{S+s}$$

then a stable equilibrium will be maintained

We have seen how the frequencies of particular genes may be maintained by selection acting against both homozygotes in suitably balanced amounts. Furthermore, these amounts of selection may be large, so that the homozygotes are partly or even completely removed by selection in each generation. This is illustrated by the sickle cell gene in man, which conveys to individuals who are homozygous for the gene the lethal disease of anaemia. Unfortunately, the alternative gene (or *allele*) to the gene is such that the homozygous individuals for it have low resistance to malaria. In those tropical regions where malaria is widespread, therefore, the sickle cell gene is maintained in the form of a balanced polymorphism, with the homozygotes, who die in each generation, being replaced by reproduction from the heterozygotes.

In populations which are not subject to selection or to other forces that reduce the numbers of individuals with particular genotypes, the frequencies of the three genotypes as we have seen will be p^2, $2pq$ and q^2. These frequencies are obtainable as a special case of the *binomial probability distribution* or, alternatively, as coefficients of terms in the binomial expansion $(pA + qa)^2$ which we may choose to write in the form

$$p^2(AA) + 2pq(Aa) + q^2(aa)$$

where the bracketed symbols refer to the genotypes. The constancy of these genotype frequencies from generation to generation arises from the *Hardy–Weinberg Law* (or Theorem) which, for the genetic system we have chosen to look at, states that if the gene frequencies for A and a are p and q then, *whatever* the genotype frequencies in the population in one generation, after reproduction the frequencies in the offspring will be as given above. These offspring genotypes are said to be in *Hardy–Weinberg equilibrium*. The name of the theorem acknowledges the independent discovery in 1908 by G. H. Hardy, the English mathematician, and W. Weinberg, the German physician. (Here we have used the notion of probability in the sense of the *proportion* of all individuals in a *large* population who would possess some characteristic. This is known as a 'relative frequency' definition of probability. A broader definition is used in Chapter 3. For a fuller account of probability consult a specialized text [1].

The importance of Hardy–Weinberg equilibrium lies in its use as a base-line for comparison. We have seen that selection and assortative mating may affect the genotype frequencies. Population data may therefore be examined to see if these frequencies are in Hardy–Weinberg equilibrium. If they are, it would be difficult to support any hypothesis suggesting that natural selection or any of a number of other factors was operating in a population.

We may extend the mathematics used in the analysis of a single gene A and its allele a to cover situations with two or more genes. Our description of the biological mechanisms has been greatly simplified, but we may tentatively say that some genes may be inherited together and are said to be *linked*. Such linkage may be only partial, and genes may only tend to be found together. The importance of the phenomenon lies in the fact that we may use the Hardy–Weinberg Law upon the combinations of the two sets of genotypes. If observed combinations do not have the frequencies predicted by the law then we call the state of affairs 'linkage disequilibrium'. This is a useful notion because it helps to explain the possession of certain attributes which appear to depend upon particular patterns of genetic inheritance. We find that the occurrence of certain genes within families may be correlated with disease susceptibility in some individuals. There are reasonable signs that with our increased ability to determine biochemically the genes we possess and our refined mathematical analysis of genetic systems, we shall profit from our knowledge of associations concerning some types of leukaemia and cancer.

We now turn to an important phenomenon which affects many human populations: *intermixture*. By this term we mean the aggregation of groups of people with different genetic characteristics and, subsequently, the random mating which might follow the bringing together of such groups. In the first stages of this population aggregation, in terms of our simple system, we suppose that two populations of equal size have the gene A with frequencies p_1 and p_2. The genotype frequencies in the separate populations will be

$$p_1{}^2, 2p_1q_1 \text{ and } q_1{}^2 \text{ and } p_2{}^2, 2p_2q_2 \text{ and } q_2{}^2$$

and in the total population aggregate

$$\tfrac{1}{2}(p_1{}^2 + p_2{}^2), p_1q_1 + p_2q_2 \text{ and } \tfrac{1}{2}(q_1{}^2 + q_2{}^2).$$

The frequency of the A gene in this aggregate population is, however, $\tfrac{1}{2}(p_1 + p_2)$, and the Hardy–Weinberg equilibrium for the aggregate will therefore be predicted as

$$\tfrac{1}{4}(p_1 + p_2)^2, \tfrac{1}{2}(p_1 + p_2)(q_1 + q_2) \text{ and } \tfrac{1}{4}(q_1 + q_2)^2$$

and these will agree with the actual population only if $p_1 = p_2$. However, after only one generation of reproduction, the offspring genotypes will have the latter set of frequencies, provided that the mating has been random, rather than assortative, and that there has been no selection.

Another problem arises when the proportions of the contributing populations to some admixed population are unknown. We assume that the frequencies of a gene are known in each of two 'parental' populations as p_1 and p_2 and in the 'hybrid' population as p. A simple estimator of the proportion m_1 of individuals contributed by the first parental population is the proportional distance of p from p_2 on the interval from p_1 to p_2, i.e., $(p_2 - p)/(p_2 - p_1)$. If p is half-way between p_1 and p_2 then $m_1 = \frac{1}{2}$. Alternatively if $p_1 = 0.2$, $p_2 = 0.5$ and $p = 0.3$ then $m_1 = \frac{2}{3}$ and this high value reflects the high proportion of the first population in the hybrid population. Use is made of this quantity by population geneticists in an interesting way to look for natural selection or selective movement of peoples. This is done simply by calculating the estimator of mixture using gene frequencies for each of several genetic systems, particularly the blood group systems known as ABO, Rhesus, MNS etc., and comparing the results. If the values we get agree well, there is no evidence for natural selection or other non-random effects. Fig. 2.4 shows how mixture rates may be presented for two and three parental populations in such analyses.

In practice, we discover some problems in such mathematical approaches to intermixture. On the one hand, there are difficulties because of the complex nature of common blood group systems like ABO and Rhesus. These each have a number of alleles arranged in a complicated way. Also, not all of the genotypes may be detectable by blood tests. When this happens we say that there is *dominance*. In the simplest case, of the three genotypes AA, Aa and aa, we may only have one test which distinguishes those individuals with the gene A from those without it. Hence two 'phenotypes' are observed: $AA + Aa$ and aa. Here we say that the A gene is *dominant*, and that its allele a is *recessive*. There are some important limitations on our ability to interpret what we see and these are often met in rare diseases which only arise in people who have the recessive gene in its homozygous form (suppose we call this *aa*). Marriage between individuals with the gene *A* does not preclude the production of children

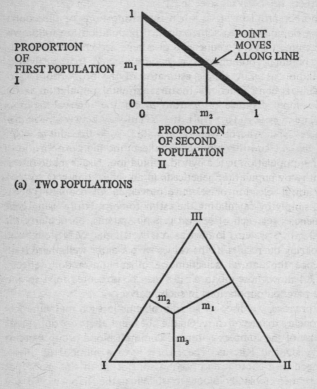

(a) TWO POPULATIONS

(b)   THREE POPULATIONS

Fig. 2.4. Graphs of mixture rates. The proportions of mixing populations in a hybrid population can be represented by a point on a simple graph. If there are two mixing populations, this point lies on a straight line as indicated in the upper graph (*a*). For three populations, this point lies within a triangle as shown in the lower graph (*b*). The *m* values are the proportions of the mixtures

with *aa*, for both parents may be of *Aa* type. This kind of problem arises in the field of genetic counselling. In the mixture situation, this adds to the difficulty of interpretation.

Another obstacle in our path in population genetics analyses of human populations comes from the small sizes of our samples. These samples have only a few hundred observations and often cover only a fraction of the actual population. There are two sources of 'error' in this situation: 'sampling' and 'drift'. The sampling error is the distortion of the sample from the actual population, while the drift is seen on a time dimension to be the change in gene frequencies from our grandparents' generation to our parents' generation to our own caused by the random shuffling of genes at reproduction. These discrepancies can be covered by our methodology and adequate procedures have been devised for intermixture analysis in recent years.

An illuminating example was provided by P. L. Workman and his co-workers in 1963. They estimated admixture rates for the American Negroes in Claxton, Georgia, in terms of the white Americans in Claxton and of West African Negroes. Their estimates, made using the methods we have been looking at, show that the American Negro comprises 10% White American and 90% African Negro. At least this was the case for all but four blood group systems which indicated a much higher 'white' contribution of around 30%. Although there are some uncertainties in using present-day descendants as representatives of their ancestors because of 'drift', we may tentatively conclude that these four systems were affected by selection for 'white' alleles after the African Negroes had migrated into the American environment.

Where there are several intermixing populations, a useful extension to the methods may be made using matrices. We need an *admixture* or *migration matrix*, *M*. At this point, however, we may profit from a digression which will take us through some useful properties of matrices.

For those unfamiliar with matrices the excellent chapters on the subject in W. W. Sawyer's books [2, 3] will be found valuable but for others the following notes may serve as a refresher.

To begin with a basic definition, a matrix is a rectangular table of numbers, usually denoted by a capital letter. Two matrices of the same size may be added or subtracted by carrying out these operations upon corresponding elements. For example, if

$$P = \begin{pmatrix} 2 & 3 & 8 \\ 4 & 2 & 1 \end{pmatrix} \text{ and } Q = \begin{pmatrix} 5 & 7 & 2 \\ 1 & 4 & 3 \end{pmatrix}$$

then $$P + Q = \begin{pmatrix} 2+5 & 3+7 & 8+2 \\ 4+1 & 2+4 & 1+3 \end{pmatrix} = \begin{pmatrix} 7 & 10 & 10 \\ 5 & 6 & 4 \end{pmatrix}$$

and $$P - Q = \begin{pmatrix} -3 & -4 & 6 \\ 3 & -2 & -2 \end{pmatrix}$$

(Note that P and Q have 2 rows and 3 columns and each may be said to be a '2 by 3' or a 2×3 matrix.)

Although it would seem natural to proceed to multiply and divide matrices using the corresponding elements, there is a much more powerful and useful definition for these operations. Matrix multiplication is only permitted when the two matrices forming the product have dimensions which match in a particular sense. If we call P and Q these two matrices, then the product PQ will be possible only if P has as many columns as Q has rows. The reason for this will be apparent when we see an actual multiplication. But the dimensions of P and Q in the last example will not allow us to form a product, for whereas P had 3 columns Q had only 2 rows. Instead we shall choose Q to be a new matrix

$$\begin{pmatrix} 6 & 4 & 3 & 2 \\ 1 & 2 & 5 & 8 \\ 4 & 7 & 1 & 6 \end{pmatrix}$$

with 3 rows and 4 columns, and Q now has as many rows as P has columns. The product PQ is possible, but we note that QP would not be possible. What then is PQ? It has for its first element in the first row

$$(2 \times 6) + (3 \times 1) + (8 \times 4) = 47$$

and we can see that this is the sum of the pairwise products of the corresponding elements in the first row of P and the first column of Q, respectively.

$$\begin{pmatrix} \underline{2 \quad 3 \quad 8} \\ 4 \quad 2 \quad 1 \end{pmatrix} \begin{pmatrix} 6 & 4 & 3 & 2 \\ 1 & 2 & 5 & 3 \\ 4 & 7 & 1 & 6 \end{pmatrix} = \begin{pmatrix} \textcircled{47} & \cdot & \cdot & \cdot \\ \cdot & \cdot & \cdot & \cdot \end{pmatrix}$$

Now the *second* element in the first row of PQ is the sum of the products of the elements in the first row of P with those in the second column of Q. The *third* element in the first row of the product uses the first row of P and the *third* column of Q and so on. The *second row* of the product is obtained using the *second* row of P and the columns of Q in turn: similarly for all other rows of the product. The reader may now verify with a pencil and paper that the product PQ has the elements

$$\begin{pmatrix} 47 & 70 & 29 & 76 \\ 30 & 27 & 24 & 38 \end{pmatrix}.$$

We note that the dimensions of the product are 2 rows and 4 columns and this arises from

$$\underset{2 \times 3}{P} \times \underset{3 \times 4}{Q} = \underset{2 \times 4}{PQ}$$

where we 'cancel' the 'inner' dimension 3 which was the number of pairs of elements used to form each new element of the product matrix. As we have seen, the product QP is not possible, but even in the special case when P and Q are square matrices of the same order, i.e., each has k rows and k columns although both PQ and QP exist they will not usually be equal.

Matrices which have only one column are called *vectors* and we shall often be using a square matrix multiplied by a vector. For example

$$\begin{pmatrix} 2 & 3 & 6 \\ 4 & 5 & 2 \\ 1 & 7 & 8 \end{pmatrix} \begin{pmatrix} 3 \\ 4 \\ 6 \end{pmatrix} = \begin{pmatrix} 54 \\ 44 \\ 79 \end{pmatrix}$$

and we notice that the product in this case is a vector and has the same dimensions as the original vector. The heart of all matrix applications is now within our grasp. It is this: for a square matrix M of order k there are usually k unrelated vectors which have the remarkable property that when multiplied by M they reproduce themselves, except for a scaling factor. Such a vector is called a *latent vector* (or *eigenvector*) of a matrix and each scaling factor is called a *latent root* (or *eigenvalue*) of the matrix.

The reader may care to verify this property in the case of a simple matrix. If we multiply the matrix $\begin{pmatrix} 4 & 2 \\ -1 & 1 \end{pmatrix}$ by the vector $\begin{pmatrix} 2 \\ -1 \end{pmatrix}$ we obtain $\begin{pmatrix} 6 \\ -3 \end{pmatrix} = 3 \times \begin{pmatrix} 2 \\ -1 \end{pmatrix}$ and 3 is called a latent root and $\begin{pmatrix} 2 \\ -1 \end{pmatrix}$ is the latent vector corresponding to that root. What happens if we multiply the matrix by $\begin{pmatrix} 1 \\ -1 \end{pmatrix}$?

We shall comment briefly upon the importance of the latent roots and vectors. For a physical or biological system whose movement can be portrayed in mathematical terms by means of a matrix equation the latent roots and vectors completely determine the behaviour of the system. The system may be stable or unstable or perhaps oscillating, and this information can be determined from an examination of the latent roots and vectors.

Unfortunately, there is not space to pursue the basics of matrix algebra in further detail, and the reader may care to explore these in W. W. Sawyer's books [2, 3] already mentioned. We shall see that the hidden properties of matrices reveal themselves most when they are multiplied repeatedly. Because a matrix multiplication corresponds to some real transformation of the state of the world, the reader would do well to check at this stage that he is clear about multiplication at least! Here are some revision examples:

$$\begin{pmatrix} 1 & 2 & 3 \\ 4 & 5 & 6 \\ 7 & 8 & 9 \end{pmatrix} \begin{pmatrix} 1 \\ 1 \\ 1 \end{pmatrix} = \begin{pmatrix} 6 \\ 15 \\ 24 \end{pmatrix} \quad \text{and} \quad \begin{pmatrix} 6 & 1 & 4 \\ 3 & 0 & 0 \\ 0 & 8 & 0 \end{pmatrix} \begin{pmatrix} 1 \\ 2 \\ 3 \end{pmatrix} = \begin{pmatrix} 20 \\ 3 \\ 24 \end{pmatrix}$$

As a preliminary word of introduction to some useful biological matrices which appear in later sections of this chapter, we now mention the *migration matrix*, M, which has the form:

$$\begin{array}{c} \text{From place} \\ A \quad B \quad C \end{array}$$

$$\begin{array}{c} \text{To} \\ \text{place} \end{array} \quad \begin{array}{c} A \\ B \\ C \end{array} \begin{pmatrix} \cdot & \cdot & \cdot \\ \cdot & \cdot & \cdot \\ \cdot & \cdot & \cdot \end{pmatrix}$$

Each dot represents the proportion moving from place to place in one unit of time. These proportions are multiplied by the numbers (of people, or whatever) in the places A, B and C and the result – the numbers in A, B and C one time unit later:

$$\begin{array}{ccccc} \text{PROPORTIONS} & & \text{NUMBERS} & & \text{NUMBERS} \\ \text{MOVING} & \times & \text{NOW} & = & \text{NEXT} \\ & & & & \\ M & \times & n & = & n' \end{array}$$

(the Matrix Equation for Migration).

If this operation is repeated several times, we can see the effects of the pattern of migration contained in M upon the numbers in A, B and C over several time periods. As time advances these latter numbers depend less and less upon their initial values and after some time the relative numbers come to depend only upon the migration matrix, M. This property is common to other types of matrix.

As a numerical illustration, we may think of $M = \begin{pmatrix} \frac{2}{3} & \frac{1}{3} \\ \frac{1}{3} & \frac{2}{3} \end{pmatrix}$ for movement between two populations with 54 and 108 individuals, so that we take $n = \begin{pmatrix} 54 \\ 108 \end{pmatrix}$. After migration, the new population sizes are the elements of n' where $n' = Mn = \begin{pmatrix} \frac{2}{3} & \frac{1}{3} \\ \frac{1}{3} & \frac{2}{3} \end{pmatrix} \begin{pmatrix} 54 \\ 108 \end{pmatrix} = \begin{pmatrix} 72 \\ 90 \end{pmatrix}$. Repeating this migration gives $\begin{pmatrix} 78 \\ 84 \end{pmatrix}$, $\begin{pmatrix} 80 \\ 82 \end{pmatrix}$ and, eventually, $\begin{pmatrix} 81 \\ 81 \end{pmatrix}$. Hence $\begin{pmatrix} 81 \\ 81 \end{pmatrix}$ is the latent vector corresponding to the latent root 1 of the matrix M. Furthermore, this has the important interpretation: any amount of migration under the symmetric pattern indicated by the elements of M will not change the numbers in the two populations once they have become equal.

We now return to the intermixture situation to make use of the matrices which we have introduced. Having dealt with mixtures of two populations we can now embrace any number.

For p populations, the matrix is square and is of order $p \times p$. The element m_{ij} in row i and column j represents the proportion of population i which has migrated in from population j. To complete the model we shall define a matrix Q containing the frequencies of all genes in all populations. Suppose that there are k genes to be considered, then we shall require the matrix Q to have k columns and p rows. Each row then describes the 'gene pool' for one of the populations. The 'gene pool' is a useful concept for large randomly mating populations and essentially includes all the available genes in a population without regard to the individuals who possess them. The effect of multiplying the matrix Q from the left by the matrix M is to produce a new matrix Q' which describes the state of the gene pools in the populations after one generation of migration and admixture.

Each column of Q we can think of as a vector which changes with migration. This is precisely what happened in the Matrix Equation for Migration. The point of having these vectors side by side in the matrix Q is sheer convenience. For now we can describe the simultaneous change in the different genes in all populations by a single equation. There is no new mathematical difficulty involved in this step. To be quite sure of what is going on here, we may suppose that for two given populations, $\frac{3}{4}$ of those in one population possess a gene A and $\frac{1}{2}$ of those in the other population have this gene. For a second gene B, let us suppose that these proportions are $\frac{1}{2}$ and $\frac{1}{4}$, respectively. Also we assume that a third of those in one population move to the other population each generation. Substitution of these values into the appropriate matrices means that

$$M = \begin{pmatrix} \frac{2}{3} & \frac{1}{3} \\ \frac{1}{3} & \frac{2}{3} \end{pmatrix} \quad \text{and} \quad Q = \begin{pmatrix} \frac{3}{4} & \frac{1}{2} \\ \frac{1}{2} & \frac{1}{4} \end{pmatrix}$$

so that after migration and mixture, the new gene frequencies are

$$Q' = MQ = \begin{pmatrix} (\frac{2}{3} \times \frac{3}{4}) + (\frac{1}{3} \times \frac{1}{2}) & (\frac{2}{3} \times \frac{1}{2}) + (\frac{1}{3} \times \frac{1}{4}) \\ (\frac{1}{3} \times \frac{3}{4}) + (\frac{2}{3} \times \frac{1}{2}) & (\frac{1}{3} \times \frac{1}{2}) + (\frac{2}{3} \times \frac{1}{4}) \end{pmatrix}$$

$$= \begin{pmatrix} \frac{2}{3} & \frac{5}{12} \\ \frac{7}{12} & \frac{1}{3} \end{pmatrix}$$

From these values we can see that the effect of the mixing is to bring the frequencies for the A gene in the two populations closer together: the higher value $\frac{3}{4}$ being reduced to $\frac{2}{3}$ and the lower value being increased from $\frac{1}{2}$ to $\frac{7}{12}$. A similar change can be seen in the case of the B gene.

The importance of the equation

$$Q' = MQ$$

lies in its use in a number of different ways in population genetics. In the first place it can be used to determine, as we have seen, the change in gene frequencies in the populations which results from known amounts of migration. Conversely, if the frequencies before and after migration are known, then from them the migration rates may be estimated.

The migration matrix M may be used to predict the relatedness between populations and the model allows this to be followed through time. The rate at which populations become more closely related is determined by that latent root of the matrix which has the largest modulus (that is, positive numerical value). By 'relatedness' is meant the proportion of ancestors which populations, or, strictly, randomly chosen individuals from populations, have in common.

Another interpretation of the migration matrix model, for use when population sizes are small, allows the determination of the expected variation between populations which will arise from genetic drift. Clearly if there is no migration the drift variance will be considerable for populations of small size. Migration tends to reduce this variance and an equilibrium state is reached when a balance is achieved between the forces of migration which strive to make populations more genetically similar and drift which tends to make them different.

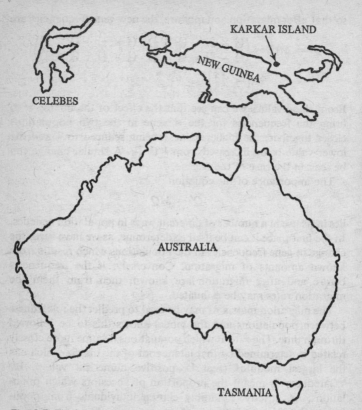

Fig. 2.5. The location of Karkar Island, New Guinea

Fig. 2.5 shows the geographical position of Karkar Island, New Guinea, for which a migration analysis has been made. On this island some fifty-nine villages are to be found arranged around the circular coast-line. The centre of the island is a volcanic cone which proves impassable and communications, therefore, take place around the circular perimeter of the island. Further barriers to movement exist in the form of areas of tropical rain forest (see Fig. 2.6(a)). From a demographic survey of the 16 000 inhabitants of the island, a migration matrix was constructed. In this case, migration involved, for the most part, a woman moving from the village of her birth to that of her husband.

The results of the migration analysis provide a contour map as shown in Fig. 2.6(*b*). These contours are times in generations for the villages to become closely related under the action of the migration currently prevailing. Each contour encloses within it villages which will be closely related at the time shown for that contour. As a main feature of this contour map, we notice that there is a north–south division of the island which persists for more than fifty generations. This division is certainly a result of the topography of the island, but also of the separation of the north and south peoples of the island into two distinct tribes locally known as Waskia and Takia, respectively.

Following the demographic analysis and predictions from the migration matrix, blood samples were taken from the islanders

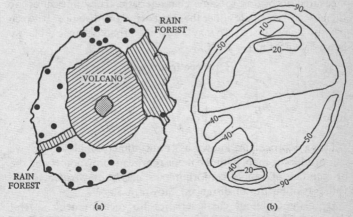

(a) (b)

Fig. 2.6. The topography of Karkar Island is shown in the left-hand map (*a*). The right-hand graph (*b*) is a contour map of the predicted relatedness between villages on Karkar Island. The numbers on the contours are times (in generations) before villages within each contour become closely related, assuming that migration continues at the present rates

and the variations of the gene frequencies determined from these samples largely confirmed the predictions made concerning the relatedness of the villages on the island. We discover from such an analysis, however, that proximity and topography may predict genetic differences better than do cultural factors such as 'tribe' or 'language'.

Another illustration of the methods of population analysis starts by examining the blood donor records for England, Wales and Scotland. From these records, A. Kopec has published tables of gene frequencies for the ABO system for all Blood Transfusion Service areas in these countries. After some computer smoothing of these data, contour maps may be produced showing the actual distribution of the blood groups in Britain. Fig. 2.7 shows one of these maps for the O gene frequency distribution.

A different type of matrix is useful here, the *similarity matrix*, S, which is quite distinct from M. Items or individuals are measured according to some characteristic. These measures are then used to calculate either the 'similarity' or 'distance' between pairs of items. Just as we may construct a road distance table, we enter distances into a matrix:

$$
\begin{array}{c}
\text{Distance from item} \\
\begin{array}{ccc} A & B & C \end{array} \\
\begin{array}{c}\text{To}\\ \text{Item}\end{array}
\begin{array}{c} A \\ B \\ C \end{array}
\begin{pmatrix} \cdot & \cdot & \cdot \\ \cdot & \cdot & \cdot \\ \cdot & \cdot & \cdot \end{pmatrix}
\end{array}
$$

Using a procedure known as 'multi-dimensional scaling' we can produce a map, with the items represented by points and the proximities of points conforming *as nearly as possible* to the values (similarities or distances) shown in S.

By using different characteristics for measurement, several maps can be produced. When compared with each other these maps reveal differences on the one hand between the items and on the other between the characteristics. Sometimes, as would be true for a map constructed from road distances alone, a comparison is possible with a geographical map.

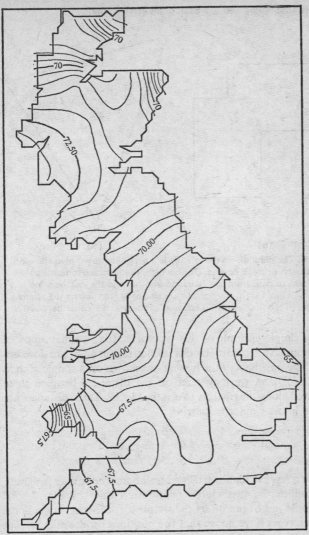

Fig. 2.7. The distribution of blood group O in Britain. The contours describe 'smoothed' gene frequencies for the O gene in the A B O blood group. These frequencies are based upon blood transfusion donors whose data have been collated and presented by A. Kopec

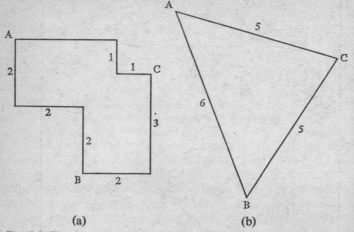

(a)   (b)

Fig. 2.8. The diagram on the left (*a*) is a simplified town plan showing the positions of three houses. On this plan, the lines are routes we would take to pass from one house to another and the numbers are numbers of blocks. On the right (*b*), a similarity map shows the relative positions of the houses taking account of the true distances between them

As a simple example of this type of matrix, we may think of three houses in separate 'blocks' of housing in an urban development. The positions of the houses are shown on a simple plan in Fig. 2.8(*a*), and we may read off the distances between them measured along the roads between the blocks. These distances are entered into a similarity matrix:

$$S = \begin{pmatrix} 0 & 6 & 5 \\ 6 & 0 & 5 \\ 5 & 5 & 0 \end{pmatrix}$$

After a few calculations, using an elaborate computer program, we produce the predictable relationship map of Fig. 2.8(*b*)! We should not be put off by the simplicity of this example. Some value emerges if we notice that the positions on the map are not as on the geographic map. Now the houses have been given relative positions and these are what matters most to us if we wished to call at the houses one after another. The connection between this example and a general and very famous problem will be discovered when the reader meets the 'travelling salesman' problem in Chapter 3. But our needs at the moment are not exactly those of a manager. In passing, and before leaving the little example, we should be aware that some of our better town planners are already using this kind of approach. What matter really are not physical measures of distance, but conceptual ones, and these are affected by subtle features of the environment. To get these measures, we ask 'do you think that B is nearer to A than it is to C?' and similar questions. The answers are often surprising. Cognitive distances can be used as those above to produce a cognitive map. From this map, our shopping, social and recreational facilities can be sited properly and the stress in coping with our environment can be reduced!

The Similarity Matrix, S, is widely used by sociologists, biologists (particularly in genetics and taxonomy), and perhaps surprisingly, by archaeologists and historians [4].

Having introduced this new matrix and having seen something of what it can do for others, we shall apply it to the British ABO data. In fact, the elements of the matrix used for this illustration were taken as negative genetic distances based upon the A, B and O genes. Using a multi-dimensional scaling method, a map results which illustrates the genetic similarity between the places. Since the scaling method draws its map in a higher number of dimensions than two, some arbitrary rotations and projections on to different planes is needed before a recognizable map emerges. In the end, however, the map produced in Fig. 2.9 leads us to the conclusion that geographic proximity generally indicates genetic similarity, for the ABO blood groups at least!

On a local scale, migration data for eight contiguous ecclesiastical parishes in the Otmoor region of Oxfordshire have been collected covering the period since 1500 to the present day. From a migration matrix, scaled maps have been produced by D. G. Kendall and also relatedness has been predicted on the basis of recent migration (i.e., post-1850); Fig. 2.10 shows the developing relatedness with time [4]. According to these pre-

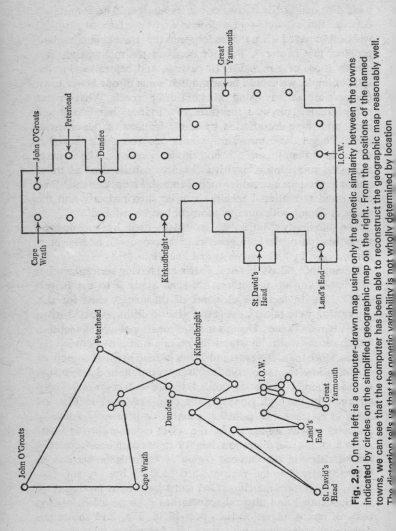

Fig. 2.9. On the left is a computer-drawn map using only the genetic similarity between the towns indicated by circles on the simplified geographic map on the right. From the positions of the named towns, we can see that the computer has been able to reconstruct the geographic map reasonably well. The distortion tells us that the genetic variability is not wholly determined by location

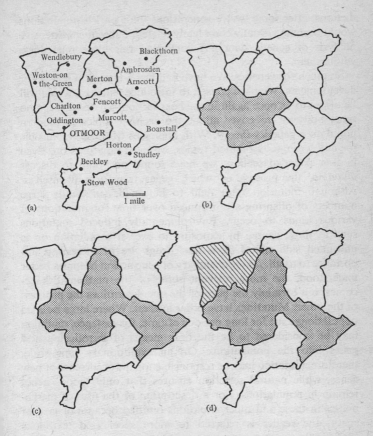

overall time = 12 generations
to complete relatedness

Fig. 2.10. Maps of Otmoor, Oxfordshire, showing the developing relatedness with time of the parishes indicated by boundaries on the top left map (*a*). The shaded parts of the other maps show the spread of localized genes over 6(*b*), 9(*c*) and 10(*d*) generations which would result from migration of the amounts observed for the period since 1850

dictions, after some twelve generations these parish populations would become virtually completely related. Subsequent extensive analysis of genetic data has confirmed this close relatedness within them.

So much for genetics. We have had just a taste of a few newer developments. We have paused to introduce matrices, which will be of use to us once again in the next section. There has been no time, however, to cover all of the exciting topics in modern population genetics theory. With apologies to those responsible for any wrongfully omitted topics, we must now pass on. Even in this extended section, we have found no room to discuss *mutation*, the process by which genes transform themselves. Although mutation rates tend to be small, sufficiently large numbers of offspring are produced over time for mutations of various kinds to occur. Environmentally induced mutations such as those caused by exposure to excessive radiation or as unwanted side effects from new drugs are of increasing importance to us all. As the chemistry of gene action becomes better understood, the more relevant some of the mathematics becomes, but certainly the pace of the one determines the progress of the other. Mutations have other interest. Where large isolated populations exist for long periods of time, for instance, mutation may be considered to be the only means of introducing new genes into the populations. On an evolutionary time scale, therefore, mutation may be responsible for the occurrence of new genes, while natural selection ensures that only 'good' genes remain in populations. For a description of the uses of mathematics in these and other situations omitted because of lack of space, the reader is referred to more specialized textbooks [5, 6, 7, 8].

3. Populations

We may use mathematics in population biology in many widely different contexts. At one extreme, we consider the response of organisms to *external* factors which originate in the physical environment. We can think of enough examples all around us.

The weather affects the growth of our crops. The discovery of rich mineral deposits can seemingly generate human population growth. At the other extreme, we could conceive of organisms which grow in number in total dependence upon other biological species, which we may choose to call *internal* factors. Although extreme situations may not exist in reality, there is often some value in excluding the less important features from a mathematical analysis. In this section, therefore, we describe some of the simpler new applications of mathematics in population studies.

The simplest situation for a single population is that where an organism grows and reproduces itself using an inexhaustible food supply. If the cycle time of the reproduction is a fixed one and if the organism lives for one reproductive cycle, we may consider the growth of the population at discrete points in time. Perhaps such a point will be the instant which terminates a generation. The size of the population after t generations, N_t, may be some multiple $1 + r$ of the reproducing population, N_{t-1}, where r is the proportional excess in the new generation (Fig. 2.11).

The size of this population changes in one generation according to the equation

$$N_1 = (1 + r) N_0, \quad N_2 = (1 + r) N_1 = (1 + r)^2 N_0$$

and we may write N_t in terms of N_0 in general as

$$N_t = (1 + r)^t N_0.$$

This model is the same as one for an investment of a capital amount N_0 at a rate of r with compound interest. In the population model, r is called the *intrinsic rate of natural increase* and a positive value means that the population grows to an infinite size, whereas a negative value, between -1 and 0, means a continuing reduction in population size towards zero. If $r = 0$ the population is said to be *stationary* as its size remains constant with time.

This unconstrained growth mathematics may be developed to deal with constrained populations. In 1847, Verhulst incorporated

POPULATION MATHEMATICS

Single population

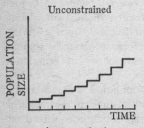

Unconstrained

increase at rate r

(a)

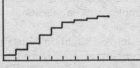

Constrained

increase at rate r
until some limit reached
then decrease in growth
at this rate

(b)

Interacting populations

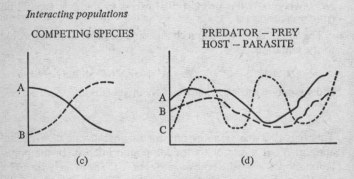

COMPETING SPECIES

(c)

PREDATOR – PREY
HOST – PARASITE

(d)

Simple examples of interaction
Very complex systems studied by simulation

Fig. 2.11. Population growth and interaction. The upper graphs (*a*) and (*b*) show two simple patterns of growth of a single population. The lower graphs (*c*) and (*d*) illustrate some possible interactions between several populations. These are explained further in the text

a time limit upon the available food supply. The effect of this is that the growth of the population increases in a 'compound interest' manner in the early stages. This happens only while the population size is small in relation to the limiting size, which is the maximum population that can be supported by available food resources. As this limit is reached, the growth of the population slows down. We can see from Fig. 2.11 that this results in an S-shaped growth function. This is sometimes called the *logistic* growth curve.

Although the notion of a limit has a certain appeal, the logistic curve seldom fits very well to real situations. There have been a few successes in describing human populations but these curves were found to have only temporary validity.

One common constraint on the growth in numbers of particular animals or plants is the existence of other species with which they have to share food resources. We can identify different forms of interaction between species, but perhaps the simplest is the direct competition between two species which results in the eventual replacement of one species by another. A mathematically general analysis of this situation is possible which allows for either of two species to win at the expense of the other; or, for equilibrium to be maintained, special conditions have to be imposed upon the reproductive values for the species in relation to their demands upon the common food resources. From such analysis we can show that the equilibrium can exist only if, for each species in turn, the inhibiting effect of one species on the other is less than its effect upon itself. J. Maynard Smith has concluded that if the food supply available to two species was in part common and in part species-specific, the stable equilibrium would be conceivable. It would not be, however, if the two species were micro-organisms, each of which released into their surrounding medium a toxic substance, for then the toxic substance could have a greater inhibiting effect upon the other species than on itself [8].

The analysis of more complex interactions between species leads to applicable results for predator–prey and host–parasite situations. In Fig. 2.11 some of the possible oscillatory results are presented. The curves labelled A and B might represent a parasite and its host, respectively. Such oscillations could then result from a host limited only by the parasite and a parasite which confines itself to one host species. Alternatively if, in Fig. 2.11, species A represents the predator and C the prey, we notice that cyclic behaviour in the prey can induce a similar pattern in the predators. These cycles are not necessarily of the same periodicity, for any one of a number of reasons. A first reason might be that the species have life cycles of different lengths. Alternatively, the feeding behaviour of the predator may change when the prey are plentiful; or it may be that the prey may be successful and avoid being caught. For detailed analysis, the reader should consult a specialized textbook [8].

We now return to the single population. The structure of a population which is usually considered to be most important is its age structure. The reason for this is the inevitable dependence of fertility and mortality upon age. Without any knowledge of the age structure for a population there is little point in refined calculations and the description and projection can be made only in crude terms. Fig. 2.12 shows the population pyramids for two actual populations which differ markedly in their age distributions.

The matrix notation introduced in the last section can help us once again. We need a tool to give us insight into changing age structure with time. Ideally built for this purpose, the *population projection matrix*, or *Leslie matrix*, has the form of the migration matrix, but instead of 'places' we use 'age classes'. After all, we move to the next age class each birthday – a kind of migration. Also population projection must take account of births and deaths. (This matrix is named after P. H. Leslie, the Oxford zoologist, who has been primarily responsible for its widespread use [9].)

$$\begin{array}{c} \text{age} \\ \text{class} \\ \text{next} \end{array} \begin{array}{c} 1 \\ 2 \\ 3 \end{array} \begin{pmatrix} f_1 & f_2 & f_3 \\ s_1 & 0 & 0 \\ 0 & s_2 & 0 \end{pmatrix}$$

age class now
1 2 3

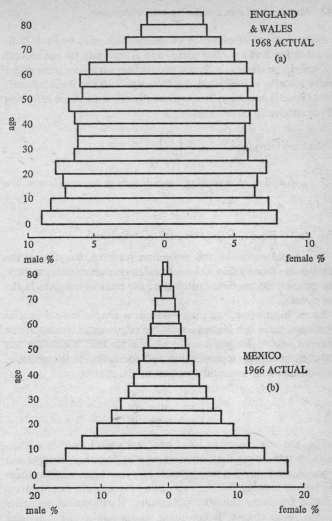

Fig. 2.12. Population pyramids of two contrasting populations. The upper graph (*a*) shows the age composition of the actual population in England and Wales in 1968 and the lower graph (*b*) that in Mexico in 1966. Mexico is a much younger population

f_1 is the number of offspring produced each year by each individual in the first age class and f_2 and f_3 similarly for the second and third age classes; s_1 is the proportion surviving from age 1 to age 2 and s_2 is that surviving from age 2 to age 3 during each year. (Strictly speaking, the f-values include only those offspring which survive to be counted.)

Next we introduce $$n = \begin{pmatrix} n_1 \\ n_2 \\ n_3 \end{pmatrix}$$

with n_1, n_2 and n_3 representing the numbers in the age classes. We have

$$n' = Pn$$

(the Matrix Equation for Population Projection).

This equation shows the projection forward one year of the population. Once again, the repeated operation reveals *stability*, this time of the age distribution (i.e., the relative numbers in the age classes).

As an illustration, we may envisage a simple animal species which has three age classes consisting of juveniles, young adults and old adults. To begin with we assume that each class has 100 members. The reproduction and mortality of the animals, we suppose, is known and specified by the matrix

$$P = \begin{pmatrix} 0.1 & 0.5 & 0.01 \\ 0.5 & 0 & 0 \\ 0 & 0.25 & 0 \end{pmatrix}$$

Taking the age classes to be of one year each in width, we can see from this matrix that, on average, a juvenile may survive and produce one-tenth of a new juvenile one year later, whereas there will be one new juvenile for every two young adults but only one juvenile for every hundred old adults. Furthermore, juveniles have a 50–50 chance of becoming young adults one year later but young adults have only a 1 in 4 chance of becoming old adults. Old adults, we suppose, just die.

After one year, some animals in each age class will have died,

thus reducing the numbers of all ages, but some will have reproduced, thus contributing new members to the juveniles class. The
aggregate effect of these vital processes is a changed population
age structure with 61, 50 and 25 animals in the three age classes.
These are the numbers which result when the matrix P is multi-
plied by the vector $v = \begin{pmatrix} 100 \\ 100 \\ 100 \end{pmatrix}$ $v' = Pv = \begin{pmatrix} 61 \\ 50 \\ 25 \end{pmatrix}$. A further year
(and one more multiplication of P by the new v) sees the numbers
change to 31, 30 and 12, respectively (rounding off to the nearest
whole numbers), and then to 19, 16 and 8 and again to 10, 9 and 4.
Although this animal species is heading for extinction we can see
that the ratios in the age classes, even on such small numbers,
do show some stability. This stability is a necessary feature
produced by the latent roots and vectors of the matrix P. (The
reader may care to verify that the largest latent root of the matrix
in this example is 0·555 and its associated latent vector is $\begin{pmatrix} 2·462 \\ 2·219 \\ 1 \end{pmatrix}$.
This tells us that each year the population reduces in size by
55·5% of its value at the beginning of the period. Furthermore,
the ratios of the numbers in the age classes will stabilize at
2·462 : 2·219 : 1 or in approximate percentages 43 : 39 : 18.)

Returning to our example shown in Fig. 2.12, we can see if this
applies to the two populations: England and Wales in 1968 and
Mexico in 1966. Clearly, we can see that the England and Wales
population cannot be stable because the numbers in the age
classes do not decrease with increasing age. Some of the age
groups (20–24 and 45–49) have more in them than do some
younger age groups and the 'bulges' and 'waist-lines' in the
population pyramid will have to move upwards with time. This
will ensure that the profile will change for many years to come.
We can reflect that if sizable migrations were allowed at certain
specific ages and times then stability could be guaranteed by
emigrating out or immigrating in as required. But this is fantasy.

Turning to the Mexico population, this seems much more
agreeable and it certainly satisfies the reducing age class condition.

We tentatively accept that it *could* be stable. But we can now make use of the mortality and fertility data provided in the form of a Leslie matrix in Table 2.1. If we proceed to calculate the stable age distribution by the method described earlier, we obtain the results presented graphically in Fig. 2.13. Surprisingly,

TABLE 2.1   Leslie matrices: *Actual populations*

The mortality and fertility rates are presented here in the form of survival probabilities from each age group to the next and reproduction rates. These values are the elements of the Leslie matrices for England and Wales and for Mexico. The n_∞ columns are the stable age distributions in the form of percentages in the age classes depicted in the central frame. These stable distributions when plotted appear as the 'ghost' outlines in the population pyramids of Fig. 2.13.

| *England and Wales 1968* | | | AGE | *Mexico 1966* | | |
s	f	n_∞		s	f	n_∞
0·997	0	8·7	0–4	0·972	0	18·4
0·999	0·0001	8·3	5–9	0·992	0·001	15·0
0·998	0·058	8·0	10–14	0·994	0·107	12·5
0·998	0·254	7·7	15–19	0·991	0·429	10·5
0·998	0·395	7·3	20–24	0·987	0·677	8·7
0·997	0·309	7·0	25–29	0·984	0·631	7·2
0·995	0·161	6·7	30–34	0·980	0·517	6·0
0·992	0·065	6·4	35–39	0·976	0·367	4·9
0·987	0·014	6·1	40–44	0·972	0·152	4·0
0·979	0·001	5·8	45–49	0·964	0·025	3·3
0·968	0	5·4	50–54	0·950	0	2·7
0·807	0	22·5	55–	0·817	0	6·8

$$P = \begin{pmatrix} f_1 & f_2 & f_3 \\ s_1 & 0 & 0 \dots \\ 0 & s_2 & 0 \dots \\ 0 & 0 & s_3 \dots \\ \cdot & \cdot & \cdot \\ \cdot & \cdot & \cdot \\ \cdot & \cdot & \cdot \end{pmatrix}$$

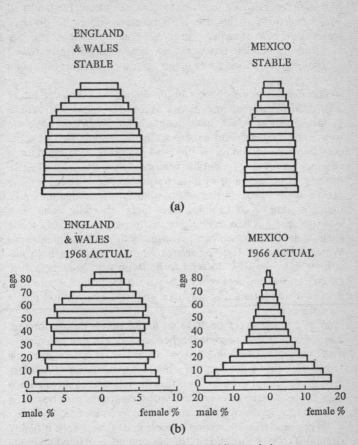

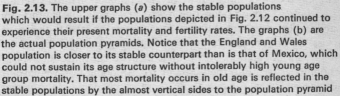

Fig. 2.13. The upper graphs (a) show the stable populations which would result if the populations depicted in Fig. 2.12 continued to experience their present mortality and fertility rates. The graphs (b) are the actual population pyramids. Notice that the England and Wales population is closer to its stable counterpart than is that of Mexico, which could not sustain its age structure without intolerably high young age group mortality. That most mortality occurs in old age is reflected in the stable populations by the almost vertical sides to the population pyramid

Mexico appears to be further from stability than England and Wales and this remains true if we take account of the different scales at the base of each pyramid. We conclude that the increase in births in Mexico has not been accompanied by increased mortality in childhood or by major migration and this instability is real.

The matrix approach can be developed to take account of additional structure factors as well as age. The relevant structural factors are those which have demographic importance. We are well aware that different subgroups in the populations of some countries have different fertility and mortality. For example, a religious group may forbid its members to use the more effective forms of contraception, with the result that family sizes for these people tend to be larger. Also, dangerous occupations like racing driving or coal mining have higher mortality rates than safer occupations like teaching or mathematics! At a very crude level of description, we can get away with some simplification. We find that high fertility groups in many societies, both advanced and primitive, do tend to be those with high mortality. Of course, we can deal separately with these groups where they exist in a non-interacting way in a country. The method above then applies.

In reality, however, we find that interaction is more common. Sons of coal miners become teachers and sons of mathematicians may even end up as racing drivers! Over several generations, this 'social mobility' will affect the growth of the different social class groups in a country. This means that the numbers in these groups may not be in balance over time and some groups may be taking over from others which may be doomed to extinction.

To see how this interesting process works, we adopt a highly simplified view of things. If we divide a population into two crude social classes, one upper, one lower, and apply lower mortality and fertility rates to the upper class and higher rates to the lower class, we can proceed to determine the long-term stability of the aggregate population. More than this, we can introduce social mobility as well into the Leslie matrix. In doing this, we have to decide which vital rates apply to those changing class in any generation. We could insist that an upwardly mobile person carried with him or her the higher mortality and fertility associated with the early environment. Alternatively, we could believe in 'instant' adaption to the new environment and give them vital rates of the destination class. A compromise, which we have used, is to apply rates to all movers which are half-way between their origin and destination rates. The other input information for this highly simplified illustration is provided in Table 2.2, together with the enlarged Leslie matrix.

TABLE 2.2   Differential fertility and mortality

2 populations with migration

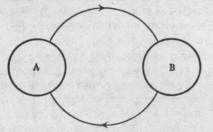

Consider females in 3 age groups *0–14, 15–29, 30–44*

1. FERTILITY
   Average number of daughters in 15 years

   for *A*   0·25   0·5   0·25
   for *B*   0·35   0·7   0·35

   for these age groups

2. MORTALITY
   Survival probabilities from one age group to next:

   for *A*   0·99   0·999
   for *B*   0·97   0·998

FERTILITY and MORTALITY both higher in *B* than *A*. Above rates apply to non-migrants, migrants take values half-way between *origin* and *destination* values.

3. MIGRATION

Suppose 20% move in each direction in 15 years. (Chosen large to exaggerate effects of differential vital rates.)

LESLIE MATRIX for A for B

$$P_A = \begin{pmatrix} 0.25 & 0.5 & 0.25 \\ 0.99 & 0 & 0 \\ 0 & 0.999 & 0 \end{pmatrix} \qquad P_B = \begin{pmatrix} 0.35 & 0.7 & 0.35 \\ 0.97 & 0 & 0 \\ 0 & 0.998 & 0 \end{pmatrix}$$

for A and B, where $m = 0.2$,

$$P_{A+B} = \left(\begin{array}{c|c} (1-m)P_A & \frac{1}{2}m(P_A + P_B) \\ \hline \frac{1}{2}m(P_A + P_B) & (1-m)P_B \end{array} \right)$$

The reader will notice that P_{A+B} is a 6×6 matrix. Writing out this matrix in full, substituting numerical values, we see that

$$P_{A+B} = \left(\begin{array}{ccc|ccc} 0.2 & 0.4 & 0.2 & 0.06 & 0.12 & 0.06 \\ 0.792 & 0 & 0 & 0.196 & 0 & 0 \\ 0 & 0.7992 & 0 & 0 & 0.1997 & 0 \\ \hline 0.06 & 0.12 & 0.06 & 0.28 & 0.56 & 0.28 \\ 0.196 & 0 & 0 & 0.776 & 0 & 0 \\ 0 & 0.1997 & 0 & 0 & 0.7984 & 0 \end{array} \right)$$

The results are presented in Table 2.3 and Fig. 2.14 and show the effects of differential vital rates upon the growth of the two classes, which increase at different rates and have different stable age distribution within the classes. In particular, 14·6% of the whole under-45 population are upper class under-15s whereas 22·4% of this age are lower class. Similar but smaller excesses in favour of the lower class exist in other age groups. Consequently, the lower class has a younger population than the upper class. More importantly perhaps, we can see that the lower class increases its share of the total with time. The end result of this is, of course, the extinction of the upper class! As we have admitted, this is a greatly simplified description, but the power of the matrix approach in dealing with complex behaviour is apparent from this example.

With the passage of time, species of all types are at risk of

TABLE 2.3

| POPULATION | | TIME (years) | | | Eventual | |
Age	0	15	30		(per cent)	
A	0–14	100	104·0	106·1	33·9	14·6
	15–29	100	98·8	109·0	33·7	14·6
	30–45	100	99·9	98·4	32·4	14·0
	Total	300	302·7	313·5	100·0	43·2
B	0–14	100	136·0	144·6	39·3	22·4
	15–29	100	97·2	125·9	32·4	18·4
	30–45	100	99·8	97·3	28·3	16·0
	Total	300	333·0	367·8	100·0	56·8
Total $A + B$		600	635·7	681·3	—	100·0

extinction and nature often compensates for this by providing for the introduction of new species. This balancing mechanism produces an equilibrium for the number of species in an environment if the immigrant species equal in number those made

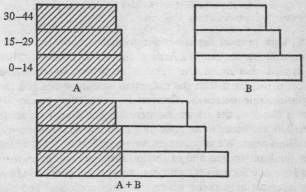

Fig. 2.14. The age composition for a stable population with two classes *A* and *B* experiencing different vital rates. Class *A* has lower fertility and mortality and this results in a population for that class which has a greater mean age

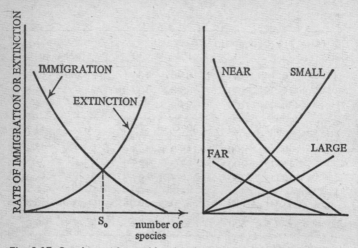

Fig. 2.15. Species number and immigration or extinction rates. The intersection of the two curves corresponds to a balanced equilibrium with optimal species number, S_0. The right-hand graph shows immigration curves for islands near or far from the mainland together with extinction curves for small and large island populations. Note the different S-values for the intersection far–small and near–large but similar values for near–small and far–large

extinct. This proposition has been formally made by F. W. Preston, and by R. H. MacArthur and E. O. Wilson. They have argued that when the number of species in an island situation is too great then the extinction rate increases and the immigration rate decreases. Conversely, when the number of species is too low, the reverse happens. This is a very plausible proposition and leads to an appealing mathematical description of such situations. We may inspect the graph in Fig. 2.15, which depicts the immigration and extinction rates as two curves, one decreasing, one increasing, with the number of species. These curves intersect at a point where the rates are equal and the island may then be seen to be, in one sense at least, in a state of balanced equilibrium. (The stability of this equilibrium is not fully demonstrated here, but the method used earlier for a balanced polymorphism could be used.)

Although there is some support for this simple description in real biological populations, the theory has to be developed a little further to justify it fully. After all, we need to take account of the varying distances travelled by the immigrants according to the position of the island. For islands near to a mainland the immigration rates would be higher than those further away. Also, the probability of extinction would depend upon the size of the population for a species, and hence on the size of the island, ultimately. The result of such changing rates allows for data to be plotted in a more structured way and compared with the right-hand curves in Fig. 2.15. We note that the effect of distance or size upon optimal species number can be considerable, but 'large–near' and 'small–far' numbers could be similar.

The concept which the biologists call a 'biotic equilibrium' is what we have just illustrated in a simple way. This is a new concept in ecology and there are insufficient data so far available to verify or reject the propositions of the above type which are now being made since more mathematicians have become interested in the biological world. Without much extended discussion, we present below some of the law-like relationships which have been discovered.

1. The number of species on an island, S, depends upon the area A of the island according to the law $S = CA^z$ where C is a constant dependent upon the organisms concerned and z is a constant dependent upon the units of area but apparently invariant to the type of organism.

2. The *extinction rate* at equilibrium, X (also equals the immigration rate at equilibrium), the expected number of species S at equilibrium and the time taken to reach 90% of this number upon colonization, t, are related by the equation $X = \dfrac{cS}{t}$ where c is a constant.

3. The evolution of species and the colonization of a habitat may be traced in terms of r, the *intrinsic rate of natural increase* introduced earlier in the chapter, and K, the *carrying capacity of the environment*. Both constants may be expressed in terms of the fitness of individuals possessing particular characteristics (and

presumably particular genes). When a species is newly colonizing it requires individuals to have high fertility and *selection* (strictly *r-selection*) but when it is established it will require its members to be highly adapted to the environment and we then say that *K-selection* operates [10].

The success of man in this environment would seem to have little to do with the imperfect understanding we have of the processes which govern the growth of human populations. We might look instead at our ability to manipulate the biological environment and control the growth of populations of other species. In agriculture, we apply our knowledge of genetics in order to improve the supply of food both in quality and quantity. This process requires an application of the principles of selection, which we have discussed in an elementary way in this chapter, to animal and plant-breeding situations. In this way we may modify those characteristics of animals and plants which depend upon particular genes. Not all species may be controlled in this way, however, and we may contrast the position of farm animals which may be raised in a closely monitored and regulated environment with fish living off-shore over which our control is considerably less. In this instance we shall, however, confine ourselves to a brief examination of the latter topic, which has proved to be an exciting and profitable newer use of mathematics. The mathematics for describing the dynamics of fish populations may be stated in outline, in a simple way. This takes the form of a fundamental equation for C, the change in the total fishing stock; in terms of G, the growth in weight; R, the weight of new recruits; F, the mortality from fishing; and M, the natural mortality (including the net migration balance):

$$C = G + R - F - M.$$

None of the four quantities on the right is particularly simple to express as a mathematical function but all may be specified in numerical form. Consequently, computations may be made on different assumptions concerning the natural population or the deployment of fishing fleets including the sizes of the fishing nets which they use. By this process, we may arrive at optimum strategies for exploiting the resources of the fishing grounds. R. J. H. Beverton and S. J. Holt have used this approach to examine North Sea trawled fish. Eventually, they achieved the *eumetric fishing curve* which describes the optimum mesh size for a given fishing intensity. This curve is depicted in Fig. 2.16(*b*). Let us hold *F* fixed and examine the mesh size/yield relationship. It appears that we may increase the yield in weight by taking smaller fish, but there is a lower limit to this fish size, beneath which the yield declines because of the depletion of stocks. The importance of choosing an optimum size becomes clear and the difficulty in doing so arises partly from the lack of knowledge concerning the natural mortality. The value of the approach is that it allows different assumptions to be made and their consequences examined by calculation.

Recent work in this field has examined different fish species which are of critical importance to our food supply, and also different strategies using variable rates of exploitation and more mobile fishing fleets. This appears to be an area to which mathematicians have only recently turned but one where the eventual benefits from their activities will be very considerable [5].

4. Medicine

When we are ill we go to a doctor, not to a mathematician! Mathematics and medicine may seem to us as incompatible as oil and water. Our family doctor may have little interest in or aptitude for mathematics. After all, he is concerned with the 'art' of medical practice, and in his world mathematics can have little place. The world is changing. More and more medical research depends upon a mathematical or statistical argument. We can see this by opening any medical journal or sometimes even by reading our newspaper or watching television. Every day some new medical discovery is reported in the popular media. Often these associate one of our favourite foods *F* with a feared disease *D*. Such proclamations are made in statistical or probabilistic terms: 'we are ten times more likely to suffer from *D* if we eat *F* than if we do not'. This is not to say that these are *newer*

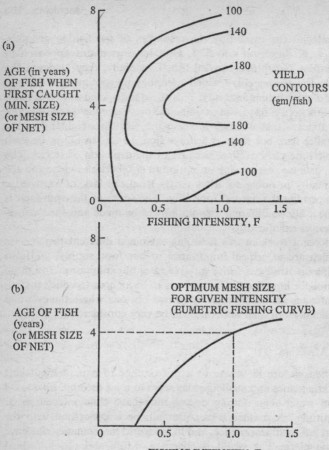

Fig. 2.16. The yields for several levels of fishing intensity have been calculated for different ages of fish caught. These yields are shown as contours in the upper graph (*a*). For each given fishing intensity, *F*, there is an optimum mesh size in order to produce a maximum yield. The curve of optimum mesh size for given intensity is shown in the lower graph (*b*). We can see, for example, that a policy of fishing 1% of the stocks produces an optimum yield if the nets have a mesh size to catch fish aged 4 years and upwards. This curve is known as the *eumetric fishing curve*

uses of mathematics. Much of this work involves the use of fairly standard statistical methods in the design and analysis of clinical studies, population health surveys, laboratory experiments, and so on. Rather more relevant to the theme of this book is the development of special mathematical arguments in response to specific problems arising in medicine which have no exact analogues in other fields of inquiry. We shall outline a few of these situations in the following pages. Some further examples are given by P. Sprent [11].

What happens when we visit the doctor? He tries to find out what is wrong. This is known as diagnosis. In order to discover which disease we may have, he will elicit signs, and this means he will take our blood pressure or temperature or an X-ray or measure the levels of concentration of substances in our blood or urine. He may impose upon us some more dreadful tests. Using these technical measurements to help him he will attempt to assess alongside other general characteristics such as our age, sex, occupation or family history. His total knowledge of us can be seen as a set of observations. If we consider all of the forbidding diseases which are candidates in his mind for this diagnosis, we could label these D_1, D_2, ... etc. Now D_1 has a characteristic pattern of values for the variables which we have called observations and our doctor is trained to recognize this pattern and to distinguish it from the patterns of the other diseases. He may know intuitively that we have D_1. Perhaps D_1 is easy to recognize and is so distinct from other candidates that he is quite certain about the diagnosis. Unfortunately for us, this is not always our experience. Our doctor sometimes says very little and prefers to wait for more information. This will come as our illness develops or as the results of further tests are known. For our part, we hope that this all happens in time for his diagnosis to be beneficial!

Sometimes, particularly in trying to distinguish between rare but closely similar diseases, our doctor may have rather little personal experience, may find it difficult to marshal all the available facts and may need recourse to documentary evidence of past cases. How best can he use past experience in arriving at a diagnosis for a particular case? This is a fairly standard type of problem in statistics, and objective methods can be described [12]. They usually involve at some stage the use of computers to do the necessary calculations, although the resulting rules of procedure may be simple enough for the doctor to apply in his surgery. There are now a number of research findings which suggest that in some such situations computer-assisted diagnosis can be rather more reliable than an intuitive judgement.

Our doctor makes use of a diagnosis as a base for further action – in particular to decide what treatment to recommend for us. His decision will be based on a judgement about the likely effect on us, whether beneficial or adverse (through unwanted side-effects). Again, the problem may be common and its solution rather clear to the doctor; in other less common situations he may know a lot about the effects of alternative treatments in similar cases to ours.

The ideas we now present are discussed much more fully in later chapters. For the moment, we think of alternative decisions with outcomes which have known risks or probabilities and, possibly, costs attached to them. Armed with this information, we can pursue some calculations and arrive at an optimal decision.

To illustrate these ideas, we use a simplified example. We shall suppose that for two diseases D_1 and D_2 there are two treatments, T_1 and T_2, available. We have a 50 : 50 chance of improvement if we have D_1 and receive treatment T_1, but a 60% chance if we have this disease and receive treatment T_2. Also we know that if we have disease D_2 and are given treatment T_1 then our improvement is certain whereas with T_2 we only have a 10% chance of improvement. Now the doctor can easily decide upon the best treatment for us if the diagnosis is certain (and if there are no great differences in costs of the treatments and in their side-effects which might affect his decision). For if we have D_1 then T_2 has the higher chance of success and would be the obvious treatment. Similarly, if we have D_2 then T_1 leads to certain improvement. His task is then easy.

If the diagnosis is not certain, however, but our doctor may only know that the symptoms which we have would indicate in nine out of ten cases disease D_1, and in the remaining case disease D_2 then what is he to do? There are three choices open to him. He may ignore the one in ten chance and immediately recommend treatment T_2 because it is best for disease D_1. Alternatively, he may recommend treatment T_1 irrespective of the diagnosis because it has the highest over-all improvement rate (i.e., 100% and 50% are greater than 60% and 10% by some impressionistic judgement!). Thirdly, he may ask a mathematician who specializes in statistical decision theory for help. At this point, the reader is urged to make his own guess at the best treatment. What would you do?

The mathematician might suggest that the doctor draws a tree diagram as in Fig. 2.17 and on this diagram we have entered the known numerical data. Moving from left to right along the tree we multiply pairs of numbers to give the chance of improvement for a particular treatment and disease. For example, the chance of our having disease D_1 and improving following the application of treatment T_1 is $\frac{9}{10} \times \frac{1}{2} = \frac{9}{20}$ whereas the chance of our having D_2 and improving after the application of this treatment is $\frac{1}{10} \times 1 = \frac{1}{10}$. The next stage in the calculation is to add these two numbers to give the chance of improvement following the application of T_1 *whichever disease we have*. This chance is $\frac{9}{20} + \frac{1}{10} = \frac{55}{100}$ or $\frac{11}{20}$.

If we repeat this calculation for treatment T_2 we get $\frac{9}{10} \times \frac{6}{10} + \frac{1}{10} \times \frac{1}{10} = \frac{55}{100}$ or $\frac{11}{20}$, the same! This means that in this situation, whichever treatment is given in the absence of certain knowledge of the disease, the chance of improvement following treatment is the same. Our doctor could justifiably toss a coin to decide which treatment to give!

In this greatly simplified analysis of a consulting room problem, we have not included costs, and if the treatment costs of T_1 and T_2 were different this would make the choice between them easy. For example, T_1 may be an expensive drug and T_2 may be 'no treatment' and in this case the cost of T_1 and the possibility of unknown side-effects would on all grounds suggest the treatment T_2. One final point, concerning the possible improvement of the diagnosis: in our example if the uncertainty of the diagnosis were removed the chance of improvement could be increased by

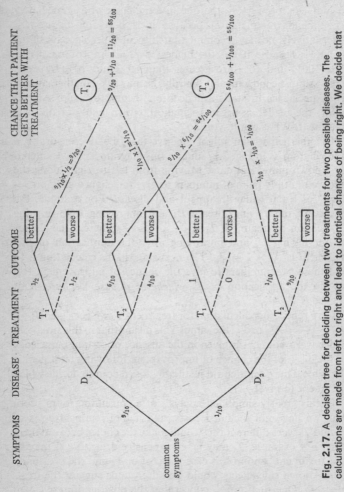

SYMPTOMS   DISEASE   TREATMENT   OUTCOME

CHANCE THAT PATIENT
GETS BETTER WITH
TREATMENT

$\frac{9}{10} \times \frac{1}{2} = \frac{9}{20}$

$1 \times \frac{1}{10} = \frac{1}{10}$

T_1 $\frac{9}{20} + \frac{1}{10} = \frac{11}{20} = \frac{55}{100}$

$\frac{9}{10} \times \frac{6}{10} = \frac{54}{100}$

$\frac{1}{10} \times \frac{1}{10} = \frac{1}{100}$

T_2 $\frac{54}{100} + \frac{1}{100} = \frac{55}{100}$

better — $\frac{1}{2}$

worse — $\frac{1}{2}$ T_1

better — $\frac{6}{10}$

worse — $\frac{4}{10}$ T_2 D_1 $\frac{9}{10}$

better — 1

worse — 0 T_1

better — $\frac{1}{10}$

worse — $\frac{9}{10}$ T_2 D_2 $\frac{1}{10}$

common
symptoms

Fig. 2.17. A decision tree for deciding between two treatments for two possible diseases. The calculations are made from left to right and lead to identical chances of being right. We decide that either treatment is best when we are uncertain about the diagnosis

using the best treatment from 55% to 60% in the case of disease D_1 and to 100% in the case of D_2. This suggests that more effort is needed in order to ascertain the disease D_2 when it exists because of the substantial improvement which could be realized. Our doctor would then be well advised to conduct further tests specially designed to detect disease D_2 in this case.

Even from this grossly simplified and rather special example we can see the way the minds of doctors and mathematicians can be fused to help patients. Some more extended decision theory applications to medical problems have been made by J. Aitchison and others and their work is very promising. There are, however, some real difficulties in getting sufficiently reliable information about the comparative effects of different treatments for specific types of patients. We shall see later that this kind of information is normally obtained from clinical trials.

Any attempt to discuss the best way of treating patients with a certain disease must depend on the natural course of development of the disease – whether it is communicable, whether in an individual patient it flares up and then disappears quickly, whether it moves through a series of easily distinguished stages, and so on. Processes of this sort can often be described and studied mathematically. There is, for instance, a considerable mathematical theory of communicable disease transmission [5, 13] which can be of use in the study of possible measures for the control of infectious diseases. The basic mathematical tools are the use of differential equations and the theory and methodology of stochastic processes. Similarly, many chronic diseases, such as cancer, mental disorders and heart disease, progress from one stage to another in a way which can be described numerically and studied mathematically. The stages may relate to methods of medical care: for example, many patients with psychiatric disorders move from home care, with various forms of medical treatment, to institutional care, again of various types. If a mathematical model can be constructed which simulates these movements, with frequencies approximating to those found in practice, it should be possible (perhaps by computer analysis) to study the effects on the system of changes in various parts of it (e.g. reducing the length of hospital stays). The technique, known as *computer simulation*, is described at some length in the next chapter, when we shall plan a doctor's appointment system by this method.

We now look at a topic which has received much recent attention. This concerns patients who have dyslexia following brain damage. If we receive a bang on the head in a road accident or grow a tumour or experience a spontaneous brain haemorrhage, it may cause damage to many areas of the brain the results of which can be very distressing. For certain localized head injuries only the language area of the brain may be affected and in this situation we will have language difficulties in reading or writing, or both, which usually show some rapid early recovery. The questions which will be of most concern to us are 'how long will the disability last?' and 'what, if any, permanent after-effects will there be?'. Recent work by Dr Freda Newcombe and her colleagues has shown that by careful regular testing and recording of our performance, these questions can be satisfactorily answered by using mathematical analysis followed by a statistical procedure which includes the fitting of a recovery curve. The importance of this work is that it has led to the collection of improved data on a regular basis from these patients. Also the mathematical structure of tests which are applied to the patients turns out to be a feature of critical importance.

A typical recovery curve for a severely affected patient is shown in Fig. 2.18. The horizontal asymptote to the curve represents the permanent reading deficit and the time taken to approach this asymptote may be estimated in advance. These two quantities and the precision of their determination are of utmost importance. The precision can be improved by good design of the testing procedure both in the choice of tests used and in the timing involved. Recovery depends upon the natural healing of damaged brain cells and the neural network between them and upon the reorganization of function which compensates when cells are affected by injury or operation. Both processes are supported by exposure to external stimuli, some of which are contained in the tests applied to the patients: hence the importance of the test procedures. If we find ourselves in the unfortunate position of being such a patient we may be gratified to know that some of the uncertainties of our predicament can be removed by mathematics.

The study of cancer gives rise to some questions of a more biological nature. If we consider cancer of a particular site, and examine from national statistical data the risks of incurring the

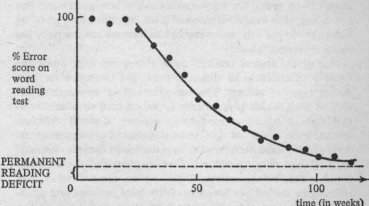

Fig. 2.18. A recovery curve for a patient with acquired dyslexia. This patient improved performance on a reading test between week 20 and week 100 but little further recovery takes place after 100 weeks. The residual deficit is shown by the asymptote to the curve and this deficit can be predicted from the earlier t est scores. Also, the time to reach this asymptote may be predicted in advance

disease at different ages, we find a remarkable consistency: the risk (or 'incidence') seems to vary in proportion to the kth power of age, where k is about 4 or 5.

The very consistency of this relationship, recently reported by R. Doll, suggests that there may be some common underlying mechanism giving rise to it. One widely discussed theory is that the observed cancer is the end-product of a number of successive changes in a cell (perhaps in its nuclear material). Mathematical theory shows that a rather more specific formulation of this idea would lead to precisely the power-law described earlier, where $k + 1$ is the number of changes required. This and similar theories have been quite successful in describing the results of experiments in which carcinogens are applied to the skins of animals and are also reasonably compatible with observations on human populations exposed to cigarette smoke and other carcinogens. For example, it is sometimes thought to be surprising that the average age of onset of lung cancer is about the same (55–60 years) for non-smokers as for heavy smokers. But this is just what would be expected if the effect of heavy smoking was to retain the kth power law but to increase substantially the factor of proportionality.

One of the central tasks of medical research is to study the relative effectiveness of different treatment strategies for particular types of patient. The basic method of approach is the clinical trial, which is essentially an experiment in which treatments are allocated at random to patients. Clinical trials are amenable to most of the general precepts of experimental design which have been familiar to statisticians for forty years or so, but they raise special problems particularly of an ethical nature. The circumstances under which clinical trials can ethically be carried out have been fairly fully explored and documented [14]. These special features of clinical trial have given rise to the idea of conducting trials sequentially, so that a trial can be stopped (as, ethically, it usually should be stopped) if a large difference between the effects of rival treatments becomes apparent during the study. Many sequential trials have, in fact, been conducted and reported in the literature. Such a tendency to examine repeatedly the accumulating records of a trial produces a rather subtle statistical difficulty. It increases the chance that strong evidence for a difference will be found at some stage, even when the treatments being compared have absolutely identical effects. This phenomenon is well known in statistics and has given rise to a branch of the subject called *sequential analysis*. It would be inappropriate to discuss this topic in detail, but one point to be made is that the special needs of clinical trials have given rise to special sequential plans, i.e., rules which tell the investigator whether, at any stage, to stop or to go on [15]. Given any such plan, how do we discover what are its properties (e.g. how it copes with the problem of repeated tests on accumulating data mentioned above)?

Such questions call for a variety of mathematical methods according to the type of data and the type of sequential plan being considered. Some approximations come from the theory of diffusion, the separate steps corresponding to the addition of an extra observation being replaced by a continuous movement analogous to the diffusion of a particle. Other exact results come from repetitive use of very simple results in probability theory. In some other problems the probability calculations can be done effectively only by the use of a computer. A different use of the computer is in simulation, which is particularly appropriate when the data are of an unusual type or if the sequential plan which seems intuitively desirable is too complex to be studied analytically.

The plans referred to above, which have been used extensively in practice, are usually specified in terms of probabilities of error. In comparing two treatments one might specify that (i) if they are really equivalent in their effect one wants only a small probability of concluding that they are different, and (ii) if they are really different (by a specified amount) one wants a high probability of being able to say so. Over the last fifteen years or so a good deal of discussion has taken place about the alternative possibility of viewing the clinical trial as a decision procedure and applying standard methods of decision theory.

Suppose N patients (the 'horizon') with a certain disease are to be treated with one of two treatments, A or B. We might compare A and B in a (possibly sequential) randomized trial, and then decide to use whichever appears to be the better of A or B on the remainder of the horizon. How many patients should be in the trial? If the trial is to be sequential, what plan should be followed? In decision theoretic problems we must make assumptions about the consequences of actions; in this case it seems reasonable to make the loss in using the wrong treatment depend on the difference between the true measures of effectiveness. We also should take account of our preliminary knowledge or beliefs about the possible magnitudes of the difference. When put into more formal mathematical terms these concepts enable answers to be obtained. The real difficulty lies in attributing numerical values to quantities such as the horizon and to the investigator's preliminary knowledge, and such problems have prevented this approach being fully applied in practice. It does, however, clarify some of the features which ideally should affect the design of a clinical trial, and it reminds the investigator that the decisions to be taken in planning a scientific investigation often depend on a judicious guess at the values of unknown quantities.

5. Pharmacology

We are a drug-consuming society. Most of us take drugs, if only to relieve the occasional headache or to suppress the occasional cough. Some of us who suffer from such chronic illnesses as diabetes and epilepsy may even need to take drugs daily in order to lead normal productive lives. Pharmacology, the science of drugs, has therefore become increasingly important and it forms an integral part in the training of physicians, pharmacists, veterinarians, dentists, nurses and others concerned with administering drugs to patients (usually man) and in the training of scientists developing new drugs. Within pharmacology, mathematics is playing an expanding rôle. We shall explore just a few examples of these newer applications of mathematics. These and others are discussed in the references [11, 16, 17].

Drug treatment involves maintaining a desired effect for a period of time. Usually, we maintain this effect by repeatedly administering a dose of drug (a tablet, capsule or teaspoonful of syrup) for the period of treatment. We call this course of treatment a *dosage regimen*. But why do we need to take some drugs every few hours and others only once a day, and why do we sometimes have to take more at the beginning of the treatment? In the past, the answers to these questions were obtained by trial and error. Namely, the physician adjusted the dose and the time between doses and followed the patient's progress. Efficacy and toxicity were noted carefully and the dosage regimen was adjusted empirically until a maximal effect with a minimal toxicity was obtained. After considerable experimentation, on a sufficiently large number of patients, reasonable dosage regimens were established. But a great deal of information had to be gathered and, in arriving at the final choice, many of the regimens may have produced excessive toxicity, or may have had no effect at all. Today, mathematics is allowing us to take a more rational approach in the establishment of dosage regimens.

As already mentioned, the objective of drug therapy is to reach and then keep the amount of drug in the body within a range: the amount must be enough to produce the desired effect but not so much as to cause excessive ill effects. Placing the right amount of drug in the body will not suffice because the effects will wear off as the liver (the chemical converter and power house) and the kidneys (the filtering system) rid the body of drug. We need to replenish continually the body drug stores. In some ways the problem is analogous to that of maintaining a level of water in a tank, when both the tap and outlet are open, but in other ways we are dealing with an infinitely more complex situation. Drugs are usually swallowed; but swallowing is one matter and onset of effect is another. It can take from minutes to hours before all the drug has passed from the stomach and intestine into the blood stream. Once in the blood stream, the drug is then distributed at different rates to all parts of the body. Patients who marvel that a drug taken by mouth can 'find' the brain to relieve their headache fail to realize that the drug distributes throughout the body *including* the brain. You might think that these and other complexities would preclude any tractable mathematical solution to our problem; but simplicity is often a virtue. So let us proceed by initially considering the body, like the tank, to be a single container (Fig. 2.19). Before proceeding, however, we should point out one major difference between the two situations. In the tank problem, we can maintain the level by adjusting either the flow from the tap, the size of the outlet, or both. In our problem we have no control over the outlet (liver and kidneys); we can only adjust the *rate of administration*.

Like the water in the tank, the rate of loss of drug from the body is proportional to the amount in the body. The proportionality constant which we will denote by k is appropriately called the *elimination rate constant*. When, as in this case, the

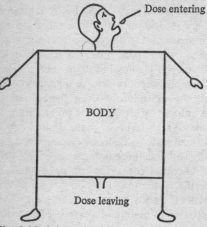

Dose entering

BODY

Dose leaving

Fig. 2.19. Schematic representation of the body as a single container with drug entering (usually via the mouth) and leaving (via the chemical conversion by liver and excretion by kidneys)

rate of the reaction is proportional to the first power of the amount present, the process is said to be first-order. First-order processes are found in many physical and biological systems. Suppose that the drug enters the body continuously at a constant rate, R_0. Then, the amount of the drug in the body, denoted by A, changes at a rate given by the equation:

Rate of change in the amount of drug in the body = Rate of administration − Rate of loss of drug from the body

(for the reader who has calculus we can write this:

$$\frac{dA}{dt} = R_0 - kA).$$

The analytical solution for this rate equation can be found in any standard textbook of calculus and its form is shown by a continuous line (Fig. 2.20). Initially, with no drug in the body, the rate of administration exceeds the rate of loss and the body level rises. As the level rises, so does the rate of loss, and eventually the rate of loss matches the rate of administration so that the rate of change equals zero. Thereafter, the level in the body remains constant so long as the rate of administration is maintained. The above equation shows that this happens when $A = R_0/k$. As this depends only on R_0 and k, by adjusting the rate of administration we can obtain any level or plateau we choose, provided the value of k is known. In practice it is more common

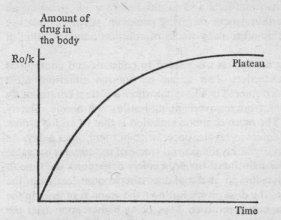

Fig. 2.20. When drug enters the body at a constant rate, the amount in the body rises until, when output matches input, a stable level is reached. The time to reach the plateau depends only on the half-life of the drug

to refer to the *half-life* of a drug rather than its elimination rate constant.

The half-life is the time it takes for the amount of drug in the body to fall by one half when the drug administration is stopped. It can be shown to be equal to $0{\cdot}693/k$ and usually it is independent of the amount of drug in the body. Each drug has a characteristic half-life which can vary from minutes to days or weeks. For example, in man, the half-life of the antibiotic penicillin is only 20 minutes, while that of quinacrine, a drug used to treat malaria, is almost 2 weeks.

Now we can see why a delay always exists between the initiation of the constant rate of administration and the attainment of the stable desired level, and why the full effects of the drug may not be felt until some time after treatment is started. But just how long is this delay? An examination of this system shows that the *sole* factor influencing the rise to the stable level is the *half-life* of the drug. Also the closer we are to the plateau, the slower the rate of rise of drug in the body. Thus, one half of the plateau is reached in one half-life but only three-quarters of the plateau is reached in two half-lives and so on. For example, the plateau is reached within 1 hour of giving penicillin, whereas a plateau is reached only after many weeks of constant administration of quinacrine.

So far the discussion has related to constant and continuous drug administration. This is an uncommon situation, most frequently encountered in a hospital where a patient continuously receives drugs, from an overhanging bottle via a needle, directly into a vein. The more common situation is that of taking a dose of drug at discreet intervals; once, twice or more times a day. As we can see in Fig. 2.21, the general shape of the curve is the same as before; the half-life of the drug solely determines the time to reach the plateau and at the plateau the amount lost from the body matches the dose given. But now the amount of drug in the body fluctuates between each dose, being highest soon after the dose and lowest just before the next dose. In drug therapy, it is important to know the magnitude of this fluctuation at the plateau because the maximum and minimum levels should lie within the desired range. We can determine these limits in the following manner. Suppose a dose D is given each time. After the first dose, the amount remaining in the body *just before* the next dose is $D.r$ ('$D.r$' means '$D \times r$'), where r is the fraction of the dose retained in the body at that time. Immediately after the second dose, the amount in the body is $D + D.r$, or $D(1 + r)$, and immediately after the third dose it is $D(1 + r + r^2)$. Thus we are dealing with nothing more than a geometric progression, with the amount after the nth dose being given by $D(1 - r^n)/(1 - r)$. After a large number of doses, n is large and since $r < 1$, r^n has become very small, so we have nearly reached the plateau at which the maximum amount is $D/(1 - r)$ and the minimum is $Dr/(1 - r)$. For example, when 100 mg of a drug is given every half-life ($r = 0.5$), the maximum and minimum amounts of drug in the body at the plateau are 200 and 100 mg, respectively. Conversely, to maintain these amounts of drug in the body, a dose of drug must be given every half-life. This explains, in part, why some drugs are given three times a day and others given only once daily.

Sometimes, because a prompt response is needed, the delay in reaching the plateau is too long. This (as you might expect) is especially true for drugs with long half-lives. In those cases, the problem is solved by giving a large first dose, which immediately attains the plateau level, followed by smaller supplementary doses which maintain the level in the body within the desired range. For example, a prescription for a drug with a half-life of 1 day might read: 'Take two tablets to start and then one tablet once daily.'

We previously stated that we have no control over the outlet

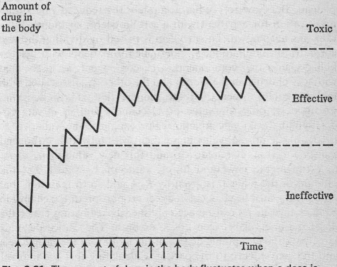

Fig. 2.21. The amount of drug in the body fluctuates when a dose is given at fixed times but the general shape of the accumulation curve is much the same as displayed in Fig. 2.20. The frequency of giving drugs should be adjusted to prevent the plateau level in the body fluctuating outside of the effective range

(liver, kidneys) from the body. This is true, but the ability to remove materials from the body can vary with dose of drug, health, age, weight and genetic background of the patient as well as other drugs the patient is receiving. Liver and kidney function are severely depressed in the new-born and deciding on appropriate dosages in the neonate is difficult. From 1 year onwards, however, successful adjustments can be made empirically on the basis of (body weight)$^{0.7}$. This experimental observation probably has some fundamental explanation since body heat production, body surface area, output of blood from the heart and blood flow to various organs are also proportional to the 0·7th power of body weight.

With the advent of the kidney machine, many people with poor or no kidneys, who would have otherwise died, live. It was known that, because of the diminished kidney function, these patients needed less drug to maintain normal amounts in the body at the plateau. But how to predict accurately the reduced rate of administration for a patient with a certain degree of kidney function was not known. The problem is solved by dividing the rate constant for elimination, k, determined in patients with healthy kidneys, into two components. One component, k_e, associated with loss of drug via the kidney and the other, k_m, associated with loss by routes other than the kidney, that is $k = k_e + k_m$. Fortunately, only the value of k_e changes and in direct proportion to the kidney function. For example, the value of k_e is 10% of the normal value in a patient with only 10% of normal kidney function. It remains to add this new value of k_e to k_m to estimate the new (and lower) value of k. Recall that the amount at the plateau is given by R_0/k and so to maintain that amount R_0 must be changed in direct proportion to the change in k. A change in R_0 can be accomplished by adjusting either the dose, the interval between doses, or both. The choice depends largely on the acceptable degree of fluctuation in the drug levels and the convenience to the patient. Calculating that the interval between doses should be changed, from the normal value of every 6 hours, to 16·5 hours is fine in theory but, in practice, expecting the patient to comply would be asking too much. It would be better to have the patient take the medicine every 12 hours and reduce the dose accordingly.

Drugs have a multiplicity of effects, some good and some bad. We are as much interested in these effects and how they vary with dose and time as we are in the amount of drug in the body, if not more so. An appreciation of these changes in effect has been approached by adding to the simple model in Fig. 2.21 (and to more complex models of drug handling) another that relates the concentration of drug, D, at the site of action to the effect produced, E. Most drugs act reversibly, some act irreversibly. One model for the reversible effect, shown in Fig. 2.22, has the form

$$\text{Effect, } E = \frac{E_{max} \cdot D}{K + D}$$

where K is an experimentally determinable constant. The importance of this equation lies in its prediction of the observed maximal effect at high drug concentration, $D \gg K$, $E \to E_{max}$ ('$\gg$' means 'very much larger than'), and proportionality between effect and concentration at low levels by the drug, $D \ll K$ ('$\ll$' means 'very much smaller than'). The maximal effect arises from the full occupancy of the limited number of sites of action in the body. We see from this equation that a condition of diminishing returns exists when $D > K$. Analysis will also show that when E lies between 20 and 80% of E_{max} then, if benefit is defined as the area under the effect time curve, greater benefit is gained by dividing a dose and giving the smaller dose at frequent intervals than is gained by giving the entire dose less frequently. Moreover, toxicity, often associated with high levels of drug in the body, might be avoided.

Cancer remains one of the most serious diseases facing man. Unfortunately there is no 'magic bullet' that kills only cancer cells. The problem in cancer drug treatment is therefore one of trying to eradicate the cancer cells from the body without permitting the toxicity of the drug to kill the patient. Cancer cells, like many other cells, go through a cycle. Broadly speaking, the cancer cell is either dividing and proliferating or resting.

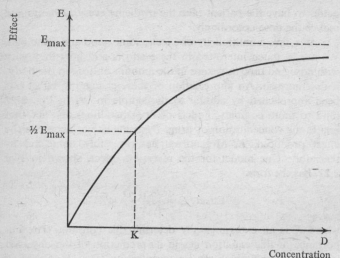

Fig. 2.22. The curve relating concentration to the effect produced by a drug. For concentration levels higher than K, the effect approaches its maximum value and then diminishing returns are seen from increasing concentrations

There are many mathematical models describing the kinetics of cancer cell growth; all account for the initial rapid growth (this is often called the logarithmic phase) and the subsequent slow down as the number of cells increase. If we examine the dynamics of the growth, we find that the fractional rate of growth decreases with an increase in cell numbers, or, stated differently, the fraction of the total number of cells which are growing decreases with increasing numbers. This is one reason why treatment of cancer with drugs is so difficult. Many of the agents used kill the cancer cell during the proliferating stage; these are known as cell-cycle-specific drugs. When detected, many tumours are well developed and so most of the cells are in the resting phase. Giving a cell-cycle-specific agent will kill the proliferating cells but as the total number of cells decreases, so the rate of proliferation increases. And, if the dosage regimen of the drug is poorly chosen, a point may be achieved where the number of new cells produced during a dosing interval equals or exceeds those killed. The host may then fail to survive. Analysis of the situation shows that eradication of the cancer is more likely to succeed if a large dose of the cell-cycle-specific agent is given initially, followed by smaller doses.

Another reason why drug treatment is difficult is that other cells in the body, particularly the blood cells developing in the bone marrow, are also sensitive to these agents. These blood cells are necessary for the body's defence and, if their number is suppressed too much, the host's defence weakens and the patient may die. A balance must therefore be sought. One approach is first to give a non-cell-cycle-specific drug which drastically reduces the total number of all cancer cells. Then, when the number of cells is small, most will be in the proliferating stage, which is a good time to 'hit them' with a large dose of the cell-cycle-specific agent. The problem facing us is to decide dose, frequency of dosing, duration of therapy and when each drug should be started. At present there is much empiricism. There are exciting possibilities, however, utilizing mathematics, in quantifying and predicting the clinical outcome of cancer treatment.

ACKNOWLEDGEMENTS

I am grateful to Professors P. Armitage (University of Oxford) and M. Rowland (University of Manchester) for generously providing material for the sections on Medicine and Pharmacology, respectively.

Fig. 2.6 is redrawn from graphs which appeared in *Phil. Trans. R. Soc. London. B*, 268, 241–249 (Harrison, G. A., Hiorns, R. W., and Boyce, A. J.) (1974).

Figs. 2.7 and 2.9 were redrawn from computer produced maps provided by Dr A. J. Boyce and Mrs P. J. Carrivick.

Fig. 2.10 is redrawn from *Annals of Human Genetics*, 32, 237–251 (Hiorns, R. W., Harrison, G. A., Boyce, A. J., and Küchemann, C. F.) (1969).

Figs. 2.12 and 2.13 were redrawn from *Population: Facts and Methods of Demography* (Keyfitz, N., and Flieger, W.), W. H. Freeman, San Francisco (1971).

Fig. 2.15 was redrawn from *The Theory of Island Biogeography* MacArthur, R. H., and Wilson, E. O.), Princeton Univ. Press (1967).

Fig. 2.16 is based on a figure in *Fish. Invest., London*, **19**, 7–533 (Beverton, R. J. H., and Holt, S. J.) (1957).

REFERENCES

1. DURRAN, J. H., *Statistics and Probability*, School Mathematics Project, CUP, 1970.

A very comprehensive text providing a sound basis in these subjects. Many good examples including some biological ones.

2. SAWYER, W. W., *A Pathway to Modern Mathematics*, Penguin Books, 1966.

An excellent introduction to some of the relevant mathematics used in this chapter. See Chapters 3 and 4 for matrices. Very well written.

3. SAWYER, W. W., *Prelude to Mathematics*, Penguin Books, 1955.

Like the 'pathway' book, this is useful supporting reading to what is here. See Chapter 8, in particular.

4. HODSON, F. R., KENDALL, D. G., and TAUTU, P. (editors), *Mathematics in the Archaeological and Historical Sciences*, Edinburgh University Press, 1971.

Contains some good articles, particularly strong on multivariate techniques, graphs and trees, computer maps by scaling techniques and seriation applied in interesting contexts.

5. BARTLETT, M. S., and HIORNS, R. W. (editors), *The Mathematical Theory of the Dynamics of Biological Populations*, Academic Press, New York, 1973.

A collection of articles discussing the uses of mathematics in ecology, demography, epidemiology and population genetics.

6. CAVALLI-SFORZA, L. L., and BODMER, W. F., *The Genetics of Human Populations*, Freeman, San Francisco, 1971.

A very comprehensive survey of this field. There is a strong quantitative bias in this large book, but the mathematics used is carefully developed. There is one appendix on statistics and probability and another on segregation and linkage analysis.

7. HARRISON, G. A., and BOYCE, A. J. (editors), *The Structure of Human Populations*, OUP, 1972.

A collection of largely expository articles in the areas of population studies including human genetics, demography, human ecology and sociology. There is some emphasis placed on the quantitative aspects of the topics covered.

8. MAYNARD-SMITH, J., *Mathematical Ideas in Biology*, CUP, 1968.

A reasonably up-to-date account of the kind of mathematics used by biologists. Written for biologists, but assumes some post O-level mathematics (calculus and difference/differential equations).

9. KEYFITZ, N., *Introduction to the Mathematics of Population*, Addison-Wesley, Reading, Mass., 1968.

A very illuminating account of the impact which mathematics has made in the last decade or so in the field of applied demography. Supplementary to the population mathematics in this chapter.

10. MACARTHUR, R. H., and WILSON, E. O., *The Theory of Island Biogeography*, Princeton University Press, 1967.

A stimulating book dealing with spatial relationships in biological populations.

11. SPRENT, P., *Statistics in Action*, Penguin Books, 1976.

This book contains some medical and pharmacological examples presented in an attractive way.

12. TEATHER, D., 'Diagnosis – Methods and Analysis', *Bulletin IMA*, Vol. 10, p. 37, 1974.

This article contains an appraisal of the methodology of computer-aided diagnosis and suggests some improvements. Several good illustrations are used and there are helpful references.

13. BAILEY, N. T. J., *The Mathematical Theory of Infectious Diseases and its Applications*, Griffin, London, 1975.

An up-to-date account of the use of mathematics in describing the mechanisms which underlie many infectious diseases. The level of the mathematics is advanced in places but the exposition is everywhere clear. The latter part of the book has good illustrations with real data.

14. HILL, A. B., *The Principles of Medical Statistics*, Lancet Publications, 9th edn, 1971.

A classic introduction to statistics for medical students. Also provides a good exposition of standardization techniques and of the Life Table.

15. ARMITAGE, P. *Sequential Medical Trials*, Blackwell, Oxford, 2nd edn, 1975.

A standard reference book to this methodology.

16. TEORELL, T., DEDRICK, R. L., and CONDLIFFE, P. G., *Pharmacology and Pharmacokinetics*, Plenum Press, New York, London, 1974.

Critical reviews of classical and modern approaches of drug kinetics to problems in pharmacology.

17. RIGGS, D. S., *The Mathematical Approach to Physiological Problems*, Williams and Wilkins, Baltimore, Md., 1963.

A primer, with problems and answers, to the use of elementary calculus and mass balance relationships in physiology and pharmacology.

3 Methods of Operational Analysis

1. Introduction

The two preceding chapters have been concerned with some of the more recent applications of mathematics in the natural sciences. The emphasis now shifts, and will remain for the rest of the book, to the social sciences; in particular, for the most part, to the broad field of management. The mathematical invasion – and effective colonization – of this hitherto largely innumerate territory was made possible by two events, both of which occurred in the 1940s – during the Second World War and the immediate post-war years. The first was the success of operational research; the second was the advent of the electronic digital computer. The impact of the computer is too well known to need further mention here [1,2]; in this chapter we shall attempt a brief survey of some of the more important mathematical methods used by the operational research worker and the management scientist. This chapter, with its emphasis on methods and examples rather than 'real-life' applications, is designed as an introduction to the chapters that follow, which deal with the uses of mathematics in particular areas of management and planning. We shall exclude conventional statistical techniques from the survey, notwithstanding their extensive use throughout the whole social science field. Such techniques are well known [3] and the more elementary ones have been available for quite a long time; they do not qualify for inclusion among the 'newer uses' of mathematics.

What, then, is operational research (or 'operations research', as the Americans call it)? Roughly speaking, it is the application of scientific methods – and thus, where appropriate, of mathematical methods – to the analysis and solution of executive problems, when it is necessary to take decisions leading to action. Nowadays most large organizations have their operational research teams, with trained mathematicians usually well to the fore.

Most management decisions exhibit one or more of the following characteristics. The first is that they are taken on the basis of information that is inadequate, or uncertain, or both. The manager is confronted with a mixture of fact, opinions and plain ignorance. If he attempts to call on mathematical assistance, the appropriate tool is more likely to be the calculus of probabilities than the calculus of Newton. The second is that the framework within which the decision must be taken (the *system*, to use the current jargon) is extremely complex. It involves a large number of factors, many of them changing all the time, which interact with each other in ways usually not fully understood. The third is that the object of the exercise is to achieve what is in some sense the 'best' solution of the problem within the limitations – of availability of money or labour, for example – imposed by the situation itself.

With this preamble, let us now give the official definition of operational research; 'official' in the sense that it appears at the front of every issue of the British journal the *Operational Research Quarterly*.

'Operational Research is the application of the methods of science to complex problems arising in the direction and management of large systems of men, machines, materials and money in industry, business, government and defence. The distinctive approach is to develop a scientific model of the system, incorporating measurements of factors such as chance and risk, with which to predict and compare the outcomes of alternative decisions, strategies and controls. The purpose is to help management determine its policy and actions scientifically.'

This definition merits careful scrutiny; it contains most of the key ideas that will form the framework of this chapter. If we wish to 'predict and compare' in quantitative terms, the 'scientific model of the system' must take the form of what is termed a *mathematical model*. This is usually a set of equations or inequalities which purports to represent the essential structure of the system and the relationships between its constituent parts. The behaviour of the model can then be studied by 'solving' the equations and interpreting the results in terms of the system being modelled. Alternatives can be examined by varying the co-efficients, and perhaps the form, of the equations. It is usually much easier to alter a model than to alter the actual situation it represents!

Mathematical models are used by the operational research worker in essentially the same way as they are by the natural scientist; by the meteorologist engaged in numerical weather forecasting, for example. Indeed, model-making is at the heart of all applied mathematics, whatever the field of application. One of its merits is that it can reveal similarities in the underlying structure of situations which, on the face of it, seem to have little in common: for example, working out a diet for farm animals or a duty rota for railway porters.

We have mentioned that the rapid growth in the influence of mathematics during recent years can be attributed in large measure to the arrival on the scene, more or less at the same time, of the electronic computer and operational research. These lusty twins have grown up together during the last thirty years. It is interesting to reflect that the origins of both can be traced to the same man. Charles Babbage (1791–1871) – for eleven years Professor of Mathematics at Cambridge – is generally accepted as the founding father of the automatic digital computer. A strong paternity claim can also be made for him in respect of operational research. He was a tireless advocate of the application of quantitative, scientific methods to the affairs of the world and he carried out a variety of investigations himself – into the workings of the Post Office, the pin-making industry and the printing trade, among others. His studies of the operations of the Post Office led him to the conclusion that most of the cost of a letter arises from what goes on in the sorting office, not during its journeys from sender to receiver. He argued, therefore, that the cost of sending a letter should be independent of the distance it has to go. Shortly afterwards, in 1840, the 'penny post' was introduced in Britain. Babbage's most successful book, *The Economy of Manufactures and Machinery*, was published in 1832; it was reprinted in the United States and translated into several European languages.

2. Linear Programming

Many managerial decisions boil down to decisions as to how to allocate limited resources to best advantage. This means, in terms of the mathematical model, that we wish to maximize something (e.g., profits) or to minimize something (e.g., costs). This something is usually expressed as a function (known as the *objective function*) of a number of factors, called *variables*. The process is usually termed *optimization*, although some operational research workers dislike the term because they believe it is virtually impossible, except in the simplest situations, to know what is the best, or the *optimal*, let alone to achieve it. In practice, the ranges of the variables of the model are subject to a number of limitations. These may arise from shortages of labour, materials or cash; from contractual obligations to customers; from legal requirements; or from many other things. Some of these constraints are so obvious as to pass unnoticed but may be of great significance in the mathematical model. Thus some variables must be non-negative, that is either positive or zero in value (e.g., the amount of material used on some project); others may be further restricted to integral (whole number) values (e.g., the number of men or machines allocated to a task).

Problems of this kind can sometimes be formulated – usually after some simplification – in such a way that they can be solved by a mathematical technique which has the somewhat unfortunate name of *linear programming*. The term 'programming' here is not used in the computing sense, but as a synonym for 'planning'; the significance of the term 'linear' will appear shortly. We will first illustrate the use of the technique by applying it to a very simple problem.

Consider a company which manufactures two kinds of cloth (*A* and *B*) and uses three different colours of wool. The material required to make a unit length of each type of cloth, and the total amount of wool of each colour that is available, are shown in Table 3.1.

TABLE 3.1 Data for the cloth manufacturing example

Requirements for unit length of cloth of type		Colour of wool	Wool available
A	B		
4 kg	4 kg	red	1400 kg
6 kg	3 kg	green	1800 kg
2 kg	6 kg	yellow	1800 kg

The manufacturer can make a profit of £12 on a unit length of cloth A and £8 on a unit length of cloth B. How should he use the available material so as to make the largest possible profit? To answer this question we must construct a mathematical model of the situation.

Let us denote by x_1 and x_2 the number of lengths of cloth A and cloth B, respectively, that are produced. Then, since only 1400 kg of red wool are available, we have

$$4x_1 + 4x_2 \leqslant 1400. \qquad (1a)$$

Similarly, considering the available green and yellow wool, we have

$$6x_1 + 3x_2 \leqslant 1800 \qquad (1b)$$

and

$$2x_1 + 6x_2 \leqslant 1800. \qquad (1c)$$

(The symbol '$\leqslant$' means 'less than or equal to'.)
The profit, P, in £ is given by

$$P = 12x_1 + 8x_2. \qquad (2)$$

We wish to choose x_1 and x_2 so as to maximize P, subject to the three constraints represented by $(1a)$ to $(1c)$ and the fact that neither x_1 nor x_2 may be negative. We may express this limitation as

$$x_1, x_2 \geqslant 0. \tag{3}$$

(The symbol '$\geqslant$' means 'greater than or equal to'.)

The situation is set out in graphical form in Fig. 3.1. Any pair of values of x_1 and x_2 corresponds to a point whose coordinates are (x_1, x_2). The first thing to note is that, in virtue of (3), we may confine our attention to the first quadrant of the (x_1, x_2) plane.

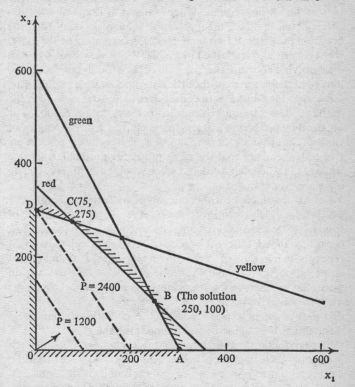

Fig. 3.1. Linear programming solution of the cloth manufacturing problem

The three full lines correspond to the three constraints (1). In the case of the first constraint, for instance, the form of the inequality in (1a) is such that we must exclude that part of the plane which lies further from the origin than the line whose equation is $4x_1+4x_2 = 1400$: similarly for the other two constraints. We see, therefore, that the admissible (or *feasible*) region of the plane is the polygon (both boundaries and interior) whose vertices are O, A, B, C and D. All that remains is to find the point in this region whose coordinates make $(12x_1+8x_2)$ as large as possible. Now all lines whose equations are of the form $12x_1+8x_2 = P$ are parallel to each other; the greater the value of P, the further the line from the origin. (Two such lines, corresponding to $P = 1200$ and 2400, are shown dashed in Fig. 3.1.) It is now clear what we must do; namely move the 'profit line' (2) as far as possible in the direction of the arrow without leaving the feasible region. This takes us to the point B with coordinates (250, 100) and solves the problem. The maximum profit comes out to be £3800; it is obtained by producing 250 lengths of cloth A and 100 lengths of cloth B. Substituting these values of x_1 and x_2 in the constraints (1a) to (1c), we see that the most profitable production scheme will use all the available red and green wool, but will leave 700 kg of yellow wool unused.

Let us now look at a more complex industrial example, for which I am indebted to Professor B. H. P. Rivett, the author of Chapter 6. A shoemaking company operates a production line consisting of six different processes through which pass eight products, two in each of four main categories, A, B, C and D: not all categories go through the same processes, and the way in which the products in the four categories are moved through the six processes is shown in Fig. 3.2. The figures in brackets indicate that four of the processes are subject to operating limitations in the number of machine hours that are available.

The production manager wishes to decide how much of each product to make in order to achieve the largest possible profit. To do this he must know what demand each unit of the eight end-products will impose on the processes through which they pass. Table 3.2 shows the number of machine hours needed to produce one unit of each product.

Finally, the production manager must consult his marketing colleague on the profitability of each product. This information is given in the last row of Table 3.2. If we denote by $x_1, x_2 \ldots x_8$ the number of units of each product that are produced, we can write down the constraint inequalities (four in this case) and the profit

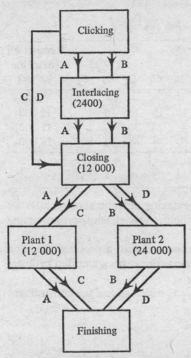

Fig. 3.2. Production flow chart for shoemaking example

equation, just as we did for our cloth manufacturing example. How can we solve this problem? Clearly, with eight variables instead of two, a graphical approach is not feasible. Needless to say, this is also the case with nearly all 'real' problems. What we need, therefore, is an algebraic method of solution – and furthermore, a method that can be expressed as an algorithm suitable for embodying in a computer program. Such programs which can solve linear programming systems with many hundreds, even thousands, of variables are now widely available and used as a matter of routine. At the other end of the spectrum, we find that linear programming is now being introduced into the school-room; it forms part of the School Mathematics Project syllabus for the GCE examination at O-level [4].

TABLE 3.2   Data for shoemaking example

	Product	A_1	A_2	B_1	B_2	C_1	C_2	D_1	D_2	Restriction (machine hours)
Processes	Interlacing	5	6	4	4	—	—	—	—	2400
	Closing	15	20	12	15	15	15	10	12	12 000
	Plant 1	30	20	—	—	20	30	—	—	12 000
	Plant 2	—	—	60	60	—	—	50	80	24 000
Profit per unit of product		5	5	4	4	5	6	4	5	

We shall return to our shoemaking problem shortly, after we have discussed the general method of solving linear programming problems.

Once a mathematician is presented with a problem, he soon wants to generalize it. The general linear programming problem can be stated as follows.

Find the values of $x_1, x_2 \ldots x_n$ which satisfy the constraints

$$\left.\begin{aligned}
a_{11} x_1 + a_{12} x_2 + \ldots + a_{1n} x_n &\leqslant b_1 \\
a_{21} x_1 + a_{22} x_2 + \ldots + a_{2n} x_n &\leqslant b_2 \\
&\ldots \ldots \ldots \ldots \ldots \\
a_{m1} x_1 + a_{m2} x_2 + \ldots + a_{mn} x_n &\leqslant b_m
\end{aligned}\right\} \tag{4}$$

and also

$$x_1, x_2 \ldots x_n \geqslant 0 \tag{5}$$

and which make the objective function

$$f = c_1 x_1 + c_2 x_2 + \ldots + c_n x_n \tag{6}$$

as large (or as small) as possible.

We thus have m constraints and n unknown variables, where usually, but not necessarily, n is greater than m. Inequalities are awkward to deal with and it is convenient to convert all the constraints into equations by introducing a further set of non-negative variables, one for each constraint inequality. They are called *slack variables* and we shall denote them by $s_1, s_2 \ldots s_m$; some will be zero if some of the constraints are already expressed as equations. The variables $x_1, x_2 \ldots x_n$ may be called *problem variables* (an alternative term is 'activities'). Our problem can now be expressed formally as:

Maximize (or minimize)

$$f = \sum_{j=1}^{n} c_j x_j \tag{7}$$

subject to

$$\left. \begin{array}{l} \sum_{i=1}^{n} a_{ij} x_j + s_i = b_i \, (i = 1, 2 \ldots m) \\ x_j \geqslant 0 \, (j = 1, 2 \ldots n) \\ s_i \geqslant 0 \, (i = 1, 2 \ldots m). \end{array} \right\} \tag{8}$$

For brevity, we have used the well-known 'summation convention'. Thus, in (7), $\sum_{j=1}^{n} c_j x_j$ is a shorthand way of writing $(c_1 x_1 + c_2 x_2 + \ldots + c_n x_n)$, and similarly for the constraint equations. The symbol '$\sum$' (sigma) is the Greek letter for S, denoting 'sum of'.

We can now see why such systems are called 'linear'. Both the objective function and the constraints are represented as linear expressions in terms of the x's. The linear form of the objective function means that a profit, cost, etc. per unit of activity can be ascribed to each of the activities and the separate contributions simply added together.

THE SIMPLEX METHOD OF SOLUTION

A number of algebraic procedures for solving the linear programming problem have been devised during the last thirty years. We shall confine ourselves to the earliest, and still the most widely used, known as the *Simplex method* (or the Simplex algorithm).

It will be convenient at this stage to introduce a few definitions. A set of values of the $(m+n)$ variables, $x_j(j = 1, 2 \ldots n)$ and $s_i(i = 1, 2 \ldots m)$, which satisfies the constraint equations (8), including the non-negative conditions, is called a *feasible solution*. A solution obtained by setting n of the variables to zero (thus leaving us with m equations in m variables) is called a *basic solution* and the set of m non-zero variables is called the *basis*. It can be shown that if a system has a feasible solution, then it also has a *basic feasible solution*. It is such solutions that we shall normally be looking for. The solution of the complete problem, including (7), we shall call the *optimal* or *final solution*. In most cases of practical interest, the optimal solution is one of the basic feasible solutions. To see how the Simplex method works, we shall use it to solve our cloth production example. The problem, after introducing slack variables s_1, s_2 and s_3, may be formulated as

$$\left.\begin{array}{l} 4x_1 + 4x_2 + s_1 = 1400 \\ 6x_1 + 3x_2 + s_2 = 1800 \\ 2x_1 + 6x_2 + s_3 = 1800 \\ x_1, x_2, s_1, s_2, s_3 \geqslant 0 \end{array}\right\} \quad (9)$$

maximize $\qquad\qquad P = 12x_1 + 8x_2. \qquad\qquad (10)$

We recall that the feasible region consists of the interior and boundaries of the polygon $OABCD$ in Fig. 3.1. Each vertex of the polygon corresponds to a basic feasible solution – for example, vertex A corresponds to the case $x_2 = 0$, $s_2 = 0$ – of (9), one of which is the optimal solution. This observation gives us a clue as to how to proceed in the general case [6].

The Simplex procedure consists, in essence, of the following steps.

(1) Find a basic feasible solution and evaluate the objective function for that solution. This solution corresponds to an *extreme point* of the feasible region (a vertex of the feasible polygon in the two-dimensional case).

(2) Examine each boundary edge of the feasible region passing

through this extreme point to see whether movement along any such edge will increase the *value* of the objective function.

(3) If it does, move along the chosen edge to the new extreme point and evaluate the objective function.

(4) Repeat steps (2) and (3) until movement along any edge no longer increases the value of the objective function. We then have the optimal solution. Note that we have not specified how the basic feasible solution required in step (1) is to be found; we shall return to this point later. Meanwhile, let us see how things work out in our example.

We have three constraint equations and five variables so we must select three of the variables and imagine that each of the others has a value of zero. If the three non-zero variables satisfy the constraint equations (9), then we shall have found a first basic feasible solution with these three variables constituting the basis. How do we make our selection? We notice that each of the slack variables appears in one equation only. This makes our choice very easy; we can choose the slack variables to form the basis. Our basic feasible solution is then given by $s_1 = 1400$, $s_2 = 1800$, $s_3 = 1800$, $x_1 = 0$, $x_2 = 0$. The next step is to solve the constraint equations for the basic variables in terms of the non-basic variables, and also to express the objective function in terms of the non-basic variables. This gives

$$\left.\begin{aligned} s_1 &= 1400 - 4x_1 - 4x_2 \\ s_2 &= 1800 - 6x_1 - 3x_2 \\ s_3 &= 1800 - 2x_1 - 6x_2 \end{aligned}\right\} \tag{11}$$

and
$$P = 12x_1 + 8x_2.$$

This solution corresponds to the point 0 in Fig. 3.1, but unfortunately there is no profit to be had from it!

We see that we can increase P by increasing either x_1 or x_2. In the Simplex procedure only one variable is increased at a time; the usual rule is to select that variable which has the largest coefficient in the objective function. In this instance this is x_1. However, x_1 must not be increased beyond 300, otherwise s_2 would become negative. (Remember that $x_2 = 0$ at this stage.) So we make $x_1 = 300$ and therefore $s_2 = 0$. We are now at the point A in Fig. 3.1. x_1 has become a basic variable and s_2 a non-basic one.

Once again we express the basic variables and the objective function in terms of the non-basic variables to give

$$\left.\begin{array}{l} s_1 = 200 \ +\tfrac{2}{3}s_2-2x_2 \\ x_1 = 300 \ -\tfrac{1}{6}s_2-\tfrac{1}{2}x_2 \\ s_3 = 1200+\tfrac{1}{3}s_2-5x_2 \\ P = 3600-2s_2+2x_2. \end{array}\right\} \tag{12}$$

The variables on the right-hand sides of the equations have zero values at this stage and so the constants give the current values of the basic variables and the objective function. It is gratifying to find that the profit has increased from 0 to £3600. Can we do still better? Yes, we can. The fact that P has a positive coefficient for x_2 shows that we can increase P by increasing x_2. The maximum permissible value of x_2 is 100, thereby making s_1 zero. So we exchange x_2 for s_1, making x_1, x_2 and s_3 the basic variables. Proceeding as before we obtain

$$\left.\begin{array}{l} x_2 = \ 100-\tfrac{1}{2}s_1+\tfrac{1}{3}s_2 \\ x_1 = \ 250+\tfrac{1}{4}s_1-\tfrac{1}{3}s_2 \\ s_3 = \ 700+\tfrac{5}{2}s_1-\tfrac{4}{3}s_2 \\ P = 3800- \ s_1-\tfrac{4}{3}s_2. \end{array}\right\} \tag{13}$$

The form of P, with negative coefficients for both s_1 and s_2, shows that any increase in either would decrease P. We have reached the final solution corresponding to the point B in Fig. 3.1. A full discussion and formal proof of the validity of the Simplex method will be found in [6].

The Simplex method may also be applied to our shoemaking example. We shall not go through the working but shall merely state the result. Here it is: the best schedule is to produce 240 units of product C_1 (x_5), 240 units of product C_2 (x_6) and 480 units of product D_1 (x_7), giving a total profit of £4560. Only three of the eight products are made and it turns out that the interlacing process is not used at all. There is thus a surplus of 2400 machine hours here. Three other processes are also subject to capacity restrictions, but the full capacity of each is used. We have four constraints and so in any basic feasible solution only four of the eight variables will have non-zero values. For the optimal solution, the basis consists of x_5, x_6, x_7 and s_1, where s_1 denotes the slack variable that must be introduced to convert the constraint on the interlacing process into an equation.

So far in this example we have dealt only with the capacity constraints on production. In practice, there are likely to be marketing constraints as well. Suppose, for example, that the maximum sales of all category C goods were estimated to be 300. This would introduce a new constraint (namely $x_5 + x_6 \leqslant 300$) and so the solution would have to be reworked. A basis must now consist of five non-zero variables and the optimal solution turns out to be $x_1 = 20$, $x_2 = 120$, $x_5 = 300$, $x_7 = 480$, $s_1 = 1580$, yielding a maximum profit of £4420. (Another exercise for the keen reader!)

In 'real life' the formal solution of one or two linear programming models is most unlikely to be the end of the story. One obvious question the management would want answered is: how sensitive are the results to small changes in the various coefficients that have been put into the model – some of them, perhaps, on pretty flimsy information? The production manager, for instance, will probably be interested in such things as the profit bonus to be obtained by making small increases of capacity in each of the critical processes. The sales manager might ask for information on the cost of forcing into production one or other of the products which do not appear in the 'optimal solution' he is presented with. Indeed, he may well want an indication of the relative lack of desirability of each of the excluded products. It must be most unusual for a first solution to be accepted right away! A good computer program would work out not only the optimal solution, but a range of marginal profits and costs, as well as other refinements. To pursue such matters here would take us beyond the scope of this chapter.

We must remember, too, that a computer executing a program, unlike even the most stupid human being, is quite unable to go beyond the letter of its instructions. This means that in constructing a computer program, account must be taken of every possible contingency, however unlikely. Several kinds of degenerate behaviour may indeed be encountered in linear programming models. The rules for carrying on may fail although we have not reached the optimal solution; applying the test for determining which basic variable is to be removed from the basis may fail to give an unambiguous answer; or some of the basic variables may be zero. These niceties are discussed in the literature [5, 6] and we shall not pursue them here except to remark that a case of this last type of degeneracy will be encountered in the next chapter. Here we have a linear programming problem with eleven variables and five constraint equations; it may be stated thus:

$$\left.\begin{array}{l} x_1+x_2+x_3 = 35 \\ y_1+y_2+y_3 = 20 \\ z_1+z_2 \quad\;\; = 15 \\ w_1+w_2 \quad = 30 \\ x_1+y_1+z_1+w_1+s = 50, \end{array}\right\} \quad (14)$$

with all variables non-negative.

Minimize
$$\left.\begin{array}{l} C = 6x_1+8x_2+7x_3+8y_1 \\ \quad +10y_2+9y_3+10z_1 \\ \quad +12z_2+12w_1+14w_2. \end{array}\right\}$$

Applying the Simplex method, we obtain a basic feasible solution given by

$$x_3 = 35, \; y_1 = 5, \; y_3 = 15, \; z_1 = 15 \text{ and } w_1 = 30.$$

Expressing C in terms of the six non-basic variables in the usual way, we get

$$C = 930+x_1+y_2+z_2+w_2+s. \quad (15)$$

The fact that all the coefficients in (15) are positive means that we have achieved an optimal solution; the minimum value of C is 930. Degeneracy arises however because the coefficient of x_1 in (15) is zero. This means that the above solution is not unique. We can interchange x_1 and y_1 in the basic set without altering the value of C. The general optimal solution can be written as

$$x_1 = t, \; x_3 = 35 - t, \; y_1 = 5 - t, \; y_3 = 15 + t, \; C = 930$$
$$\text{where } 0 \leqslant t \leqslant 5. \quad (16)$$

Most types of degeneracy can be illustrated geometrically by making simple alterations to one or more of the equations (9) and Fig. 3.1. The reader is invited to experiment on these lines for himself.

FINDING A STARTING POINT

We must now return to the question of how to find a first basic feasible solution to enable the Simplex process to get started. Perhaps, indeed, there may not be one! It is apparent from our wool example that an initial basic feasible solution can always be found when all the constraints are 'less than or equal to' inequalities having positive numbers on the right-hand side. What do we do if this is not the case?

The procedure is to introduce a number of artificial non-negative variables, as many as we need, one on the left-hand side of each constraint which either has no slack variable (because it is an equality) or where the slack variable does not immediately contribute a basic variable (because the inequality is the 'wrong way round'). There is now no difficulty in constructing an initial basic feasible solution for the modified system. However, we do not want to have the artificial variables in the final solution, so we introduce them into the objective function but multiplied by a large negative coefficient, $-M$. (If the objective function were to be minimized, the coefficient would be $+M$.)

The purpose of this strategem is to ensure that any objective function containing M will not have its maximum value. Thus if there is a basic feasible solution with all the artificial variables having zero values, we shall eventually reach it. A further point is that once we remove an artificial variable from the basic set, we can forget about it; we have reached a basic feasible solution not involving that variable.

All this looks rather complicated, so let us illustrate how the process works by making a simple modification to our 'cloth' example. Let us suppose that the manufacturer insists that all his yellow wool must be used. This means that the '$\leqslant$' symbol in (1c) must be replaced by '$=$' and we only need two slack variables,

s_1 and s_2. We therefore introduce one artificial variable, z say, into the third constraint equation and modify the objective function as explained above. The equations now become

$$\left.\begin{array}{l} 4x_1 + 4x_2 + s_1 = 1400 \\ 6x_1 + 3x_2 + s_2 = 1800 \\ 2x_1 + 6x_2 + z = 1800 \\ x_1, x_2, s_1, s_2 \text{ and } z \geqslant 0 \\ P = 12x_1 + 8x_2 - Mz. \end{array}\right\} \qquad (17)$$

The initial basic feasible solution is $s_1 = 1400$, $s_2 = 1800$, $z = 1800$, $x_1 = x_2 = 0$. Corresponding to (11) we have

$$\left.\begin{array}{l} s_1 = 1400 - 4x_1 - 4x_2 \\ s_2 = 1800 - 6x_1 - 3x_2 \\ z = 1800 - 2x_1 - 6x_2 \\ P = 12x_1 + 8x_2 - M(1800 - 2x_1 - 6x_2) \end{array}\right\} \qquad (18)$$

but

since we must express the basic variable z in terms of the non-basic variables x_1 and x_2.

The large size of M means that the terms of P which contain M are over-riding, so the rule for selecting which non-basic variable to bring into the basic set indicates x_2, not x_1 as in the original example. Furthermore, x_2 is to be exchanged for z, not s_2. Applying the usual procedure, we express s_1, s_2, x_3 and P in terms of x_1 and z. We have now reached the happy situation where an artificial variable (the only one in this example) has been removed from the basis and can be discarded, as also can the terms in M in the objective function. The situation we have now reached corresponds to the point D in Fig. 3.1.

The profit equation turns out to be $P = 2400 + \frac{28}{3}x_1$, and the fact that we still have a positive coefficient in P means that its value can be further increased. We find that x_1 must be exchanged with s_1, and reach the final solution, namely $x_1 = 75$, $x_2 = 275$ and $P = 3100$. This corresponds to the point C in Fig. 3.1. The manufacturer's insistence on using all his yellow wool has reduced his profit by £700 and has left him with 525 kg of unused green wool.

THE TRANSPORTATION PROBLEM

An important special type of linear programming problem is the *transportation problem*, so named because, in its classic form, the objective is to minimize the over-all cost of moving a quantity of goods from a number (m) of dispatch points to a number (n) of destinations. We can formulate the basic transportation problem as follows. Let x_{ij} be the number of units of goods sent from dispatch point (i) to destination (j) and let c_{ij} be the cost of sending a unit of goods along this route. The suffix i can take values from 1 to m, and the suffix j values from 1 to n.

If the dispatch point (i) can send a_i units, then

$$\sum_{j=1}^{n} x_{ij} = a_i \text{ for } i = 1, 2 \ldots m. \tag{19}$$

Similarly if destination (j) requires b_j units, then

$$\sum_{i=1}^{m} x_{ij} = b_j \text{ for } j = 1, 2 \ldots n. \tag{20}$$

These equations imply that

$$\sum_{i=1}^{m} a_i = \sum_{j=1}^{n} b_j. \tag{21}$$

One of the equations (19) and (20) is therefore redundant and we have ($m + n - 1$) independent constraint equations. The total cost of the operation, C, is given by

$$C = \sum_{i=1}^{m} \sum_{j=1}^{n} c_{ij} x_{ij} \text{ (a total of } mn \text{ terms).} \tag{22}$$

The problem is to minimize C while satisfying the constraint equations (19–21) and the usual non-negative conditions.

As a numerical example, fully discussed in [7], we consider the case of a company which has four warehouses and three customers. The data of the problem, with the transportation costs (in £) given in the central matrix, are set out in Table 3.3. The reader is invited to solve this problem for himself; we shall merely state the result that the minimum over-all transport cost works out to be £298.

TABLE 3.3 Data for transportation example

			Figures in matrix are costs		
			Stock in warehouses		
	$i \rightarrow$	1	2	3	4
j ↓	Total 55	15	16	11	13
Customer's requirements	1	17	8	9	6	3
	2	20	6	11	5	10
	3	18	3	8	7	9

The pattern of coefficients in the constraint equations of the transportation problem is very simple; most of them are zero and all the rest have the value 1. It is not surprising, therefore, that there are quicker methods of solving such problems than by using the Simplex method. Hitchcock first opened up the subject in 1941 and several other methods have been devised since. We shall not discuss them here; they are well covered in the literature [5, 6, 7, 8]. The transportation problem can also be formulated as a network problem, and we shall be meeting it in this guise in the next chapter, where it is referred to as the 'Hitchcock problem'.

Many linear programming problems which are not obviously related to the movement of goods can be formulated as transportation problems. There are also several important variants of the basic problem which go by such names as 'transportation with restrictions' or 'the assignment problem'. Here again, we must refer the reader to the literature for further information [5, 6, 7].

Before leaving the subject we will mention another kind of transportation problem to illustrate a quite different approach – the use of what might be called an 'analogy model' (once again, I am indebted to Professor B. H. P. Rivett). Suppose we have only one warehouse to supply known demands for goods at a number of towns. We wish to determine the best place at which to site the warehouse so as to minimize the total transport cost. This is the sum of the costs to the separate destinations, each of which is taken to be the product of the distance from warehouse to destination and the total tonnage of goods carried to that destination.

The problem is not very nice to solve algebraically, but a rapid solution may be obtained by using a physical analogy. We represent each of the towns by its appropriate position on a map, which must be mounted on a rigid, horizontally held, base. We then bore a hole at each position and suspend weights on a string through each of the holes. The weight hanging from a hole is proportional to the demand at the corresponding town. These strings are then joined together and the knot will take up its position of equilibrium under the action of gravity. This equilibrium point corresponds to the optimum position of the warehouse. The problem is solved! Fig. 3.3(*a*) illustrates the situation for four towns.

There is no need for all the towns to be destinations: the flow of goods may be in either direction. Thus, for example, one of the towns might be the port of entry for imported goods which are stored in the warehouse before being transported to the other towns.

The physical arrangements need slight modification when all the towns are situated on a straight line. In this case the optional location of the warehouse is at *one* of the towns; which one depends on the tonnages to be delivered to each. In fact this location depends only on the relative tonnages and not at all on the distances. This paradoxical result is illustrated in Fig. 3.3(*b*). *A*, *B* and *C* represent the positions of three towns in a line, having the requirements shown. The optimal location of the warehouse is at *A*, regardless of the position of *B* and *C*. In fact the warehouse should be sited at *A* whenever the tonnage required at *A* exceeds 50% of the total.

In constructing our string and weight model, we have tacitly assumed that (apart from the 'straight line' case of Fig. 3.3(*b*)) there is no sharing of routes, i.e., that the same piece of road never forms part of the routes to two or more towns. It turns out that if this restriction is removed, it is sometimes possible to achieve a small saving in costs. The simplest example of this is when we have four destinations situated at the corners of a rectangle, with

4 customers at A, B, C and D require delivery of W_1, W_2, W_3 and W_4 tonnage of goods, respectively.

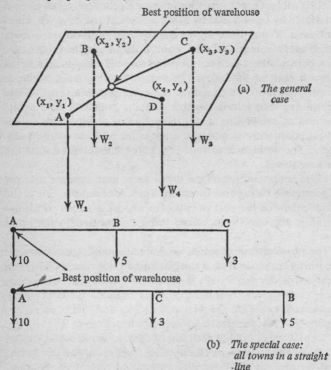

Fig. 3.3. The warehouse positioning problem

each requiring the same tonnage of goods. The reader is invited to work out this case for himself. An analogy may be very useful, but is seldom perfect.

APPLICATIONS OF LINEAR PROGRAMMING

Linear programming is a post-war subject. The development of general methods of solution can be said to date from Dantzig's work on the Simplex method in 1947. His classic paper on the subject was presented at a conference in Chicago in 1949 and published in 1951. We have seen that the first treatment of the transportation problem – a special case of linear programming – was rather earlier – by Hitchcock in 1941. One other early study is worth mentioning. In 1945 Stigler wrote a paper called 'The Cost of Subsistence' in which he discussed the cheapest diet that would suffice to meet some prescribed minimum nutritional standard. He considered twenty-seven food items and arrived at the cheapest combination by trial and error methods. 'The procedure', he said, 'is experimental because there does not appear to be any direct method of finding the minimum of a linear function subject to linear conditions.' This was true in 1945 but ceased to be so soon afterwards. When the new techniques were subsequently applied to Stigler's problem, it was found that he had got extremely close to the optimal solution – if this term may be allowed in so sombre a context [6].

For some years after Dantzig's work began to attract attention, linear programming was regarded as a subject for scientific papers rather than as an effective tool for solving 'real' problems. Gradually, however, the creation of efficient computer programs brought the subject to a state where it could be of direct help to managers, planners and policy makers. Speaking in broad terms, one can say that recognition of this fact took place first in the oil industry, then in chemicals and food. It has now spread through industry, agriculture and government. The variety of problems discussed in [5] illustrates the wide scope of the subject.

Sets of pre-prepared computer programs (known as packages) are now available for all the larger computers. Linear programming systems involving many thousands of constraints can now be solved as a matter of routine. Some of these packages implement the Simplex method, but many are designed to take advantage of special features of the model. The transportation problem is an obvious example; there are many others. Indeed, many industries generate linear programming models having distinctive patterns. This in turn makes it worth-while to develop special computer programs.

A broad classification of the main industrial usages of the technique, based on [9], is outlined below.

(*a*) *Blending problems*. The objective is to choose the most economical mixture of ingredients – crude oils, iron ores, foodstuffs, etc. – while taking account of constraints on the physical or chemical composition of the mixture, and perhaps also of the availability of some of the materials needed.

(*b*) *Production problems*. The typical situation is that there are alternative ways of making one or more products using a number of scarce resources. The object is to achieve the most profitable production programme. Our cloth manufacturing illustration provides a simple example.

(*c*) *Distribution problems*. The basic requirement is to organize the movement of material from a number of sources to a number of destinations so as to minimize either the distribution costs or the time taken, or some combination of both. This, in its simplest form, is the transportation problem. In practice, distribution problems often lead to more general linear programming models, but they usually exhibit special features that can be turned to advantage when devising a method of solution.

(*d*) *Combined problems*. When products can be made at more than one place, it may be worth-while combining production and distribution in a single model. Similarly, production and storage problems may be combined with the object of minimizing stock holding costs over a period of time. (Storage may be treated in much the same way as distribution, but with the material being moved in time rather than in space.)

Comprehensive 'over-all policy' models dealing with production, distribution and storage over a number of time periods are becoming increasingly practicable and useful. Many of the large linear programming models embodying several thousand different constraints that are now being solved on a production basis are of this type.

INTEGER LINEAR PROGRAMMING

If the variables of a linear programming problem represent numbers of machines, persons or any other indivisible entity, then only integral values of such variables make sense. In the particular case of the transportation problem, if all the marginal totals (i.e., the figures bordering the cost matrix in Table 3.3) are integral, then all basic feasible solutions will also be expressed in integers. This is not true, however, in general.

Consider the following example taken from [5]:

$$\left.\begin{array}{r}10x_1 - x_2 \leqslant 40 \\ x_1 + x_2 \leqslant 20 \cdot 5. \\ \text{Maximize } C = 11x_1 - x_2 \\ \text{with } x_1, x_2 \geqslant 0 \text{ and restricted to integral values.}\end{array}\right\} \quad (23)$$

If we ignore the integral requirement for the moment and apply the Simplex procedure in the usual way, introducing two slack variables s_1 and s_2, we obtain the optimal solution $x_1 = 5\cdot5$, $x_2 = 15$, $C = 45\cdot5$. This solution is unacceptable, however, since x_1 is not a whole number. The nearest integers are 5 and 6, so it is reasonable to consider the pairs (5, 15) and (6, 15). We find that the former pair yields a value of 40 for C, while the latter pair is not feasible – the constraints are not satisfied. It turns out, however, that even if we limit ourselves to integral values of the variables, we can achieve values of C larger than 40. The 'geometrically remote' pair (5, 10) produces $C = 45$ and this is in fact the optimal integral solution of the problem. (The reader may like to examine the situation graphically.) This example – admittedly contrived as a fairly extreme case – should serve to make the point that integral programming problems should be treated with respect and not dismissed with the recipe: 'round off the Simplex result to the nearest integers'.

A number of elegant methods for solving various types of integer programming problem were devised, notably by R. E. Gomory, around 1958–60, and are now in common use [6]. One interesting by-product of this work is that it enables decisions which require an unequivocal 'yes or no' answer to be represented in a linear programming model. Such logical variables are only allowed to assume the values 0 or 1. More complex logical relationships can be handled similarly.

OTHER DEVELOPMENTS

During recent years other extensions of linear programming have come into practical use and we will mention two of the most important. The first is called *parametric programming*. It is designed to take account of the fact that in many industrial applications the data to be fed into the model are not known exactly. The technique of parametric programming enables us to trace how the solution varies with changes in one or more of the coefficients, either of the constraints or of the objective function. The second is called *stochastic programming*. Here some of the coefficients of the model are not assigned fixed values, but are treated as *random variables*. This means that the chance of such a coefficient having a specified value is determined by a probability distribution. The optimal value of the objective function is then also a random variable and it is the distribution of this variable that we would like to determine. Analysis of probabilistic linear programming problems offers considerable scope for mathematical skill, but the subject lies outside the scope of this book [6].

The term 'stochastic', which will appear several times hereafter, requires some explanation. It is used broadly as a synonym for 'probabilistic', in contrast to 'deterministic' where all the operative quantities are assumed to take fixed values. It is sometimes used in a more specific sense to describe processes or situations in which the chance variability occurs over a period of time. An example would be the number of cars passing a given point on a busy road in, say, 1 minute.

This completes our review of linear programming. It is natural to wish to escape from the linear straitjacket and to consider problems in which either the constraints, or the objective function, or both, cannot realistically be formulated in linear terms. This forms part of the topic known as nonlinear optimization, which will form the subject of the next section.

3. Nonlinear Optimization

The most direct extension of linear programming into the nonlinear field is known as *quadratic programming*. The objective function is now a quadratic expression; in the general case when there are n variables we can represent it by

$$f(x_1, x_2, \ldots x_n) = \sum_{i=1}^{n} \sum_{j=1}^{n} a_{ij}\, x_i x_j \tag{24}$$

where it is convenient to let $a_{ij} = a_{ji}$. All the constraints are still linear and the non-negative conditions are also retained. An example of a quadratic objective function of two variables is $f(x_1, x_2) = 6x_1^2 + 14x_1x_2 + 3x_2^2$. The quadratic situation is well understood provided that the objective function can be guaranteed to have a unique maximum or minimum. Several satisfactory procedures for locating it have been devised; one of the most widely used, due to Beale, has many similarities with the Simplex method [6].

MATHEMATICAL PROGRAMMING

The enlargement of linear programming to include nonlinear models was signalized by a change of name; the subject was dignified by the title of *mathematical programming* [5, 6, 9]. Much of the early work – in the 1950s – was in the linear tradition, the prime objective being to solve operational research type problems in a large number (e.g., several hundred) of variables. It sought to develop the Simplex and other linear programming techniques to cope with a moderate amount of nonlinearity in the objective function and in some of the constraints, while preserving the non-negative conditions. Some success has been obtained by the use of special strategems; for example, reformulating a nonlinear problem in linear terms, either by a change of variables or by judicious approximation. However, the introduction of nonlinearities beyond that covered by quadratic programming usually gives rise to serious complications and direct algebraic methods (e.g., of the Simplex type) have yielded comparatively small dividends. An alternative – and essentially computational – approach known as *dynamic programming* has been much more successful and will form the subject of a later section of this chapter.

HILL CLIMBING

Yet another approach to the difficult problem of optimizing complex nonlinear systems was made during the 1950s. It has been termed *hill climbing*, for reasons which will become clear shortly. The motivation was to solve practical problems, mainly in the engineering and chemical industries. Here, in contrast to the typical mathematical programming model, one is usually dealing with relatively few variables (say between five and ten), the non-negative conditions may be absent, but the objective function is highly nonlinear. The constraints, if there are any, may also be nonlinear. In fact, however, one may well be concerned with minimizing (or maximizing) a complicated objective function of several variables in the absence of constraints, and it is to this topic that we now turn [10, 11, 12].

It will be convenient to present the optimization problem as one of maximization. This entails no loss of generality since the minimum of a function, F, is equal to the negative of the maximum of the function $(-F)$. Our general problem, then, will be to locate a local maximum of a function $F(x_1, x_2 \ldots x_n)$ of n variables.

Most methods of solution involve an iterative procedure of some kind. We start from an initial estimate and generate a sequence of new – and, hopefully, of steadily improving – estimates. Having reached a certain point we search about, so to speak, for a direction in which to move in order to get closer to the desired solution. Each step of the process entails making two decisions: first, to choose a direction, and second, to decide how far to move along the chosen direction. Once the first decision is made, the problem is reduced from a multi-variable to a single-variable optimization (sometimes called a *linear search*) – a much more comfortable situation. The usual procedure is to move to a 'point' where the new single-variable function attains (or closely approaches) its maximum value, and then to embark on the next step.

A number of methods for dealing with what may be called the nonlinear unconstrained optimization problem have been developed. They differ essentially in the iterative strategy to be adopted; that is, in the rules to be used in making the two basic decisions. To compare the various methods in detail would involve delving into technicalities of computational efficiency, convergence, and so forth, which are beyond the scope of this book. Most methods, however, fall into one of two main classes, known as *gradient methods* and *direct search methods*. The essential difference between them resides in the method of making the first of our two basic decisions.

GRADIENT METHODS

We have seen that the general approach is known as 'hill climbing' so let us consider a walker who wishes to climb a local hill. He is equipped with a map and a compass, but a mist suddenly descends on him. How is he to proceed? One policy he could adopt would be to start walking straight up the hill, that is in the direction of steepest ascent. He knows that this direction is at right angles to the contour lines on his map at the point where he is. His rate of climb in this direction is known as the *gradient*: it is the maximum possible rate. In hilly country the contours usually wander about a good deal and so our walker would be unlikely to reach the top just by carrying on in the same direction. He could decide to maintain his course until he found he was no longer climbing. When he had flattened out he would be walking in the direction of the tangent to the local contour line. He could then re-orientate himself by changing course to that of the new direction of steepest ascent, and then continue as before. Eventually, if there were were no subsidiary peaks in the area, he would reach the desired summit. The procedure is illustrated in Fig. 3.4.

The climber's approach is essentially the same as that adopted in gradient methods of optimization, but we must now think of a function, not of two variables (as height is a function of 'northings' and 'eastings'), but of n variables. Of course we cannot draw contours, but the direction of steepest ascent and the magnitude of the gradient can be found by a mathematical technique involving partial differentiation. In practice this 'hill climbing' procedure is likely to increase the value of the function F quite rapidly during the first few iterations and then to become increasingly unsatisfactory in the later stages. A number of methods have been put forward which seek to avoid these difficulties by using more sophisticated rules for determining the direction of search at each iterative step [11, 12].

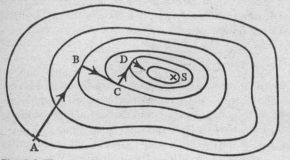

Fig. 3.4. The gradient method of hill climbing. *ABCD* shows the iterative path to the summit *S*

DIRECT SEARCH METHODS

To use a gradient method it must be possible to differentiate the function with respect to each of its n variables. It is not always possible to do this, and even when it is mathematically possible, the resulting expressions may be awkward to handle. Not only that, but the function may be specified by a numerical table, it may have discontinuities or it may be otherwise ill behaved. In such cases a gradient method cannot be used and we must resort to a direct search method. The essential feature of all such methods is that the entire procedure for approaching the maximum (both the direction of search and the step size) is determined by a set of test evaluations of the function F itself. How is this to be done? One obvious procedure is to adjust each of the variables in turn so as to maximize the value of the objective function. With this method, called 'one at a time search', a single iterative step consists of a sequence of n linear searches, each searching in one variable only. It will be seen that with this procedure, the first basic decision – to select the direction along which to move – has been effectively by-passed. A disadvantage of this method is that it is likely to be very slow when there is a strong interaction between the variables in their effect on the value of the function. A number of ingenious strategies have been devised to speed up the process [10, 11, 12].

We have already remarked that in both gradient and direct search methods the second decision – how big a step to take – can be made to depend on the results of a linear search. This means that the step size can be determined by maximizing the value of a function of a single variable. This is exactly what the hill climber did when he applied the gradient method. The same procedure can be adopted to determine the step size in a direct search method. There is, however, an alternative strategy, which is simply to take a step of fixed length in the specified direction. This is faster and simpler than making a linear search and is employed in two widely used direct search methods, known as the *Simplex search* and *pattern search* [12].

The various gradient and search methods must be judged in the end on their practical merits; in particular, on their ability to utilize the power of the digital computer. Indeed, the first accounts of many of the best current methods are to be found in the pages of the *British Computer Journal*, mostly between 1964 and 1966.

LINEAR SEARCH METHODS: BISECTION AND FIBONACCI SEARCHES

We have seen that a linear search (for the maximum value of a function, $f(x)$ say, of a single variable x) plays a central rôle in most optimization techniques. How should such a search be conducted? One method would be to evaluate $f(x)$ at regular intervals, say 0·001, until the maximum is reached. However, in practice function evaluation is often a lengthy business, and such a 'brute force' method would be very inefficient, even with the aid of a computer. We are thus led to seek the best search strategy, where we use the term 'best' in the sense of enabling the maximum to be located to a prescribed level of accuracy with the fewest possible evaluations of the function.

To make the problem precise, let us suppose that a function $f(x)$ of a single variable is known to have a unique maximum somewhere in the interval (a,b). We wish to locate this maximum to within an uncertainty of $(b - a)/k$. (This specifies the level of accuracy required.) To do this, we shall need to make a certain number, N say, of evaluations of $f(x)$. We define the best search procedure as the one which requires the smallest N for a given k; or, alternatively, the largest k for a prescribed N.

One obvious method, well suited to a computer, is to keep on halving the interval of uncertainty. Let us call this the *bisection method* of search. To halve the interval we must evaluate the function at two points on either side and near to the midpoint. The purpose of this is to find if the function is increasing or decreasing in value at the midpoint. Fig. 3.5 illustrates the two possibilities. We see that, for the bisection method, six function evaluations are needed to locate the maximum to within one-eighth of the original range. In general, we require $N = 2n$ evaluations to achieve an accuracy given by $k = 2^n$.

Can we do better than this? In fact we can. There is a remarkable theorem which not only specifies an alternative and better search strategy, but tells us that, under certain quite general conditions, it is indeed the best possible strategy. This strategy is known as the *Fibonacci search* because the positions of the points at which the function is evaluated are related to the terms of the famous Fibonacci sequence.

Leonardo of Pisa, known as Fibonacci (son of good nature),

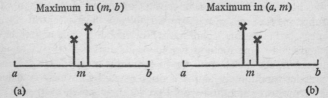

Maximum in (m, b) Maximum in (a, m)

(a)   (b)

Fig. 3.5. The bisection method of search

was a distinguished medieval mathematician who played a leading part in popularizing the Arabic number system in Western Europe. His best-known book, *Liber Abaci*, appeared in 1202 and contained a problem about the breeding of rabbits which led to the sequence that bears his name.

The defining property of the Fibonacci (or $F-$) sequence is that each term (after the second) is the sum of the two preceding terms. If we conventionally take the first and second terms as 1, the sequence is

$$1, 1, 2, 3, 5, 8, 13, 21, 34, 55 \ldots$$

The theorem that establishes the primacy of the Fibonacci search can be stated as follows. 'If k is the $(N + 1)$th term of the $F-$ sequence, the maximum of the function can be located to the required level of accuracy (as defined above) by evaluating the function $f(x)$ for N different values of x, provided that these values are correctly chosen. This is the "best" search procedure in the sense already explained.' A good discussion of the Fibonacci search – and also of a related procedure known as the *Golden Section* search – will be found in [12].

The Fibonacci search works by successively reducing the interval of uncertainty by computing function values at correctly chosen points within the current interval. By way of illustration, let us consider the case of $k = 13$, the seventh $F-$ number. The theorem tells us that *six* evaluations of the function, provided that the positions at which the evaluations are made are correctly chosen, suffice to locate the maximum to within 1/13th of the original interval.

We assume, then, that the function $f(x)$ has a unique maximum somewhere in the interval (a,b). To simplify the description, let us suppose that the values of x are scaled linearly to make $a = 0$ and $b = 13$. We first evaluate $f(5)$ and $f(8)$, 5 and 8 being the two members of the $F-$ sequence immediately preceding 13. A crucial feature is that the four points (0, 5, 8, 13) exhibit central symmetry about the current range (0, 13). We shall see that this feature persists throughout. The two possible outcomes of the evaluations are shown in Fig. 3.6 (a) and (b). In case (a), $f(5)$ is the

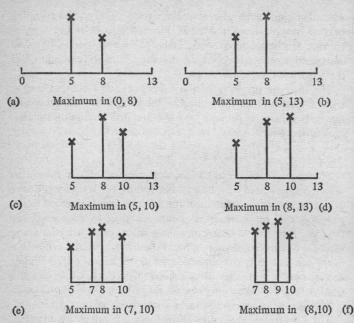

(a) Maximum in (0, 8)   Maximum in (5, 13) **(b)**

(c) Maximum in (5, 10)   Maximum in (8, 13) **(d)**

(e) Maximum in (7, 10)   Maximum in (8,10) **(f)**

Fig. 3.6. The Fibonacci search

larger, so the maximum must lie in the range (0, 8). In case (*b*), $f(8)$ is the larger and the maximum must lie in (5, 13). In either case, the range of uncertainty has been reduced from 13 to 8; in general, from F_n to F_{n-1}, where F_n is the *n*th member of the *F*— sequence.

Let us assume that (*b*) holds and consider the next step. (The transformation $X = 13 - x$ will turn case (*a*) into a situation analogous to (*b*), so the method will deal with either case equally well.) We next evaluate $f(10)$; the two possibilities are illustrated in (*c*) and (*d*). The choice of the value 10 is dictated by the need to maintain central symmetry within the current range, namely (5,13). The fact that this symmetrical pattern can always be found follows at once from the defining property of the *F*— sequence. Let us assume that (*c*) holds, so the maximum must lie in the range (5, 10). The fourth evaluation is of $f(7)$, as shown in (e), where we have assumed that $f(7)$ is less than $f(8)$. The range is thereby narrowed to (7, 10). The fifth evaluation is of $f(9)$, as shown in (f). We have assumed that $f(9)$ is greater than $f(8)$, thus reducing the range to (8, 10).

The final evaluation, of $f(9{\cdot}01)$, is of a different character, more like those needed in the bisection method. We can think of the symmetry feature as still being preserved but with the two interior points coalescing at the midpoint of the range. The purpose of this last evaluation is to test whether $f(x)$ is increasing or decreasing when $x = 9$. If it is increasing, we can say that the maximum must lie in (9, 10); if decreasing, in (8, 9). In either case we have achieved our objective of locating the maximum within 1/13th of the original range by means of six function evaluations.

In practice, the procedure would usually be taken a good deal further, with, as usual, a computer looking after the arithmetical chores. The generalization of our example should now be clear. Each evaluation (after the first) narrows the range from F_n to F_{n-1}. Eventually, the range is reduced to a width of $F_3 = 2$ and one further evaluation near the midpoint completes the process by reducing the range of uncertainty to $F_2 = 1$ unit.

The margin of advantage of the Fibonacci over the bisection search increases steadily with more evaluations, as is shown in Table 3.4. Some applications of this search method are discussed in [16]. The technique can be extended to search for the maximum of a function of more than one variable; it is then called the generalized Fibonacci search.

OPTIMIZATION WITH CONSTRAINTS

We must now briefly mention the case where the nonlinear function $F(x_1, x_2 \ldots x_n)$ to be maximized is subject to a set of constraints which themselves may be nonlinear functions of the n variables. One approach to this difficult problem is to ignore the constraints for as long as possible. We would eventually reach a situation such as that illustrated in Fig. 3.7. Suppose that the point B, on or near the boundary of a constraint, has been reached by

TABLE 3.4 Comparison of bisection and Fibonacci search methods

Number of function evaluations (N)	Accuracy factor (k) for	
	Bisection search	Fibonacci search
6	8	13
8	16	34
10	32	89
12	64	233
14	128	610

using some kind of unconstrained iterative procedure. The next such step might be expected to give a direction of search pointing more or less towards the unconstrained maximum, U, so the path would enter the forbidden, non-feasible, region. This will not do; instead we want the search to be directed towards the constrained maximum, C. The problem, therefore, is to incorporate appropriate information about the constraints into the rules for

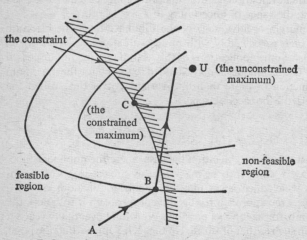

Fig. 3.7. Constrained optimization

deciding on the direction of search. With nonlinear constraints, this can be difficult and time-consuming.

One technique is to modify the function to be maximized by adding penalty terms which come into effect when, and only when, a constraint is violated. The general idea is similar to that already discussed for obtaining an initial basic feasible solution of a linear programming problem. This method works in some cases but not in all. As with most nonlinear problems, the best method of attack depends very much on the particular situation. No general recipes can be offered and the whole subject bristles with difficulties [10, 12].

4. Statistical Decision Theory

All of us, all the time, have to make decisions where the outcomes are uncertain. Shall I take an umbrella this morning? Should I make a dental appointment or wait a bit in the hope that the pain will go away? *Statistical decision theory*, a comparatively recent application of the theory of probability, is available to analyse such situations, provided that numerical measures can be attached to the consequences of the various possible courses of action and to the probabilities of occurrence of the various uncertain events. The extent to which it is sensible to do this is, and is likely to remain, a matter of controversy.

The basic purpose of the theory is the same as that of mathematical programming or hill climbing – to optimize an objective function. In this case, however, the value of the function is uncertain, so it is the *statistical expectation* of the objective function, rather than the function itself, that is maximized or minimized.

Decision theory is becoming widely used in industry and commerce, primarily to analyse situations that can be evaluated in monetary terms (e.g., marketing, investment planning, oil drilling and mineral prospecting). Its range of use is now being extended to other fields of activity, to pharmaceutical testing and screening, for instance, where the objective function might not be profit or cost, but perhaps the chance of spotting an effective new compound when the resources available for the search are limited.

Let us first consider a very simple example, based on [14]. Suppose I have travelled by car to a meeting that is expected to last about 2 hours. I am faced with the decision whether to put my car on a parking meter or in a nearby car park. The costs of the various possibilities are shown in Table 3.5.

TABLE 3.5.   Table of car parking costs

Action	Cost if meeting lasts less than 2 hours	Cost if meeting lasts more than 2 hours
Put car on meter	20p	20p + £5 penalty
Put car in car park	70p	70p

Let us suppose that I assess the chance of the meeting being over within 2 hours as 80%. Then by putting my car on the meter I am accepting an 80% chance of only paying 20p and a 20% chance of having to pay 520p. If I put the car on the meter, I pay 70p whatever happens. How do I decide what to do? One way – the way prescribed by the theory – is to work out the expected costs of the two decisions and choose the one with the lower cost. If I put my car on the meter, the expected cost is 20p × 0·8 + 520p × 0·2 = 120p. If I put it in the car park, the cost is simply the parking fee of 70p. I should, therefore, use the park. Furthermore, we can easily see that the meter would have been preferable if, and only if, the chance of the meeting finishing within 2 hours had been assessed as 90% or more.

The notion of probability as the relative frequency of occurrence of an uncertain event, such as throwing a 'six' with a die, has been introduced in the last chapter. Another and more subjective approach is to regard probability as a measure of *degree of belief*, conventionally expressed on a scale stretching from 0 (complete disbelief) to 1 (certainty). It is in this latter sense that probability will be used in this section of the chapter. Thus when I 'assess the chance of the meeting being over within 2 hours as 80%', I am making a 'degree of belief' statement. It means that if I were a betting man I would accept odds of up to 4 to 1 on the specified outcome, but would not be prepared to accept longer odds. The two usages are intimately linked because we normally like to base our degree of belief on our knowledge of what has happened in the past. We are prepared to modify it, as we shall see later, should new information become available.

DECISION TABLES

Table 3.5 is a simple example of a *decision table*. In its general form, exhibited in Table 3.6, such a table contains two lists. The first is a row-list of m exclusive and exhaustive decisions (by this we mean that one of them *has* to be taken and that not more than one *can* be taken) which we can denote by $d_1, d_2, \ldots d_m$. The second is a column-list of n exclusive and exhaustive uncertain events (i.e., events that may or may not occur) which we denote by $e_1, e_2, \ldots e_n$. With each event, e_j, is associated a probability, p_j, where $j = 1, 2, \ldots n$, that the uncertain event e_j will indeed occur. The fact that the set of events is exhaustive implies that

$$\sum_{j=1}^{n} p_j = 1.$$

The decision-maker's task is to select a single item from the first list without knowing which member of the second list is true (i.e., will occur). Let us suppose that a particular decision d_i is selected and that event e_j turns out to be true. This joint situation will give rise to certain consequences. In order to use statistical decision theory it is necessary to attach a numerical measure to each of the possible consequences ($m \times n$ in all). In many cases this is straightforward as the consequence is a direct monetary gain or loss. This is not always the case, however, so let us use a more general term and call the value of a consequence, whether monetary or otherwise, its *utility*. The body of a decision table is occupied by a set of $m \times n$ utility-numbers and we shall denote by u_{ij} the utility of selecting decision d_i when event e_j occurs. It

TABLE 3.6 General decision table. The u_{ij}'s are utilities

		List of uncertain events $e_1 \quad e_2 \ldots \qquad e_j \ldots e_n$			Expected utility of the decision
	d_1	$u_{11} \; u_{12} \cdot \cdot$		u_{1n}	$\bar{u}_1$
	d_2	$u_{21} \; u_{22} \cdot \cdot$		u_{2n}	$\bar{u}_2$
List of possible decisions	d_i		u_{ij}		$\bar{u}_i$
	d_m	$u_{m1} \; u_{m2} \cdot \cdot$		u_{mn}	$\bar{u}_m$
Probabilities of occurrence of uncertain events		$p_1 \; p_2 \cdot \cdot$	p_j	p_n	

should be remembered that a utility is a numerical measure of the attractiveness of a consequence. When we are considering losses instead of gains, we may think of the u_{ij}'s as 'negative utilities'.

The procedure, which should now be apparent from what has already been said, is as follows.

1. Compute a set of m 'expected utilities' denoted by $\bar{u}_i$ $(i = 1, 2 \ldots m)$, one for each possible decision. This is done by weighting the utility of each outcome of a decision by the appropriate probability and adding the numbers so obtained. We have, therefore,

$$\bar{u}_i = \sum_{j=1}^{n} u_{ij} p_j. \qquad (25)$$

2. Determine the largest member of the set $\bar{u}_i$. The corresponding decision will be the 'best' one within the framework of the theory. (If we are dealing with negative utilities, we would select the smallest $\bar{u}_i$.)

UTILITY AND MONEY VALUES

In our car parking example, the negative utilities shown in Table 3.5 are simply money payments, and we have tacitly assumed that the utility of an outcome is directly proportional to its expected monetary consequences. Is this, we may ask, a reasonable assumption; does it represent how most of us actually behave? Do we, in fact, habitually choose that course of action which yields the highest expectation of gain or the lowest expectation of loss? Of course the consequences of many of our actions are not naturally quantifiable (shall we go to the theatre tomorrow?) although they may have a monetary element (the price of the theatre ticket). To simplify matters, we shall restrict ourselves to situations where the consequences can be expressed entirely in monetary terms. The question is, then: what is the relationship between utility and money?

Let us consider a simple and highly fanciful example. Suppose you are offered a choice between two alternatives, (A) and (B). In (A) a coin is tossed; if heads come up, you receive £100; if tails, you pay out £10. In (B) you always receive £40. Table 3.7 shows the decision table for this situation. Most people, I suspect, would choose (B), although a rich man might choose (A) as this would offer him larger expectation of gain.

TABLE 3.7   Decision table for example discussed

| Decision | Payment if tossed coin shows | | Expectation |
	Heads	Tails	
(A)	£100	−£10	£45
(B)	£ 40	£40	£40
Probabilities	$\frac{1}{2}$	$\frac{1}{2}$	

This little example suggests that the utility (or attractiveness) of a sum of money is not directly proportional to its monetary value, particularly if large sums are involved. It seems likely that most people's utility–money curve would look something like Fig. 3.8, falling off in accordance with the principle of 'the diminishing marginal utility of money'. Here $u(x)$ represents the utility of an amount of capital, x. For a fixed gain in capital, a, the increase in

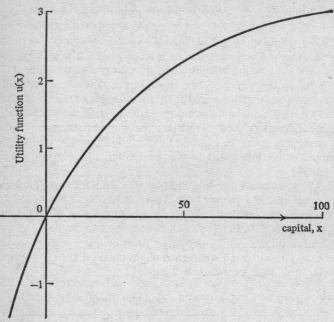

Fig. 3.8. The utility of money

utility, $u(x + a) - u(x)$, is a diminishing function of the initial capital, x.

The utility–money relationship drawn in Fig. 3.8 includes the 'points' listed in Table 3.8(*a*) (on an arbitrary scale). The corresponding decision table is shown in Table 3.8(*b*); (*B*) is now clearly the preferred choice.

Different people will have different utility curves according to their assets and their temperaments (in particular, their attitude to risk). The rich or intrepid will have straighter and flatter utility curves for a larger range on either side of the origin. The appropriate curve for any individual can, in theory, be built up by presenting him with a number of different choices and analysing his replies. Try it on your friends! Utility functions clearly tell us something about human behaviour but there are a number of

TABLE 3.8 The diminishing marginal utility of money

money x	utility $u(x)$
−10	−1
0	0
40	2
100	3

(a) A money–utility relationship

	Utility if tossed coin shows		Utility
Decision	*Heads*	*Tails*	*Expectation*
(A)	3	−1	1
(B)	2	2	2
Probabilities	$\frac{1}{2}$	$\frac{1}{2}$	

(b) The decision table

difficulties in working with them instead of with hard money. One is that many people are not entirely rational in analysing risks, another is that their utility functions are continually changing.

'CONTINUOUS' DECISION THEORY

The decision table approach assumes that we have a finite number of discrete decisions to choose between and a finite number of uncertain events, with each of which can be associated a specific probability. Many practical situations are not like this. The decision-maker wishes to select the 'best' value of a continuously varying quantity, with some at least of the uncertain events characterized not by definite probabilities but by continuous probability distributions [3].

Fig. 3.9(a) shows an example of such a distribution, which is called a *probability density distribution*. For our present purpose, its most important property is that the probability that the value of a variable quantity x will be between a and b is given by the shaded area on the right of the figure. The total area under the curve – which may extend from $x = -\infty$ to $x = +\infty$, or from $x = 0$ to $x = +\infty$, or some other range – will, of course, be 1, the probability number corresponding to certainty.

Fig. 3.9(*b*) shows another curve that will be needed later. It can be derived from the density distribution and is known as the *cumulative probability distribution*. For any value of the random variable x, say A, the height of the curve gives the probability that the value of x is *not greater than A*. Thus the height of PN in (*b*) is equal to the cross-hatched area on the left of figure (*a*). A cumulative distribution curve must rise steadily from 0 to 1 as x increases over the complete range of its possible values.

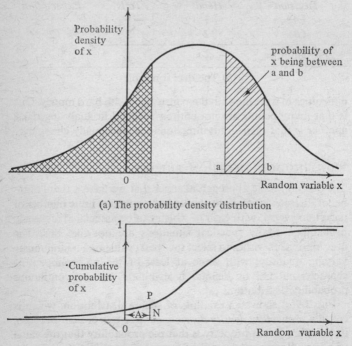

(a) The probability density distribution

(b) The cumulative probability distribution

Fig. 3.9. Probability distributions

By way of illustration of continuous decision theory, we will consider a simple example. Suppose we have a machine that produces large numbers of some article or component – light bulbs, for instance. A proportion, P, of the articles are satisfactory to the customer, but a proportion $(1 - P)$ are not and, if sent out, must be replaced. We shall assume that the defective articles turn up completely at random during a production run. It is not practicable to subject each article to the full test that a customer would make, but the quality control department can make a much simpler test. This test is not able to discriminate infallibly between good and bad articles, but it gives a figure of merit, x, for each article such that good articles usually, but not invariably, score higher x-ratings than bad ones. The ways in which the x-values are distributed for both good and bad articles are known from past experience. They are shown as continuous probability density distributions in Fig. 3.10.

What the quality control department seeks to do is to select a critical value of x, say X, such that all articles having a figure of merit greater than X are sent off to the customer and all articles with a figure of merit less than X are scrapped. The question is: how should the value of X be selected to minimize the over-all cost to the manufacturer?

Fig. 3.10 shows the situation once the choice of X has been made. The probability that a good article will be scrapped (at a cost of C_1 per article) is $p_g(X)$, represented by the shaded area under the upper curve. The probability that a defective article will be sent to a customer (at a cost of C_2 per article) is $p_b(X)$ represented by the shaded area under the lower curve. If X were to be increased, more good articles would be scrapped but fewer bad ones would be sent out. If X were to be decreased, the opposite would happen. Statistical decision theory provides a way of determining the 'best' balance between these two opposing factors.

Now as a result of a test on a particular article, one of four uncertain events (inclusive and exhaustive) will occur. They are shown along the top of Table 3.9, which is drawn up like a decision table.

The *extra* costs due to the imperfections of the tests are shown in the centre part of the table, with the corresponding probabilities below. Each of these probabilities is a product of two factors. Thus the probability of scrapping a good article is: (probability of producing a good article) × (probability of scrapping the article), or in our notation, $P . p_g(X)$: similarly for the other three probabilities. The expected cost, C, is obtained in the usual way. We have

$$C = C_1 . P . p_g(X) + C_2 . (1 - P) . p_b(X). \qquad (26)$$

We can arrive at the minimum value of C by appealing to a general principle known to economists as the principle of mar-

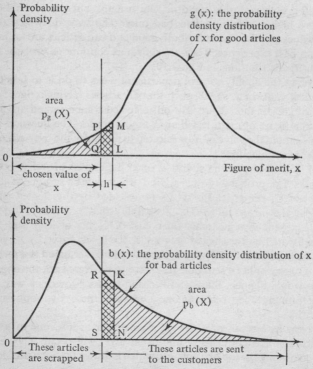

Fig. 3.10. The quality control example

TABLE 3.9 The quality control decision table

	Uncertain events			
	Good article sent out	Good article scrapped	Bad article sent out	Bad article scrapped
Extra costs to manufacturers	0	C_1	C_2	0
Probability of uncertain events is product of	P and $1 - p_g(X)$	P and $p_g(X)$	$(1 - P)$ and $p_b(X)$	$(1 - P)$ and $1 - p_b(X)$

ginal utility. Suppose we increase X by a small amount h. We have already pointed out that two consequences will follow.

(a) More good articles will be scrapped and some extra cost will be incurred and

(b) a compensatory saving will be made because fewer defective articles will be sent out to the customers.

Our principle tells us that the total cost will be minimized when the marginal cost of (a) is equal to the marginal saving of (b). The *extra* probability that a good article will be scrapped is represented by the cross-hatched area in the upper curve. The extra probability that a defective article will be sent out is represented similarly in the lower curve. Now we have chosen h to be very small, so these areas are very nearly equal to the areas of the rectangles $PQLM$ and, $RSNK$ which are simply $g(X) \cdot h$ and $b(X) \cdot h$, respectively. Applying the same argument as before, we see the marginal cost of (a) is $C_1 P \cdot g(X) \cdot h$ and the marginal saving of (b) is $C_2 \cdot (1 - P) \cdot b(X) \cdot h$.

To achieve the minimum over-all cost, we must equate these two, giving

$$C_1 P \cdot g(X) \cdot h = C_2(1 - P) \cdot b(X) \cdot h.$$

Cancelling the h's, we get

$$b(X) = \frac{C_1}{C_2} \frac{P}{(1 - P)} \cdot g(X). \qquad (27)$$

To see what this means, let us put some numbers in our example. Suppose that 96% of the articles are satisfactory and that 4% are defective (i.e., $P = 0.96$), and that it costs 60p to replace a bad article that has been sent to the customer and 20p to scrap a good article (i.e., $C_1 = 20$, $C_2 = 60$). Then (27) gives $b(X) = 8g(X)$. This means that the expected extra costs incurred by faulty diagnosis are minimized by choosing a value of X at which the height of the lower curve in Fig. 3.10 is eight times that of the upper curve (i.e. $RS = 8PQ$).

DECISION TREES

A serious limitation of the 'decision table' method of analysis is that it takes no account of the passage of time. A decision is rarely binding for all time; circumstances change and policy needs periodical review. Furthermore, all is not usually gained or lost on a single decision; further decisions can be made. Indeed, the possibility of 'second thoughts' may well influence the initial decision. Many decision problems can be broken down into a sequence of sub-problems that follow one another in some natural order, usually in time. This approach leads to an alternating sequence of decision, outcome, decision, outcome, and so on until an end-point is reached. The technique of analysis by means of *decision trees*, which we shall now discuss, enables us to take account of such sequential features. We shall illustrate the procedure by considering the following investment problem, adapted from [15].

An investor has £1000 capital. He can either leave it in the bank or invest it, in which case he stands to win £100 if the stock appreciates or lose £200 if it depreciates. (A highly simplified investment hypothesis!) The chance of the stock appreciating is estimated to be 60%. These assumptions are set out in Table 3.10. We shall now introduce a new feature; the advice of a broker is available for a fee of £f. (The value of f is left open for the present.) Is it worth employing him? The answer to this question clearly depends on the investor's assessment of the broker's ability. In order to be able to use decision theory, this judgement must be expressed in terms of numerical probabilities.

TABLE 3.10   The investor's decision

	Stock appreciates (e_1)	Stock depreciates (e_2)
Invest (d_1)	£1100	£ 800
Leave in bank (d_2)	£1000	£1000
Probabilities	0·6	0·4

To fix ideas, let us assume that the broker is assessed as being able to spot a winner (i.e., a stock that will rise in value) on 70% of occasions, and a loser (a stock that will fall) on 80% of occasions. The investor must decide whether or not to buy the advice and whether or not to invest, either with or without advice. For simplicity, we shall assume that utility is directly proportional to money value.

The decision tree for this problem is shown in Fig. 3.11. These curious trees grow horizontally with the base placed conventionally on the left. Initially the investor has three choices available to him. These are represented by the three branches on the left of the tree labelled b (broker consulted), d_1 (invest without advice) and d_2 (leave money in bank). Let us consider the branch labelled 'b'. The investor is advised either to invest or not to invest, so this branch splits into two, labelled X_1 (the advice is to invest) and X_2 (advice not to invest). Whichever of these branches is followed, the investor has finally to choose between investing his money or leaving it in the bank. Each of the X_1 and X_2 branches therefore splits into d_1 and d_2 branches. At the end of each d_1 or d_2 branch we must add two others labelled e_1 (the stock in fact appreciates) and e_2 (the stock depreciates). The structure of that part of the tree emanating from the initial branches d_1 and d_2 follows exactly the same pattern.

The complete tree consists, then, of a number of branches, each branch corresponding either to an action resulting from a decision or to an outcome of an uncertain event. The points at which the branches split are called *nodes* and are of two distinct types. Thus the first node (on the left) leads to three branches with the

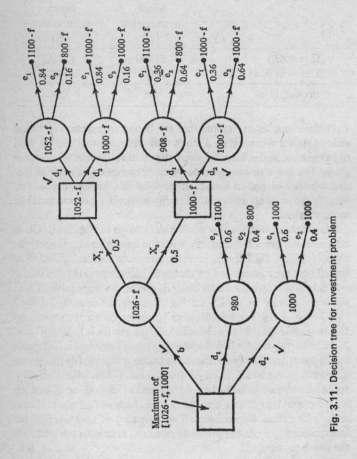

Fig. 3.11. Decision tree for investment problem

choice between them depending on the decision-maker. This node is therefore called a *decision node* and is represented by a square. Following the branch labelled '*b*', we reach a second node from which emanate the branches X_1 and X_2. Here the situation is quite different; the decision-maker has no control over which branch is selected (in this case the broker makes the choice). We have pointed out that the decision-maker must attach probabilities to the reliability of the advice he receives, so we call this node a *chance node* and represent it by a circle. Following either of the branches X_1 or X_2 we come to a decision node where the investor chooses either d_1 or d_2. Finally, each of the d_1 or d_2 branches leads to a chance node, the investor having no control over whether or not the stock appreciates.

In general, we can say that a typical decision tree consists of a number of branches emanating from nodes which are of two types. These two types alternate as we move along the tree by any route, beginning with a decision node and ending with a chance node from which emanate what may be called *terminal branches*.

Analysis of the tree involves two kinds of numerical quantity, probabilities and utilities. The first thing to do is to calculate the probabilities associated with each of the branches emanating from the chance nodes. This often needs some care and a good working knowledge of the basic laws of probability and how to use them. We must always bear in mind that a decision tree develops over time from left to right. This means that at any node we have all the information back from that node to the base of the tree, but no information from that node to the tips of the tree; that is still in the future. The probability associated with a branch depends on how much is known at the time; in technical terms, it is a *conditional probability*.

Let us return to our example and Fig. 3.11. If we begin at the base and proceed along either the d_1 or the d_2 branch, we reach a chance node, from which emanate two branches labelled e_1 and e_2. Since no advice has been obtained, the probabilities are simply the initial ones, namely 0·6 for e_1 and 0·4 for e_2. The branches labelled e_1 and e_2 are endorsed accordingly.

Now let us consider the third branch from the base, the one

labelled *b*. Here advice is sought and the branches emanating from the first chance node reached relate to what that advice is. We require the probabilities associated with the branch X_1 (advice to buy) and X_2 (not to buy). At this stage we do not know whether e_1 or e_2 is true, so we have a difficulty. We have already explained that the investor must assess the reliability of his broker; we have assumed that he believes that the probability of being advised to buy if the stock will in fact appreciate is 0·7, and the probability of being advised not to buy if the stock will in fact depreciate is 0·8. Using the standard notation for conditional probabilities, we can write $p(X_1|e_1) = 0·7$ and $p(X_2|e_2) = 0·8$. This also implies $p(X_2|e_1) = 0·3$ and $p(X_1|e_2) = 0·2$ since X_1 and X_2 form an exhaustive set. The vertical bar simply means 'given that'. Thus $p(X_1|e_1)$ means 'the probability of X_1 given that e_1 is true'. However these are not the appropriate probabilities because we do not know which of e_1 and e_2 is in fact true. All we have is our initial knowledge and so the probabilities we need are $p(X_1)$ and $p(X_2)$. Fortunately, we can calculate them by invoking the basic theorems governing conditional probabilities.

Since either e_1 or e_2 must be true (the set is exhaustive) we can express $p(X_1)$ and $p(X_2)$ in terms of the probabilities conditional on e_1 and e_2 as follows:

$$\left.\begin{aligned} p(X_1) &= p(X_1|e_1) \cdot p(e_1) + p(X_1|e_2) \cdot p(e_2) \\ &= 0·7 \times 0·6 + 0·2 \times 0·4 = 0·50 \\[6pt] p(X_2) &= p(X_2|e_1) \cdot p(e_1) + p(X_2|e_2) \cdot p(e_2) \\ &= 0·3 \times 0·6 + 0·8 \times 0·4 = 0·50. \end{aligned}\right\} \tag{28}$$

and

The fact that the computed values of $p(X_1)$ and $p(X_2)$ add to 1 provides a useful check on the working. These values are inserted in the appropriate branches emanating from the chance node.

Continuing, we come to the decision nodes, where the investor must decide what to do, having received his professional advice. We next arrive at the chance nodes corresponding to the uncertain events e_1 and e_2. By way of illustration, let us consider the node at the upper right-hand corner of the tree. This corresponds to the advice to invest (X_1) and the decision to accept the advice

(d_1). The correct probabilities to attach to the two branches emanating from this node are therefore $p(e_1|X_1)$ and $p(e_2|X_1)$; that is to say, the probabilities of the stock rising or falling, given all the information back to the base of the tree. This includes the investor's decision to buy advice, and the advice that was tendered to him. We do not have these probabilities directly, but can calculate them by means of Bayes' Theorem. The results we need are

and
$$\left.\begin{array}{l} p(e_1|X_1) = p(X_1|e_1) \cdot p(e_1) \,/\, p(X_1) \\[6pt] p(e_2|X_1) = p(X_1|e_2) \cdot p(e_2) \,/\, p(X_1). \end{array}\right\} \tag{29}$$

Thomas Bayes, FRS, was an obscure English clergyman who died in 1761. His work on inverse probabilities was published posthumously in 1763. His Theorem, one of the most important and most controversial theorems of probability theory, is concerned with how we should modify the strength of our beliefs to take account of new information. Thus, for example, the *prior probability* (representing the strength of our initial belief) that the stock will appreciate is $p(e_1)$, but this must be modified in the light of advice from the broker that he believes the stock will indeed rise. This *posterior probability*, as it is called, is denoted by $p(e_1|X_1)$ and, with a good adviser, we would expect it to be larger than $p(e_1)$.

Putting in the numerical values, we have
$$p(e_1|X_1) = 0 \cdot 7 \times 0 \cdot 6 \,/\, 0 \cdot 5 = 0 \cdot 84$$
and
$$p(e_2|X_1) = 0 \cdot 2 \times 0 \cdot 4 \,/\, 0 \cdot 5 = 0 \cdot 16.$$

The assessment of the likelihood of the stock appreciating has risen from 60% to 84%. Similar calculations can be made for the branches emanating from the other chance nodes and the results are entered on the tree (Fig. 3.11).

It should be noted that we have chosen to assess the three probabilities $p(e_1)$, $p(X_1|e_1)$ and $p(X_2|e_2)$ and to derive all the others from them. The decision-maker might, however, have chosen to proceed differently. This he is quite entitled to do provided that his probability assessments and derivations are internally consistent. In fact with the three initial probabilities used here there is no possibility of inconsistency, but this is not true in general.

Having attached probabilities to all branches emanating from chance nodes, we must next consider the utilities, which in this case are simply money values. We start with the terminal branches and associate with each of them the appropriate value of the outcome. Thus the top terminal branch relates to a situation where advice X_1 has been received, the decision d_1 to invest has been taken, and the uncertain event e_1 has occurred. The utility of this outcome is therefore the capital value of the appreciated stock, namely £1100, less the fee, £f. The utilities associated with all the other terminal branches may be computed similarly and are entered on the tree.

We can now complete the analysis by working back to the base. At each chance node we calculate an expected utility and at each decision node we select the branch with the highest expected utility. Starting, once again, at the top right-hand corner, we have a chance node with two branches: one yielding a utility of £$(1100 - f)$ with a probability of 0·84 and the other a utility of £$(800 - f)$ with a probability of 0·16. The expected utility is therefore given by £$[0·84 \ (1100 - f) + 0·16 \ (800 - f)] =$ £$(1052 - f)$, and this is entered at the chance node. The chance node immediately below clearly yields an expected utility of £$(1000 - f)$. Moving back, we find that the branches which terminate at these two chance nodes originate at the same decision node. The choice here is between decision d_1 with an expected yield of £$(1052 - f)$ and decision d_2 with an expected yield of £$(1000 - f)$. We therefore choose d_1 as giving the larger expected yield; £$(1052 - f)$ can be entered at the node and decision d_2 forgotten. Similar calculations are made for all the other branches and nodes and the results are entered on the tree. We find that the decision made at the base of the tree offers a choice between an expectation of £$(1026 - f)$ if the broker's advice is sought, and either £980 or £1000 if it is not. Clearly the second choice may be rejected, while the first is to be preferred to the third if the broker's fee is less than £26. We can regard this sum as the value to the investor of the advice that the broker is able to provide.

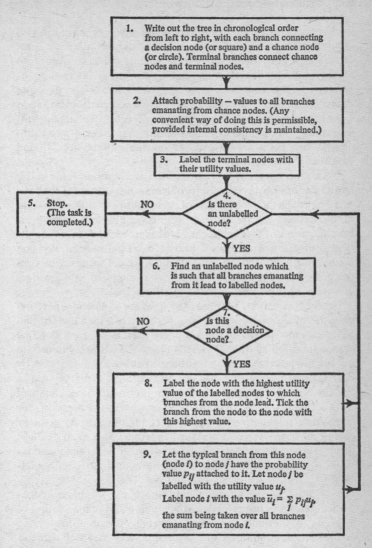

1. Write out the tree in chronological order from left to right, with each branch connecting a decision node (or square) and a chance node (or circle). Terminal branches connect chance nodes and terminal nodes.

2. Attach probability — values to all branches emanating from chance nodes. (Any convenient way of doing this is permissible, provided internal consistency is maintained.)

3. Label the terminal nodes with their utility values.

4. Is there an unlabelled node?

5. Stop. (The task is completed.)

NO

YES

6. Find an unlabelled node which is such that all branches emanating from it lead to labelled nodes.

7. Is this node a decision node?

NO

YES

8. Label the node with the highest utility value of the labelled nodes to which branches from the node lead. Tick the branch from the node to the node with this highest value.

9. Let the typical branch from this node (node i) to node j have the probability value p_{ij} attached to it. Let node j be labelled with the utility value u_j. Label node i with the value $\bar{u}_i = \sum_j p_{ij} u_j$, the sum being taken over all branches emanating from node i.

Fig. 3.12. Flow chart of procedure for analysing a decision tree

Other examples of decision tree analysis will be found in [14] and [15]. The procedure for determining the best decision is set out in Fig. 3.12 in the form of a flow chart. It will be convenient to regard terminal branches as leading to nodes, referred to as *terminal nodes*, thus bringing them into line with all other branches. Steps 1–3 are specific to the particular problem; steps 4–9 are common to all problems. The decision tree technique is well suited to computer 'mechanization'; indeed only two arithmetical processes are involved – maximization (step 8) and weighted averaging (step 9).

It is all too apparent that some of the assumptions made in our treatment of the investment problem – for example, that the stock either has a fixed chance of appreciating by a fixed amount or of depreciating by another fixed amount – are gross over-simplifications of what happens in the real world. A more sophisticated investment analysis would invoke 'continuous' decision theory and would entail constructing a model in terms of continuously varying quantities, such as changes in stock values governed by probability density distributions.

In another respect, however, the investment example is representative of a more general type of situation in which the decision tree technique is proving increasingly useful. We may describe such a situation as follows. The 'state of the world' is uncertain but further information can be obtained about it – at a price – by experimentation, by further study, or by seeking expert advice. In our investment example, this further information was obtained from a broker. In the case of an oil or minerals prospector, for instance, it might be obtained by carrying out seismic or other geological tests. Are such tests worth-while? Statistical decision theory should be able to tell us, provided it is possible to construct a mathematical model of the situation that is both sufficiently realistic and amenable to analysis. This may not be an easy task and the subject affords plenty of scope for applied mathematical research. In particular, the selection and manipulation of a variety of internally consistent probability distributions – prior and posterior, conditional and unconditional – may well demand considerable mathematical insight and skill.

5. Dynamic Programming

We now consider another approach to the general problem of nonlinear allocation and planning. It is known as *dynamic programming* and is largely the creation of one man – Richard Bellman, whose first book on the subject was published in 1957.

A LAND MANAGEMENT EXAMPLE

To show the flavour of the dynamic programming approach, we shall look at a simple example, adapted from [13]. A landowner has a piece of woodland containing z tons of timber. In a certain year he fells x tons and sells it for £$f(x)$. The balance of y tons is left to grow and will yield cy tons in a year's time (where $c > 1$). The owner wishes to sell out (with all the timber sold) after N years. The question is: what felling policy should he adopt year by year in order to maximize his total income over the N-year period? (This is, of course, a dressed-up version of one of the basic economic problems – what proportion of one's income should be saved rather than spent.) With an obvious notation, we may represent the situation thus:

Year	Sells	Leaves	Available next year
1	x_1	y_1	cy_1
2	x_2	y_2	cy_2
.			
.			
$N-1$	x_{N-1}	y_{N-1}	cy_{N-1}
N	x_N	0	0

We thus have

$$\left. \begin{array}{l} z = x_1 + y_1 \\ cy_1 = x_2 + y_2 \\ cy_2 = x_3 + y_3 \\ \dots\dots\dots\dots\dots \\ \dots\dots\dots\dots\dots \\ cy_{N-2} = x_{N-1} + y_{N-1} \\ cy_{N-1} = x_N. \end{array} \right\} \qquad (30)$$

We can eliminate $y_1, y_2 \ldots y_{N-1}$ from those N equations to give

$$zc^{N-1} = x_N + cx_{N-1} + \ldots c^{N-1}x_1. \tag{31}$$

The landowner's total income over the N years is

$$P = f(x_1) + f(x_2) + \ldots + f(x_N). \tag{32}$$

We wish to determine the policy over the N-year period which will maximize P, subject to the constraint (31) and the non-negative conditions

$$x_i \geqslant 0 \text{ for } i = 1, 2 \ldots N. \tag{33}$$

This is a nonlinear constrained optimization problem in N variables. If the form of the function $f(x)$ is linear or concave upwards, the problem is trivial, as clearly the best policy is to leave all the timber to grow until the final sell-out. If, however, $f(x)$ is of any other form – for example, a 'diminishing marginal utility' curve like that of Fig. 3.8 – then the problem is very difficult to solve either by classical methods or by any of the techniques we have discussed so far.

Let us therefore apply Bellman's approach to the problem. This will involve some mathematical sophistication, but should not present serious difficulty. Let us define $g_n(q)$ as the total income the owner will receive if he starts with q tons of timber and adopts an optimal policy of timber management over an n-year period. Now, clearly, $g_1(q) = f(q)$ since if the owner is in business for 1 year only he must sell all his timber at the end of that year. We now consider $g_2(q)$. If the optimal policy is such that x tons should be felled and sold at the end of the first year, then we have

$$g_2(q) = f(x) + \{\text{the return from } c(q - x) \text{ tons over the one remaining year}\}. \tag{34}$$

Now since $g_2(q)$, by definition, refers to the optimal policy over a 1-year period, the remaining $c(q - x)$ tons must clearly be utilized to best advantage. This implies that the value of the expression in curly brackets must be $g_1\{c(q - x)\}$, yielding $g_2(q) = f(x) + g_1\{c(q - x)\}$ if x is optimal. Now x will be optimal when the 1-year return is as large as possible, and so we have

$$g_2(q) = \max_{0 \leqslant x \leqslant q} [f(x) + g_1\{c(q - x)\}]. \qquad (35)$$

Now our argument in passing from a 1-year to a 2-year optimal policy can clearly be applied in going from any period of time, say $(m - 1)$ years, to m years. This gives

$$g_m(q) = \max_{0 \leqslant x \leqslant q} [f(x) + g_{m-1}\{c(q - x)\}]. \qquad (36)$$

We have seen that $g_1(q) = f(q)$, so this equation enables us to proceed step by step and determine in turn $g_2(q), g_3(q) \ldots$ up to $g_N(q)$.

The point to note is that we have reduced an optimization problem in N variables to a sequence of simple maximization problems, each involving only one variable. The price we have to pay for this very substantial bonus is that we have to work out the values of the sequence $g_m(q)$ for $m = 1, 2, 3 \ldots$ etc. not only for a single value but for a range of values of q. This is because we do not know at the outset how much live timber the owner has at the start of each year after the first. The dynamic programming approach thus involves a lot of computation. Once again, we have an instance of the creation of a new branch of mathematics which it would have been quite impossible to put to much practical use without the electronic computer to look after the arithmetical chores.

Although dynamic programming is, then, essentially a computer-dependent technique, it may be interesting to work through the woodland example algebraically. In order to do this without too much labour, we shall take a very simple – and therefore unrealistic – form for the price function $f(x)$, namely $f(x) = k\sqrt{x}$. This satisfies the basic requirements that $f(x) = 0$ when $x = 0$ and the slope of $f(x)$ steadily decreases as x increases while always remaining positive. We then have

$$\left. \begin{array}{l} g_1(q) = f(q) = k\sqrt{q} \\ g_2(q) = \max_{0 \leqslant x \leqslant q} [k\sqrt{x} + k\sqrt{c(q - x)}]. \end{array} \right\} \qquad (37)$$

In this particular case, the value of x which minimizes this expression may be found by differentiating the expression and equating the derivative to zero. This yields

$$\frac{k}{2}\left[\frac{1}{\sqrt{x}} - \frac{\sqrt{c}}{\sqrt{q-x}}\right] = 0 \text{ giving } x = q/(1+c). \qquad (38)$$

This value of x lies within the permitted range $(0, q)$ and it can be shown that $g_2(q)$ does in fact attain its maximum value for this value of x. We obtain

$$g_2(q) = k\sqrt{\frac{q}{1+c}} + k\sqrt{\frac{c \cdot qc}{1+c}} = k\sqrt{q(1+c)}. \qquad (39)$$

Proceeding similarly, we find that

$$\begin{rcases} g_3(q) = \max_{0 \leqslant x \leqslant q} \ [k\sqrt{x} + k\sqrt{c(1+c)(q-x)}] \\ \quad = \ k\sqrt{q(1+c+c^2)} \text{ which is attained when} \\ x = \ q/(1+c+c^2). \end{rcases} \qquad (40)$$

The general pattern is now apparent, namely,

$$\begin{rcases} g_n(q) = k\sqrt{q(1+c+c^2+\ldots+c^{n-1})}, \\ \text{attained when } x = q/(1+c+c^2+\ldots c^{n-1}). \end{rcases} \qquad (41)$$

We are now able to evaluate $g_m(q)$ for all m over a range of values of q. At this stage, however, we do not know the amount of timber (i.e., what value of q) that the owner will start with each year. Fortunately there is one value of q that we do know, namely the initial amount, z, of timber in the wood. We also know the value of N and so we can evaluate $g_N(z)$. This gives the maximum return over the N-year period.

To obtain the optimal timber felling policy for each year which will yield this maximum return we have to use the intermediate functions $g_1(q)$ to $g_{N-1}(q)$ but in reverse order. The policy decision for the first year is governed by the equation

$$g_N(z) = \max_{0 \leqslant x \leqslant z} \ [f(x) + g_{N-1}\{c(z-x)\}]. \qquad (42)$$

If x_1 is the value of x which satisfies this equation, then x_1 is the amount of timber to be felled and sold in the first year of the

N-year period. So in the second year the owner starts with $c(z - x_1) = z_2$ (say) tons of timber in his wood. The policy for the second year is therefore governed by the equation

$$g_{N-1}(z_2) = \max_{0 \leqslant x \leqslant z_2} [f(x) + g_{N-2}\{c(z_2 - x)\}]. \quad (43)$$

If x_2 is the solution of this equation, then the owner should sell x_2 tons during the second year and leave himself with $c(z_2 - x_2) = z_3$ tons for the third year. The procedure for progressing through the years is now apparent. To show how it works out, let us put some numerical values into our problem.

Suppose $N = 4$, $z = 10\,000$ and $c = 1 \cdot 1$. Since k is merely a scale factor, we may take it as 1. We start with

$$g_4(z) = \sqrt{z(1 + c + c^2 + c^3)} = 215.$$

This gives the value of the best return over the 4 years. We see further, that

$$x_1 = z/(1 + c + c^2 + c^3) = 2150,$$

so the owner should sell 2150 tons in the first year. This leaves him with $z_2 = c(z - x_1) = 8635$ tons at the start of the second year. We must now solve the $g_3(q)$ equation with $g = z_2$. This yields

$$g_3(z_2) = \sqrt{z_2(1 + c + c^2)} = 170,$$
$$x_2 = z_2/(1 + c + c^2) = 2610 \text{ tons}$$

and

$$z_3 = c(z_2 - x_2) = 6625 \text{ tons}.$$

Proceeding similarly, we evaluate $g_2(z_3)$, x_3 and z_4, and finally $g_1(z_4)$. The value of x_4 must be equal to z_4 since the maximum of $g_1(z) = k\sqrt{z_4}$ over the range $(0, z_4)$ is attained when $x = z_4$.

The results are summarized in Table 3.11. The figures in the fourth column provide a check on the arithmetic; they should add to the optimal return over the 4-year period, $g_4(10\,000)$, which has already been calculated. There are similar checks for the various shorter periods (cols. 4 and 5).

TABLE 3.11 The optimal 4-year timber management policy

(1)	(2)	(3)	(4)	(5)
Year (i)	Starts with z_i tons of wood	Fells and sells x_i tons	Annual return ($\sqrt{x_i}$) in £	Optimal return over remaining years in £ $\{g_{5-i}(z_i)\}$
1	10 000	2150	46	215
2	8635	2610	51	169
3	6625	3160	56	118
4	3800	3800	62	62
		11 720	215	

THE PRINCIPLE OF OPTIMALITY

The crucial step in the argument of the last section was when the expression in curly brackets in (34) was set equal to $g_1\{c(q-x)\}$; that is, to the *optimal* income for a 1-year period. In general, if the landowner wishes to maximize his profits with k years remaining, he must not only make the best immediate decision, but he must follow this up a year later with the best decision with respect to the remaining $(k-1)$-year period and the timber available at that time. This is the basic idea behind dynamic programming. It is known as 'The Principle of Optimality' and Bellman states it thus: 'An optimal policy has the property that whatever the initial state and initial decision are, the remaining decisions must constitute an optimal policy with regard to the state resulting from the first decision.' A rigorous proof of the Principle can easily be constructed; some may regard it as a self-evident proposition. Fig. 3.13 presents an intuitive geometrical demonstration. The continuous curve represents the optimal policy for proceeding from an initial state A to a final state N, with B, C, D ... denoting intermediate states resulting from successive decisions. Now let the optimal policy for proceeding from B to N be represented by the dashed line, which we shall assume is different from the full line. This implies that the policy represented by B, C', D' ... N is better than that represented by B, C, D ... N. Hence the policy represented by A, B, C' D' ... N is better than that represented by A, B, C, D ... N. But this latter policy is defined to be the optimum policy, so our assumption that the Principle of Optimality is violated has led to a contradiction. We therefore conclude that the Principle is true.

The method of dynamic programming can be used in any situation that can be formulated as a *multi-stage decision process*; that is to say, a process where a sequence of choices has to be made. Each such decision leads to a certain outcome, yields a certain

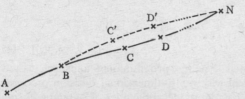

Fig. 3.13. The principle of optimality

return, and it is assumed that the total return is simply the sum of the individual returns. Usually, but not necessarily, successive decisions follow each other in time; hence the term 'dynamic'.

We pointed out in the last section that it is usually very difficult – and, indeed, often impossible – to optimize a system over N dimensions, even when N is only moderately large. The great merit of the Principle of Optimality is that it enables a problem in many dimensions to be reduced to one in far fewer; often, indeed, to a single dimension as in our woodland example.

Another important feature of the dynamic programming approach is that a specific problem is not solved in isolation, but by 'embedding' it in a set of similar problems. Thus if we had solved our timber management problem numerically, we would necessarily have obtained the solution for varying numbers of years and for different amounts of timber. This may give valuable insight into the nature of the solution as well as exhibiting the effect of varying the 'parameters' of the situation.

The more general form of equation (36) may be written as

$$g_N(x) = \max_{0 \leqslant x_N \leqslant x} [F_N(x_N) + g_{N-1}(x, x_N)]. \tag{44}$$

An equation of this kind is known as a *functional equation*. It also defines a *recurrence relation* and so provides a method of obtaining the sequence $g_N(x)$ recursively, once $g_1(x)$ is known.

In our woodland example we were concerned with only one type of resource, namely timber. In practice, the situation might well be more complex. Account might have to be taken, for example, of several different resources (e.g., labour, productive capacity, raw materials). The two-dimensional counterpart of (44) is

$$g_N(x, y) = \max_{0 \leqslant x_N \leqslant x} \max_{0 \leqslant y_N \leqslant y} [F(x_N, y_N) +$$
$$g_{N-1}(x, y, x_N, y_N)]. \tag{45}$$

While the formal dynamic programming approach can be carried over to deal with multi-dimensional decision processes, the technical and computational difficulties of obtaining a solution in a particular case may be formidable. To overcome these difficulties Bellman and his school have invoked a number of sophisticated mathematical techniques, such as the use of Lagrange multipliers and successive approximations. We must refer the reader to the literature [16] for further details.

A SHORTEST-ROUTE EXAMPLE

In our timber example, the variables, x_i, z_i are continuous; they may take any values within a presented range. Dynamic programming is also applicable to discrete problems when the decisions are limited to a finite number of alternative choices, A or B or C, etc. Consider, for example, the following problem, adapted from [7].

A person wishes to travel from a city P to a city Z by the shortest route. The alternative routes and the inter-city distances are shown (not to scale) in Fig. 3.14. Let $g_N(AB)$ denote the length of the shortest route from A to B, passing through $(N-1)$ inter-

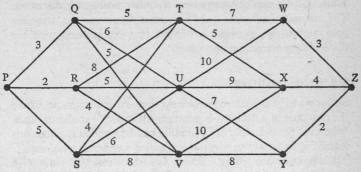

Fig. 3.14. The shortest-route problem

mediate cities. If $N = 1$ the journey is a direct one for which the distance is known. We wish to determine $g_4(PZ)$. The principle of optimality gives

$$g_4(PZ) = \text{minimum of} \begin{cases} g_1(PQ) + g_3(QZ) \\ g_1(PR) + g_3(RZ) \\ g_1(PS) + g_3(SZ). \end{cases}$$

In the same way we have
$$\text{(46)}$$

$$g_3(QZ) = \text{minimum of} \begin{cases} g_1(QT) + g_2(TZ) \\ g_1(QU) + g_2(UZ) \\ g_1(QV) + g_2(VZ) \end{cases}$$

with similar equations for $g_3(RZ)$ and $g_3(SZ)$.

The two-stage journeys (i.e., those starting from T, U and V) can be treated similarly except that there are only two choices of route from T to Z or from V to Z. Once again we proceed in reverse order. All the one-stage distances are known, so the two-stage distances and hence the three g_2-functions can be computed directly. We can then calculate the 3-stage distances and hence the g_3-functions, and finally the 4-stage distances and the desired $g_4(PZ)$. The shortest route turns out to be $PRVYZ$, a total length 16 units. The reader is invited to work through the process for himself.

Fig. 3.14 is an example of what is called a *network*. This is the subject of the next chapter where a similar problem is discussed. The method of solution there presented, known as the 'cheapest path algorithm', provides an interesting comparison with the dynamic programming approach.

SELLING A HOUSE

So far we have considered problems that are deterministic, where the outcome of a decision is precisely known. We shall now illustrate how the principle of optimality can be applied to situations involving uncertainty, to what are called *stochastic problems*.

Suppose we are selling a house. Let us assume that we receive one offer each day, which we can either accept (thus terminating the process) or reject, in which case we receive a new offer next day. Let us also assume that we know that on the tenth day, if we are still in the market, we cannot delay any longer and must accept the offer received on that day. What strategy should we adopt in order to maximize the *expected return* on the sale? (We use the term 'expected return' in the sense of 'statistical expectation', as we have already done in the section on decision theory.)

Clearly the value of this expected return will depend on two things: the number of days that remain during which decisions may be taken, and the probability distribution of the daily offer. For simplicity, we shall assume that this distribution does not differ from day to day and that it takes a discrete form in which bids are restricted to exact multiples of £1000. Suppose the distribution is given by Table 3.12.

TABLE 3.12   The distribution of bids

Value of a bid in £(s)	Probability of receiving a bid of £s ($p(s)$)
20 000	0·4
21 000	0·2
22 000	0·2
23 000	0·2

Proceeding on what are now familiar lines, let us define $f_n(s)$ as the expected return when (a) a bid of £s has been received, (b) the seller adopts an optimum policy, and (c) there are n days during which further offers may be received. Now suppose an offer of £x is received. If the seller accepts it, he will receive x; if he rejects it, his expectation of return from later bids is given by

$$\Sigma f_{n-1}(s)p(s),$$

summed over the set of all possible values of s. Thus we can write, for brevity,

$$f_n(x) = \max [x; \sum_s f_{n-1}(s)p(s)]. \tag{47}$$

We also have an 'end condition'. If the seller waits until the last possible day, he forfeits his freedom of action and must accept whatever bid is offered on that day. This means that

$$f_0(x) = x. \tag{48}$$

We can now compute the values of the sequence $f_n(x)$ for $n = 1, 2, 3 \ldots$, in turn, as far as necessary. We find that

$$f_1(x) = \max [x; \{0{\cdot}4f_0(20\,000) + 0{\cdot}2f_0(21\,000) + 0{\cdot}2f_0(22\,000) + 0{\cdot}2f_0(23\,000)\}]$$

$$= \max [x; \{0{\cdot}4(20\,000) + 0{\cdot}2(21\,000) + 0{\cdot}2(22\,000) + 0{\cdot}2(23\,000)\}]$$

$$= \max [x; 21\,200].$$

This means that, with only one day left, the seller should accept that day's bid if it is more than £21 200 and reject it if it is less. We can tabulate the decisions thus:

x	$f_1(x)$	Decision
20 000	21 200	Reject
21 000	21 200	Reject
22 000	22 000	Accept
23 000	23 000	Accept

Similarly we find that

$$f_2(x) = \max [x; \{0.4f_1(20\ 000) + 0.2f_1(21\ 000) + 0.2f_1(22\ 000) + 0.2f_1(23\ 000)\}]$$

$$= \max [x; \{0.4(21\ 200) + 0.2(21\ 200) + 0.2(22\ 000) + 0.2(23\ 000)\}]$$

$$= \max [x; 21\ 720].$$

giving:

x	$f_2(x)$	Decision
20 000	21 720	Reject
21 000	21 720	Reject
22 000	22 000	Accept
23 000	22 000	Accept

We next compute $f_3(x)$ to be max $[x; 22\ 032]$, which means that only the maximum possible bid (of £23 000) should be accepted. It is now apparent that this decision should also be taken for all higher values of n.

We are now able to formulate the complete decision rule in Table 3.13.

TABLE 3.13   House seller's optimal decision policy

x	$n = 0$	1	2	3 or more
20 000	Accept	Reject	Reject	Reject
21 000	Accept	Reject	Reject	Reject
22 000	Accept	Accept	Accept	Reject
23 000	Accept	Accept	Accept	Accept

This example has been drastically simplified in order to bring out the salient features of the dynamic programming approach to stochastic problems. In particular, we have assumed that a bid is limited to one of a set of specified amounts; that is to say, the problem is a discrete one, like the shortest route example. We can readily generalize the method to deal with the continuous case; that is, where the value of a bid is a random variable having a probability density distribution of the kind illustrated in Fig.

3.9(a). This means, for those familiar with calculus, that the probability of a bid being between s_1 and s_2 may be expressed by $\int_{s_2}^{s_1} g(s)ds$. All we have to do is to replace the summation $\Sigma f_{n-1}(s)p(s)$ in (47) by an integral of the form $\int_0^\infty f_{n-1}(s) \cdot g(s)ds$. The end condition (48) is unchanged. It is perhaps worth noting that

$$f_1(x) = \max\ [x; \int_0^\infty f_0(s) \cdot g(s)ds] = \max\ [x; \int_0^\infty s \cdot g(s)ds]$$

which may be written as

$f_1(x) = \max\ [x; m]$ where m is the *mean* of the density distribution.

THE RANGE OF DYNAMIC PROGRAMMING

Dynamic programming is a flexible technique of considerable power, always provided that adequate computing resources are available. Bellman and his school have shown great skill and ingenuity in applying the technique to a very wide variety of problems [16]. The variables may be continuous or discrete; the process may be deterministic or stochastic; the optimization may be required over a single variable or over several variables; the number of consecutive decisions may or may not be known in advance (in the latter case the objective may be to minimize costs over an indefinite future).

Most books on operational research contain at least one chapter on dynamic programming. Among the problems discussed [in 7, 13, 16, 17] are: allocation of limited resources between activities or customers, production control and scheduling, inventory management, transportation of goods and routing of vehicles, replacement and maintenance of equipment, warehousing and storage of goods, sales and purchasing policy, loading and packing of cargo, advertising policy, bidding at auctions, industrial quality control, standardization of production, farming management and catering.

We have already had an example of the close connection between dynamic programming and network analysis; the links with mathematical programming are equally close. Indeed, any linear programming problem can be formulated in dynamic programming terms. The linear programming technique, when applicable, will usually be much simpler to use, but the dynamic programming formulation has the advantage of not being restricted by linearity conditions. Finally, the connection between dynamic programming and statistical decision theory, both of which are concerned with multi-stage processes, will also be apparent.

6. Queueing and Simulation

Everyone, nowadays, is only too familiar with queues – at the supermarket, the post office, the doctor's waiting room, the airport, or on the factory floor. Queues occur when the service required by customers is not immediately available. Customers do not arrive regularly and some take longer to serve than others, so queues are likely to fluctuate in length – even to disappear for a time if there is a lull in demand. Once again we are concerned with the allocation of limited resources, but with the new feature that the conflict of interest between the supplier and the customer is now accorded explicit recognition.

The shopper leaving the supermarket, for example, desires service; the store manager wants to see his cashiers busy most of the time. If customers have to wait too long, some will decide to shop elsewhere; if the queue gets too long, it will impede access to the goods. Thus the manager might consider employing another cashier. He realizes that this would increase costs, so he would like to know what effect it would be likely to have on such things as average waiting time or average queue length. Queueing theory should be able to help him. In other cases it may be the type, rather than the total quantity, of service that is being examined. Is it better, for instance, for the various kinds of service provided at a post office to be available at separate positions along the counter or for all services to be available at all positions?

The essential feature of a queueing situation, then, is that the number of customers (or units) that can be served at a time is limited so there may be congestion. This state of affairs occurs in a variety of circumstances and its existence is not always obvious. Typical industrial queueing situations are a stream of articles passing along a conveyor belt for eventual packing into boxes, machines that stop from time to time and need attention from an operator who can attend only to one machine at a time, finished products that accumulate in a dispatch bay to be sent away in batches in lorries, or the level of stock to be held in conditions of variable supply and demand.

Queueing problems lend themselves to mathematical treatment and the theory has been extensively developed during the last seventy years. The earliest systematic work on the subject was done by Erlang of the Copenhagen Telegraph Company, whose first paper on congestion in telephone exchanges was written in 1909.

The raw materials of queueing theory are mathematical models of queue-generating systems of various kinds. The objective is to predict how the system would respond to changes in the demands made on it; in the resources provided to meet those demands; and in the rules of the game, or queue discipline as it is usually called. Examples of such rules are: 'first come, first served'; 'last come, first served', as with papers in an office 'in-tray'; service in an arbitrary order; or priority for VIPs or disabled persons.

To analyse queueing problems, we need information about the *input* (the rate and pattern of arrival of customers), the *service* (the rate at which customers are dealt with either singly or in multiple channels), and the *queue discipline*.

ELEMENTARY QUEUEING THEORY

One of the simplest queueing models postulates an input of identical units (e.g., customers), an unlimited queue that is served in order of arrival and a single service channel, with the patterns of arrival and service times specified by probability distributions. In many cases it is reasonable to assume that the rates of arrival and of completion of service (assuming that a customer is already being served) are constant, *on the average*, and are independent both of the current time and of the current state of the system. Such arrivals and service times are said to be *completely random*. What this means is that if k is a constant representing the average rate of arrival of customers, the probability of one arrival in a short interval of time may be taken as equal to k times the length of the interval. For example, if $k = 1/6$ arrivals per second, then the probability of one arrival in any period of 1/10 second (a short interval of time!) may be taken to be 1/60. The reason for stipulating a short interval is to allow us to assume that the probability of two or more arrivals during the interval will be so small as to be negligible. Furthermore – and this is the essence of the idea of randomness – what happens during this time-interval is assumed to be totally unaffected by what happens at any other time – either before or after. We must always remember that k specifies an average arrival rate; actual arrivals will be bunched or spread out in an irregular manner. In Fig. 3.15 the arrivals of

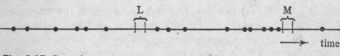

Fig. 3.15. Part of a random sequence of arrivals

customers are represented by points along a line. L and M are two short intervals of equal duration so the probability of an arrival during either interval is the same. A random sequence of arrivals is in fact a very special pattern of arrivals; there is nothing vague about it notwithstanding the popular connotation of the word 'random'. It is a good approximation to reality when the customers are drawn from a large group who all behave independently. Another of its attractions is that it is easy to handle mathematically!

With these assumptions, the probability, p_n, of n arrivals in a period of time, T, depends on the value of a single quantity, kT, which is the *average* number of arrivals during the time T. (The average interval between two successive arrivals is $1/k$. Thus if $k = 1/6$, the average interval would be 6 seconds.) This distribu-

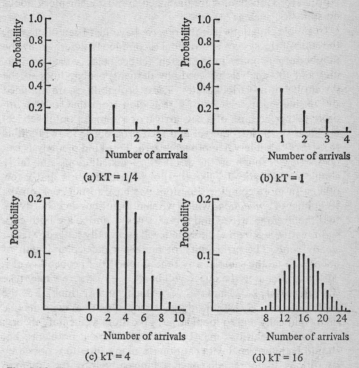

kT = average number of arrivals in the fixed time, T

(a) $kT = 1/4$

(b) $kT = 1$

(c) $kT = 4$

(d) $kT = 16$

Fig. 3.16. The Poisson distribution of the number of completely random arrivals in a fixed time

tion of p_n with n is known as the *Poisson distribution*, after the versatile French mathematician, S.-D. Poisson (1781–1840). The formula for p_n will be found in [3, 7, 13 and 18]. Fig. 3.16 gives some idea of the form of this famous distribution for a few values of kT.

If the average service time is $1/s$, similar expressions, with s in place of k, apply to completion of service, provided, of course, that the service channel is busy. The Poisson distribution – and also the two other basic distributions, the Binomial and the Normal (or Gaussian) – are discussed in most elementary books on statistics, such as [3].

In simple situations like the one we have just been discussing, the state of the system at any time is completely described by the number of customers in the system (either being served or waiting) and the analysis proceeds by deriving probabilities of the system being in various states. These probabilities are obtained by considering the chance of a transition from one state to another (either because of a new arrival or a completion of service) during a small time-interval and then proceeding to the limit as the size of this interval approaches zero. This kind of analysis can be carried through if the probability distributions are fairly simple mathematically, like the Poisson distribution and a few others. In more complex situations, we have to fall back on the technique of *simulation*, which we shall be discussing shortly.

If the average arrival and service rates, k and s, are both constant, with k less than s, the system will eventually settle down to a steady state. The probability of finding a particular length of queue will be the same at any time. In fact, the probability of n customers being in the queue will be $r^n(1 - r)$ where n may take the values 0, 1, 2 . . . and $r = k/s$, which we may think of as the traffic intensity [7, 18]. Steady states are much easier to handle mathematically and so loom large in the more elementary theory. Steady-state solutions may be useful in practical problems, but should be interpreted with caution. In many situations operating conditions do not remain constant long enough for the system to become steady.

Mathematical analysis on these lines can be applied to a considerable range of fairly simple models involving a number of different probability distributions of arrival and service times. We shall not attempt to survey the field here, as both the basic theory and its applications are well covered in the literature [7, 13, 17, 18]. More recently other techniques, notably the use of integral equations and Markov chain analysis, have been applied to the analysis of queueing problems [13, 18], but these lie outside the scope of this book.

DIGITAL SIMULATION

We have already remarked that the range of queueing models whose behaviour can be studied by analytical methods is severely limited. Where mathematical analysis fails, approximate solutions, sufficiently accurate for practical purposes, can often be obtained by *numerical experiment*. Constructing and studying the behaviour of a mathematical model which in some sense mirrors or 'simulates' the 'real' system is central to all applications of mathematics; examples will be found in every chapter of this book.

In this chapter we shall be concerned with a particular kind of simulation, sometimes called *digital simulation*, the purpose of which is to study the behaviour of models in which *variability* with time plays a major rôle. The technique is to create, and trace through over a period of time, a typical 'life history' of the system under prescribed conditions. In the case of customers seeking service, for example, one would work out, step by step, what happens to each customer as he arrives and passes through the different parts of the system. To do this, one would need to know the rules of operation and other characteristics of the system, and to have numerical information on such variable factors as the times between arrivals in the queue and the duration of service. The effect of variability on the system as a whole is reproduced by *sampling* from the appropriate probability distributions. These may be derived from frequency distributions, observations (or 'counts') of how the system has behaved in the past, or may be distributions specified directly by mathematical functions, or a mixture of both. This rather 'brute force' approach necessarily entails a lot of calculation and data processing and, once again, we have a situation where the use of computers is essential to the effective exploitation of a technique.

One of the great merits of simulation is that it enables the effects of changes in the system to be assessed by experimenting with the model instead of with the system itself. This not only yields obvious advantages in terms of cost and convenience, but in many cases it is the only practicable way to proceed. A manager cannot choose a sales plan for next year by trying out all the possibilities this year. An airport controller cannot determine the effect of a 25% increase of traffic by direct experiment. There are many situations in which a decision-maker must either rely on his 'hunch' or seek assistance from digital simulation experiments conducted on a simplified model of the actual system.

We have seen that the purpose of sampling 'at random' from a distribution is to produce a typical sequence of times at which some event occurs. One such event might be the arrival of a customer; another the completion of the service supplied to him. The interaction between these two sequences can be worked out and hence the time of arrival and departure of each customer computed.

How is the sampling process carried out? The procedure is basically the same as spinning a roulette wheel where the next number to be called is chosen at random; each number of the set is equally likely to be selected. For this reason, procedures designed to ensure random sampling of values from a probability distribution are often termed *Monte Carlo* methods, although some writers restrict the term to more refined variants of the simple sampling procedure.

The results obtained from any form of Monte Carlo method are no more than estimates; they are subject to what are termed 'sampling errors', just like the results of a public opinion poll. This means that a large number of repeated samplings must be made to get a reasonably close estimate of the desired result. In general, the accuracy of the estimate is proportional to the square root of the number of repetitions. We are thus handicapped by the operation of a law of diminishing returns. To double the accuracy, we must make four times as many 'runs'; to double it again, sixteen times as many. The advent of the computer has reduced, but not removed, the adverse consequences of this troublesome fact. Indeed, a good deal of research – using some very nice mathematics – has been devoted to refining the basic Monte Carlo procedure in order to increase the accuracy of the results. One such method, known as *importance sampling*, seeks to concentrate the distribution of the sample values in the parts of the range that are of greatest importance, instead of spreading them out evenly. Another approach is to use *stratified sampling*, a technique which will be familiar to those concerned with opinion polls and social surveys. We cannot pursue these matters here and must refer the reader to the more specialized literature [19].

It is, of course, not sensible to use a roulette wheel to obtain our sample values; it is much more convenient to make use of *random numbers*. Indeed, one writer has it that 'Monte Carlo methods comprise that branch of experimental mathematics which is concerned with experiments on random numbers' [19]. Several tables of random digits have been published since the beginning of the century and they can be used to construct sequences of random numbers of any desired length. A full discussion of the question of how we know that a given sequence of digits does in fact form part of a random sequence would lead us into deep water. Suffice to say that a sequence is accepted as 'sufficiently random' for use in sampling experiments if it passes the recognized statistical tests for randomness.

The various published tables of random numbers were designed for human computers engaged in 'pencil and paper' calculations. The electronic computer, by contrast, works so fast that it is better to arrange for it to generate its own supply of random numbers as required. Some of the early computers did this by utilizing some convenient physical process, such as the generation of 'electrical noise'. Such noise, which is due to random fluctuations of current, occurs in all electronic circuits. This procedure is open to one serious practical objection: a calculation using such numbers cannot be repeated and so cannot be checked. For this reason, it is now general practice to use sequences of what are called *pseudo-random* digits; they are generated by the computer itself by means of a repeatable arithmetical process.

We use the term 'pseudo-random' rather than 'random' because a deterministic (and so repeatable) arithmetical process operating on a finite set of digits cannot generate an infinitely long random sequence. Eventually the process will either come to a halt (i.e., yield a string of zeros) or get into a loop (i.e., the pattern of digits will repeat). Fortunately, however, it is possible to generate such sequences which satisfy all the appropriate criteria for randomness. The important thing is not where the numbers come from, but whether they are satisfactorily distributed. We shall shortly return to the question of how to construct such sequences; meanwhile, let us assume that a satisfactory supply of either random or pseudo-random digits is available to us.

SIMULATION OF A MEDICAL APPOINTMENTS SYSTEM

The way digital simulation works is best illustrated by an example. The value of the technique in medical studies has been mentioned in the last chapter, so let us choose a medical example. In 1962 the Operational Research Society and the Ministry of Health arranged a discussion on the subject of 'Appointments Systems in Hospitals and General Practice'. One of the papers given on that occasion dealt with the design of an appointments system for a GP's morning surgery which ran theoretically from 9 to 11 a.m. Before there was an appointments system, patients began to arrive at about 8 a.m. and a queue of up to ten people had usually formed outside the house by the time the surgery doors were opened at 8.30. By 9 a.m. the queue had increased to up to thirty people, overflowing the waiting room. This, together with the fact that the doctor was often late, meant that patients had to wait for a very long time – about an hour and a half on the average.

One of the unfortunate patients, an operational research man, secured the doctor's cooperation in an attempt to relieve the situation. A scheme was devised which, when reported on, had been operating satisfactorily for several years. The average patient's waiting time was reduced to about 15 minutes, apart from emergencies.

When a GP is working on his own with a part-time secretary, the administration of an appointments system clearly poses some thorny problems. Any chosen system would be likely to need modification in the light of experience. We shall not pursue these matters here, but shall concentrate on the queueing and simulation aspects. What we have then, in queueing terms, is a single

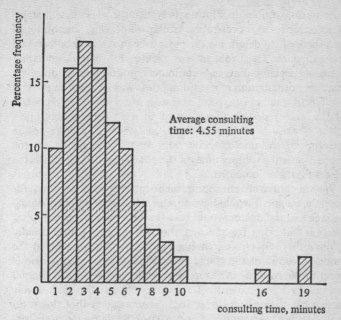

Fig. 3.17. Histogram of doctor's consulting time

service station with a 'first come, first served' queue discipline. We have already explained that, in order to specify such a system, we need to know

(a) the distribution of the doctor's consulting time, and
(b) the nature of the arrival pattern of the patients.

As to (a), observations over a period of a month covering some 800 consultations yielded a frequency distribution shown in Fig. 3.17. The average consulting time was found to be 4·55 minutes per patient. For convenience we have expressed the frequencies as percentages and grouped the observations as a histogram with a tabular interval of 1 minute. Thus 16% of the consultations took between $3\frac{1}{2}$ and $4\frac{1}{2}$ minutes and are treated as if they occupied 4 minutes each.

As to (*b*), with an appointments system the patients will arrive at prescribed – and preferably regular – intervals. The object of the exercise is to determine the best value of the interval between appointments. The problem was tackled by digital simulation using a specially designed computer program. The observed frequency distribution of consulting time was sampled to provide a sufficient number of typical examples of service time to cover a 2-hour surgery period. The effect of different appointment intervals was investigated by repeated 'runs' and the results assessed in terms of the doctor's idle time on the one hand and the average patient's waiting time on the other. Fig. 3.18(*a*) shows the results that were obtained.

We see that with an appointment interval of 5 minutes, for example, a patient would have to wait for about 6 minutes on the average and the doctor would be without a patient for 8–10% of the time, allowing for the fact that the surgery would usually continue beyond 11 a.m. In Fig. 3.18(*b*) we have expressed the results in terms of the traffic intensity, *r*, which in this case is equal to the ratio of the average consulting time to the interval allowed between patients. To ensure that neither doctor nor patient has to wait too long, it is recommended that this ratio should be held between 0·85 and 0·95. A 5-minute interval (corresponding to $r = 0·91$) would be about right in this case. A final point of interest is that the patients' queue does not become unduly large when the traffic intensity exceeds 1. This is because we do not have an infinite queueing system, each surgery being limited to 2–2½ hours.

To give an impression of how such a simulation works, we will trace through a typical sequence of events during a single surgery. The times at which successive patients arrive are determined by their appointments, the times at which they are seen by the doctor by sampling from the observed frequency distribution (Fig. 3.17 or columns (2) and (4) of Table 3.14).

To sample from this distribution we use a sequence of two-digit random numbers (i.e. the digits are to be taken in pairs) running from 00 to 99. Each number is equally likely to be chosen and will appear at random. The selected numbers are used to

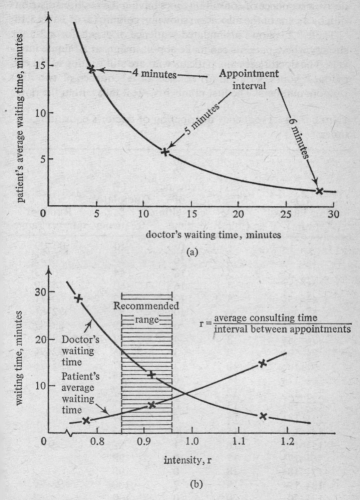

(a)

(b)

Fig. 3.18. Waiting times at doctor's surgery (average consulting time: 4·55 minutes)

derive a sequence of consulting times having the required distribution by means of the allocation shown in column (5) of Table 3.14.

Table 3.15 shows a simulated sequence of events for a 2-hour surgery where patients can make appointments at 5-minute intervals. The doctor sees any patients who are still waiting at the end of the 2 hours. Column (3) consists of a sequence of two-digit random numbers. The first number is 72; it falls within the range

TABLE 3.14   Frequency distribution of doctor's consulting times

(1) Time interval (min.)	(2) Mean time (min.)	(3) % frequency of consulting time	(4) Cumulative % frequency	(5) Random number range
½– 1½	1	10	10	00–09
1½– 2½	2	16	26	10–25
2½– 3½	3	18	44	26–43
3½– 4½	4	16	60	44–59
4½– 5½	5	12	72	60–71
5½– 6½	6	10	82	72–81
6½– 7½	7	6	88	82–87
7½– 8½	8	4	92	88–91
8½– 9½	9	3	95	92–94
9½–10½	10	2	97	95–96
10½–11½	11	0	97	
11½–12½	12	0	97	
12½–13½	13	0	97	
13½–14½	14	0	97	
14½–15½	15	0	97	
15½–16½	16	1	98	97
16½–17½	17	0	98	
17½–18½	18	0	98	
18½–19½	19	2	100	98–99
19½–20½	20	0	100	
		100		

Average consulting time: 4·55 minutes

TABLE 3.15   Simulation of a GP's surgery. Appointment interval 5 minutes

(1) Patient number	(2) Appoint-ment time	(3) Random number	(4) Consult-ing time (min.)	(5) Consultation starts at	(6) Consultation ends at	(7) Doctor's waiting time (min.)	(8) Patient's waiting time (min.)
1	9.00	72	6	9.00	9.06		
2	9.05	16	2	9.06	9.08		1
3	9.10	61	5	9.10	9.15	2	
4	9.15	01	1	9.15	9.16		
5	9.20	82	7	9.20	9.27	4	
6	9.25	97	16	9.27	9.43		2
7	9.30	14	2	9.43	9.45		13
8	9.35	63	5	9.45	9.50		10
9	9.40	50	4	9.50	9.54		10
10	9.45	86	7	9.54	10.01		9
11	9.50	23	2	10.01	10.03		11
12	9.55	83	7	10.03	10.10		8
13	10.00	18	2	10.10	10.12		10
14	10.05	42	3	10.12	10.15		7
15	10.10	56	4	10.15	10.19		5
16	10.15	02	1	10.19	10.20		4
17	10.20	08	1	10.20	10.21		
18	10.25	77	6	10.25	10.31	4	
19	10.30	43	3	10.31	10.34		1
20	10.35	70	5	10.35	10.40	1	
21	10.40	58	4	10.40	10.44		
22	10.45	98	18	10.45	11.03	1	
23	10.50	75	6	11.03	11.09		13
24	10.55	34	3	11.09	11.12		14
						12	118

Duration of surgery: 132 minutes
Doctor's waiting time: 12 minutes or 9%
Average patient's waiting time: 4·9 minutes

72–81 of Table 3.14, column (5), and so the consulting time for the first patient is taken as 6 minutes. The next number selected is 16 which falls in the range 10–25 of Table 3.14. The consulting time for the second patient is therefore taken as 2 minutes. His appointment is for 9.05, but he must wait until 9.06 before he can be seen. He is disposed of by 9.08 and so now the doctor has to wait for 2 minutes until his next patient arrives at 9.10. The construction of the remaining columns of Table 3.15 should now be clear. The results for the first hour of the surgery are presented in a somewhat different way in Fig. 3.19.

In this particular simulation the average time that a patient has to wait comes out to be just under 5 minutes, while the doctor is

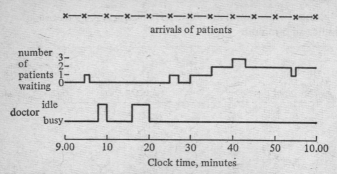

Fig. 3.19. Simulation of the first hour of a doctor's surgery

without a patient for 12 minutes (9% of his time in the surgery). This result agrees quite well with those of Fig. 3.18, which are based on a large number of such simulations and are therefore more accurate. We have ignored any waiting due to patients arriving early for their appointments as this has no direct bearing on the efficiency of the appointments system, and we have assumed that no patients arrive late.

Table 3.16 gives the results of a similar simulation but with a 4-minute interval between appointments. Here a patient must wait about 11·5 minutes on the average, but the doctor is idle for less than 3% of the time. A comparison of the two tables confirms what we saw from Fig. 3.18, namely that the system is highly sensitive to quite small changes in the appointment interval. Although in practice patients would have to be scheduled to the nearest 5 minutes, variants of the pattern can readily be devised to match any observed value of a doctor's average consulting time.

It would be interesting to know how many GPs do in fact operate appointments schemes. It has been estimated that if every patient in the UK were to wait for no more than 15 minutes on the average each time he visits his doctor, then more than five million man-days per year would be lost. (This does not include time spent waiting in hospital.) This time is more than that lost because of strikes in an average year.

TABLE 3.16   Simulation of a GP's surgery. Appointment
interval 4 minutes

(1) Patient number	(2) Appoint-ment time	(3) Random number	(4) Consult-ing time (min.)	(5) Consultation starts at	(6) Consultation ends at	(7) Doctor's waiting time (min.)	(8) Patient's waiting time (min.)
1	9.00	42	3	9.00	9.03		
2	9.04	11	2	9.04	9.06	1	
3	9.08	57	4	9.08	9.12	2	
4	9.12	90	8	9.12	9.20		
5	9.16	04	1	9.20	9.21		4
6	9.20	10	2	9.21	9.23		1
7	9.24	68	5	9.24	9.29	1	
8	9.28	96	10	9.29	9.39		1
9	9.32	77	6	9.39	9.45		7
10	9.36	51	4	9.45	9.49		9
11	9.40	23	3	9.49	9.52		9
12	9.44	88	8	9.52	10.00		8
13	9.48	60	5	10.00	10.05		12
14	9.52	19	2	10.05	10.07		13
15	9.56	45	4	10.07	10.11		11
16	10.00	82	7	10.11	10.18		11
17	10.04	65	5	10.18	10.23		14
18	10.08	07	1	10.23	10.24		15
19	10.12	36	3	10.24	10.27		12
20	10.16	51	4	10.27	10.31		11
21	10.20	40	3	10.31	10.34		11
22	10.24	99	19	10.34	10.53		10
23	10.28	44	3	10.53	10.56		25
24	10.32	01	1	10.56	10.57		24
25	10.36	49	4	10.57	11.01		21
26	10.40	77	6	11.01	11.07		21
27	10.44	31	3	11.07	11.10		23
28	10.48	81	7	11.10	11.17		22
29	10.52	04	1	11.17	11.18		25
30	10.56	68	5	11.18	11.23		22
						4	342

Duration of surgery: 143 minutes
Doctor's waiting time: 4 minutes or 2·8%
Average patient's waiting time: 11·4 minutes

SAMPLING FROM A CONTINUOUS DISTRIBUTION

In our simulation example all the distributions from which we
sampled were 'discrete' distributions obtained from frequency
observations. We have already remarked that it may be necessary
to sample from a continuous probability density distribution,

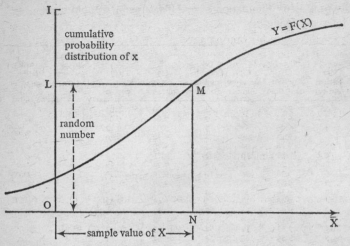

Fig. 3.20. Sampling from a continuous distribution

specified either by a mathematical function or graphically by a continuous curve. The procedure for sampling from such a distribution is as follows.

We first plot the *cumulative probability distribution function*, $F(X)$, the construction of which has already been explained (Fig. 3.9). We then choose a random number between 0 and 1 (to as many decimal places as desired), represented by the ordinate OL in Fig. 3.20. We then project the point L horizontally and downwards to obtain point N. Then ON is taken as the sample value of X. This procedure is the 'continuous' analogue of the method used for sampling from the 'discrete' frequency distribution of our medical example.

PSEUDO-RANDOM NUMBERS

We must now return to the question we posed earlier, namely, how can we produce a satisfactory sequence of pseudo-random numbers? In practical terms, what is needed is a procedure to enable a computer to generate a sequence of numbers x_0, x_1,

x_2 ... (with x_i as the general term) going on for as long as required and such that the numbers satisfy the appropriate statistical tests for randomness.

One of the earliest proposals was the 'mid square method', in which the term x_{i+1} of the sequence consists of the middle digits of x_i^2. This method was found to be unsatisfactory and what are called *congruential methods* are now widely used instead. The basic recurrence relation of such methods is

$$x_{i+1} = ax_i \text{ (modulo } m\text{)}, \tag{49}$$

which can be generalized to

$$x_{i+1} = ax_i + c \text{ (modulo } m\text{)}, \tag{50}$$

where a, c and x_i are integers between 0 and $(m - 1)$. It is often convenient to arrange matters so that each number lies within the range $(0, 1)$. To ensure this, we only have to divide each x_i by m. The expression 'modulo m' in (49) that means x_{i+1} is the remainder when ax_i is divided by m. Thus 8 (modulo 3) = 2 and 8 (modulo 4) = 0. Similarly for (50). In practice m is a large integer, usually a power of 2, and is determined by the design of the computer. Nearly all pseudo-random number generators exploit the fact that a computer has a fixed word length.

Now since x_i in (49) or (50) can take only m different values, any sequence generated by a 'modulo m' recurrence relation must repeat itself after at most m steps; that is to say, the sequence must be periodic with a period that cannot exceed m. For example, if $m = 16$, $a = 3$, $c = 1$ and $x_0 = 2$, the sequence generated by (50) is 2, 7, 6, 3, 10, 15, 14, 11, 2, 7 . . ., with a period of 8. We cannot avoid such periodicity, but we must ensure that the period is much longer than the number of random numbers required for any single simulation experiment. The value of m is usually large enough to meet this criterion (typically between 2^{30} and 2^{40}), but some elegant results in number theory can be invoked to allay any anxiety.

It can be shown that, if we use the recurrence relation (50), the full period of m can always be achieved provided that certain requirements are met [19]. For the usual case when m is a power of 2, these requirements are that (1) c is odd, and (2) a is one greater than a multiple of 4. To see what this means, let us consider a few simple examples.

1. $x_{i+1} = 5x_i + 1$ (modulo 8) with $x_0 = 2$. The sequence is 2, 3, 0, 1, 6, 7, 4, 5, 2, 3 . . . with the maximum possible period of 8.

2. $x_{i+1} = 3x_i + 1$ (modulo 8) with $x_0 = 2$. The sequence is 2, 7, 6, 3, 2, 7 . . . with a period of 4 only.

If on the other hand, we use the relation (49) to generate our sequence of numbers, again with m a power of 2, then the maximum possible period is only $m/4$, which is achieved if (1) x_0 is odd, and (2) a differs by 3 from the nearest multiple of 8.

Here are a few simple examples to show what can happen.

1. $x_{i+1} = 3x_i$ (modulo 8) with $x_0 = 3$. The sequence is 3, 1, 3 . . . with a period of 2, which is the greatest possible.

2. $x_{i+1} = 3x_i$ (modulo 8) but with $x_0 = 4$. The sequence is 4, 4, 4, . . . with a period of 1.

3. $x_{i+1} = 2x_i$ (modulo 8) with $x_0 = 3$. The sequence is 3, 6, 4, 0, 0, . . . and the process comes to a stop at the fourth term.

4. $x_{i+1} = 3x_i$ (modulo 16) with $x_0 = 1$. The sequence is 1, 3, 9, 11, 1, 3 . . . with the maximum possible period of 4.

The results we have been discussing tell us about the maximum possible period of the sequence; they do not provide any assurance about its randomness. A sequence may have the full period but be quite unsatisfactory. Consider an example quoted by Page [20] of the relation

$$x_{i+1} = 33x_i + 49 \text{ (modulo 128)},$$

for which the full period of 128 is guaranteed by the theory. Taking $x_0 = 1$, the sequence is 1, 82, 67, 84, 5, 86, 71, 88, 9, 90 . . . We observe two very odd things. First, that the terms with odd suffices steadily increase by a constant, 2, until the remainder, modulo 128, returns to zero; and, secondly, that all terms four apart in the sequence differ by 4 until the remainder returns to 0, 1, 2 or 3, when the process starts over again. Few sequences could be less random than this!

Finally, here is another sobering result [20]. The sequence

$$x_{i+1} = (2^{18} + 3)x_i \text{ (modulo } 2^{35})$$

satisfies most of the tests for randomness and was used quite happily for several years. It was then noticed (and can easily be proved) that any three successive terms are connected by a simple recurrence relation, namely

$$x_{i+2} = 6x_{i+1} - 9x_i \text{ (modulo } 2^{35}).$$

One cannot be too careful!

Digital simulation is now widely used as an aid to decision-making in industry, commerce, transport, defence and much else [21]. The lead in this field was taken about twenty-five years ago by the heavy industries – notably coal and steel – and by some defence establishments. Most books on operational research contain some discussion of this important topic [7, 13, 17]. Various applications are described in [20] which also contains several papers of theoretical and computational interest. Computational feasibility is an essential requirement of simulation, as it is of the other techniques discussed in this chapter. Indeed, a simulation model has been defined 'as the representation of a dynamic system in a form suitable for manipulation by a computer' [20]. The construction of such models, certainly of fairly complex systems with many interacting components, can be a difficult and lengthy business. To ease the process, a number of special computer programming languages – known as *simulation languages* – have been developed and publicized. Some of them are in wide use and have proved highly successful in practice [20].

7. Combinatorics

THE NATURE OF THE SUBJECT

Combinatorial analysis has a long history as a dignified and successful branch of pure mathematics. In recent years it has acquired a new importance in virtue of its practical applications in operational research, statistics and computing. Combinatorics is concerned with collections of discrete entities and with such processes as enumerating, ordering, matching, permuting and partitioning the elements of such collections. The relationships between the elements may be exhibited as arrays (or matrices), or graphically as a network of lines. Indeed, combinatorics and graph theory are very closely linked. However, we shall not pursue this aspect here because network analysis and graph theory form the main topics of the next chapter.

The applications of combinatorics in operational research have to do with such matters as the scheduling and sequencing of industrial operations, the routing of vehicles and the division of material [23]. Many of these applications give rise to permutation problems. Suppose we have N elements, $e_1, e_2, \ldots e_N$; then a particular ordering of these elements may be represented by the permutation $P = (p_1, p_2, \ldots p_N)$, where p_j denotes the element in the jth position of the permutation. With any permutation, P, we may associate an objective function, $F(P)$, which measures the 'cost', in some sense, of the permutation. In many combinatorial problems, the objective is to find the permutation, from among all feasible permutations, which minimizes $F(P)$.

The famous *travelling salesman problem* is of this type. A salesman wishes to select a route that will minimize the total distance he travels while visiting N towns, each once only, and then returning to his starting point. If we label the towns (in any order) as $1, 2, 3, \ldots N$, then any permissible route can be specified by a permutation, P, of the integers, $1, 2, \ldots N$. If d_{ij} is the distance from town i to town j we wish to find a set of N distances d_{pq}, $d_{qr}, d_{rs}, \ldots d_{uv}, d_{vp}$, where $p, q, r \ldots u, v$ is some permutation of $1, 2, 3, \ldots N$ such that the sum of the distances (which is equal to the objective function in this case) is a minimum. There are $N!$ permutations of the first N integers, but since the route is a closed circuit, any one of the N towns may be regarded as the starting point. There are thus $(N-1)!$ essentially different routes among which we must search for the solution. ($N!$ stands for the product of the first N integers.) The labour of direct enumeration becomes prohibitive for quite moderate values of N, even with the aid of a computer. The method of dynamic programming may be helpful in certain circumstances, but no satisfactory method for attacking the general case has yet been devised.

MACHINE SEQUENCING

Many industrial operations consist of a series of tasks which must be carried out one after the other, and where the effectiveness of the operation depends on the order in which the tasks are performed. Sequencing problems of this kind fall into two broad groups. In the first group we have N tasks to perform, each of which requires processing on some or all of M different machines. There are therefore $(N!)^M$ theoretically possible different sequences, although many may not be technologically feasible; we wish to select the 'best' ordering. Problems of the second kind are typified by the model of a workshop with a number of machines and a list of tasks to be performed. Each time a machine completes the job on which it is engaged a decision must be made as to which task is to be started next. The list of tasks may change as fresh orders or other information is received. Both types of problem pose serious difficulties. At present solutions have been found only for some special cases of the first type; for example, when there are only two machines and all the N jobs are to be processed on both machines in the same order, or when there are only two jobs and each job is to be processed through the M machines in a prescribed order. For the second type of problem there seems to be no general theory at all, although some empirical rules can be tested by simulation.

COLLEGE TUTORIALS

Combinatorial reasoning can be applied to the design of school and college timetables and we conclude with a discussion of a simple problem of this kind, based on [22]. In a college there are M lecturers, $L_1, L_2, \ldots L_M$, and N undergraduates $U_1, U_2 \ldots U_N$. The lecturer L_i is to take the undergraduate U_j for a total of q_{ij} 1-hour tutorials during the year. The college is generously staffed and a lecturer need never coach more than one undergraduate at a time. We can express these requirements by means of a $M \times N$ matrix, Q, whose elements are the q_{ij}'s. The problem is to determine the least number of hours, h, into which the tutorials can be fitted.

Now lecturer L_i needs to give $l_i = \sum_{j=1}^{N} q_{ij}$ tutorials altogether so h must be at least as large as the largest of the l_i's for $1 \leqslant i \leqslant M$. Similarly, the undergraduate U_j has to attend $u_j = \sum_{i=1}^{M} q_{ij}$ tutorials altogether so h must be at least as large as the largest of the u_j's for $1 \leqslant j \leqslant N$. It follows that h must be at least as large as the larger of these two quantities. We can write this as $h \geqslant S(Q)$ where $S(Q)$ denotes the largest *line sum* of the matrix Q. By a 'line sum' we mean the sum of all the elements on a line of a matrix, where the term 'line' means either a row or a column. Thus our $M \times N$ matrix will have $(M + N)$ line sums; $S(Q)$ is the largest of them.

Now we have shown that our timetable necessarily requires at least $S(Q)$ hours; the interesting result is that $S(Q)$ are indeed sufficient. The proof is a straightforward deduction from some of the standard theorems of combinatorics. One of these theorems tells us that the matrix Q can be expressed as a sum of $S(Q)$ matrices of a special type, known as Z matrices; that is

$$Q = Z_1 + Z_2 + \ldots Z_s, \tag{51}$$

where s has been written for $S(Q)$ and $Z_1, Z_2 \ldots Z_s$ are Z-matrices. A Z-matrix is a matrix in which at most one element in each line (row or column) is 1 and all the others are 0.

We can interpret any such matrix, say Z_k, as the schedule for the kth hour of our timetable. Lecturer L_i is to coach the undergraduate U_j in that hour if, and only if, the (i,j)th element of Z_k is equal to 1. The definition of a Z-matrix makes such an allocation compatible with the academic requirements. The total number of tutorials given by L_i to U_j is equal to the number of 1's among the (i,j)th elements of the set of matrices. Equation (51) shows that this is equal to q_{ij} and so all our requirements can be satisfied in

$s = S(Q)$ hours. The trivial lower bound for h turns out to be the actual value of h.

As a simple numerical illustration, we consider the case of three lecturers and five undergraduates. The matrix Q specifying their obligations is given by

	U_1	U_2	U_3	U_4	U_5	row sums
L_1	2	0	3	1	0	6
L_2	1	1	3	0	2	7
L_3	3	0	1	2	0	6
Column sums	6	1	7	3	2	

Thus the lecturer L_1 is to give 2 tutorials to undergraduate U_1, 3 to U_3, 1 to U_4 and none to U_2 or U_5. The largest line sum is seen to be 7 (occurring twice) and our theorem assures us that the Q-matrix can be expressed as the sum of seven Z-matrices. One such decomposition – there are many others – is

$$\begin{vmatrix} 2&0&3&1&0 \\ 1&1&3&0&2 \\ 3&0&1&2&0 \end{vmatrix} = \begin{vmatrix} 1&0&0&0&0 \\ 0&0&1&0&0 \\ 0&0&0&1&0 \end{vmatrix} + \begin{vmatrix} 1&0&0&0&0 \\ 0&0&1&0&0 \\ 0&0&0&1&0 \end{vmatrix} + \begin{vmatrix} 0&0&1&0&0 \\ 0&1&0&0&0 \\ 1&0&0&0&0 \end{vmatrix}$$

$$+ \begin{vmatrix} 0&0&0&1&0 \\ 0&0&1&0&0 \\ 1&0&0&0&0 \end{vmatrix} + \begin{vmatrix} 0&0&1&0&0 \\ 1&0&0&0&0 \\ 0&0&0&0&0 \end{vmatrix} + \begin{vmatrix} 0&0&1&0&0 \\ 0&0&0&0&1 \\ 1&0&0&0&0 \end{vmatrix} + \begin{vmatrix} 0&0&0&0&0 \\ 0&0&0&0&1 \\ 0&0&1&0&0 \end{vmatrix}.$$

This decomposition shows that the tutorials can be completed in 7 hours and provides an actual timetable for each person. Thus during the fifth hour, L_1 is to coach U_3, L_2 is to coach U_1 while L_3 can take a nap and U_2, U_4 and U_5 can work in the library or relax in the Students Union! Although this example is too simple to be of much practical value, it should give some idea of how combinatorial arguments may be used to solve 'real' problems.

8. Catastrophe Theory

Mathematical creativity is not just a thing of the past, a dead activity neatly encapsulated in the historical record. New mathematical ideas are being formulated and new theorems are being discovered all the time. There is usually a substantial delay, however, before such creations find application in the 'real' world; sometimes, indeed, they never do. Some of the esoteric mathematical discoveries of our own day will assuredly be put to practical use by our children or grandchildren.

One of the most exciting manifestations of contemporary mathematics is known as *catastrophe theory* [24, 25]. Its creator is one of the world's outstanding mathematicians, René Thom, whose remarkable book *Stabilitié Structurelle et Morphogénèse* appeared in 1972. Thom's theory is essentially a mathematical theory of discontinuous processes. It shows that discontinuities can arise in continuous systems in ways that can be described mathematically; that changes in form can often be predicted, at any rate in principle, in qualitative terms. Models involving the ideas of catastrophe theory have already provided useful insights into a variety of 'real life' situations: in physics (for example, the breaking of a water wave), in physiology (the action of a heart beat or a nerve impulse), and in the social sciences (a recent article by E. C. Zeeman, the main British protagonist of the theory, is entitled 'On the Unstable Behaviour of Stock Exchanges'). The potential applications of the theory – perhaps most of all in biology – are enormous, but the subject is too new and too technical to be pursued further here.

9. Miscellaneous Topics

We are now nearing the end of our review. What significant topics in management mathematics remain to be discussed? Apart from *network analysis*, which forms the subject of the next chapter, readers of the literature will detect several other omissions: the theory of search, stock control, theory of games, replacement and renewal theory, economic forecasting, among others. Some of these topics do not offer as much mathematical interest as those we have already discussed while others are perhaps of less practical importance, but we shall say a few words in conclusion about each.

TREASURE SEEKING

Suppose we wish to locate an object whose exact position is unknown, but where something is known about the likelihood of it being in various positions. If only a limited amount of searching effort (or time) is available, how should this effort be deployed so as to give the greatest chance of finding the object? It is with questions of this kind that the theory of treasure seeking – as we shall call the subject to avoid confusion with the search methods discussed earlier – is concerned. We shall limit ourselves to a brief account of a single application that is suitable for some kinds of survey work. It uses a clustering technique and entails two stages of search [13].

The objects sought (referred to as 'prizes') are assumed to be distributed at random over a large area. The first, or preliminary, search is 'cheap and nasty': it entails covering the entire area several times, but can produce false indications, i.e. can indicate possible prizes where there are none, and can fail to detect some prizes that are there. The second, or detailed, search provides perfect detection but is difficult, expensive and slow. It is therefore desirable to concentrate it in those parts of the total search area that are indicated as promising by the preliminary search. A two-stage search of this kind may be appropriate in mineralogical explorations, for instance, particularly where the preliminary search can be conducted rapidly from a moving vehicle (aircraft, truck or ship) equipped with detection devices. How are the results of the preliminary search to be interpreted?

The basic idea is that signals caused by a prize are likely to be recorded at or near the actual location of the prize, while spurious signals will be distributed at random over the survey area – they will show no tendency to cluster. We can make this precise by assuming that a first-stage detection of a prize is recorded within a known range, R, of its actual position. Let us suppose that the first-stage survey consists of n coverings of the total area and that a circle of radius R is centred about each contact point. We shall adopt the policy that those portions of the survey area that are within the intersections of at least s of the circles are to be subjected to detailed examination during the second-stage search. We assume also that all prizes within the area of the second-stage search will in fact be found. Fig. 3.21 illustrates the situation for $n = 5$, $s = 3$. The best values of n and s to choose in practice will

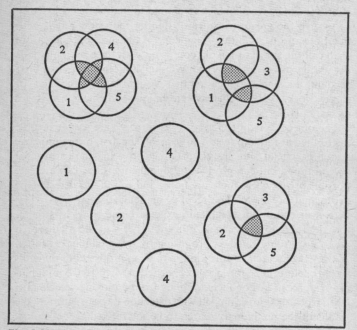

Fig. 3.21. The two-stage cluster surveying technique. Preliminary search entails five coverings of the complete area. Detailed search made of shaded areas

depend on such matters as the relative costs of the two stages of the survey, the number and average value of the prizes to be expected, and the chance of obtaining more than s false contacts within a range R of any point in the area. Mathematical analysis of the situation enables the effect of the various factors to be sorted out.

STOCK CONTROL

The reason why stocks of goods must be held is that supply and demand are not equal over periods of time. The object of stock control (or *inventory management* as it is sometimes called) is to maintain the correct balance between the cost of holding stock (capital cost, deterioration, insurance charges, etc.) and its benefits (the ability to produce in large quantities and to meet large orders, or to satisfy customers during periods of high demand, etc.). What should be the buffer stock level, the re-order level or the interval of time between ordering? Various mathematical models of the stock control process have been developed in order to answer these questions [7, 13, 17].

The simplest model is based on the assumption of a known, constant demand for goods with stock delivered immediately it is required so that shortages do not occur. We wish to determine the optimum provisioning policy: that is, the one which minimizes the total cost of holding and ordering stock. Let us assume that the stock is replenished by ordering a fixed quantity, Q, whenever the level falls below a specified buffer level B. The average stock holding will therefore be $(B + Q/2)$. If the demand per annum is D, then the number of orders placed in a year will be D/Q. Let us denote by C_1 the cost of holding a unit of stock for a year, and by C_2 the cost of placing an order for a quantity, Q. Then the total expected cost per annum is given by

$$C = C_1(B + Q/2) + C_2 \cdot D/Q. \tag{52}$$

The situation is illustrated in Fig. 3.22, from which we see that there is a value of Q which minimizes C. Now we can write (52) as

$$C = C_1 B + \sqrt{2C_1C_2D} + \left\{ \sqrt{\frac{C_1Q}{2}} - \sqrt{\frac{C_2D}{Q}} \right\}^2. \tag{53}$$

We see at once that the minimum value of C is obtained when the quantity in the curly brackets is zero. This occurs when

$$Q = \sqrt{2C_2D/C_1}$$

and the minimum value of C is given by

$$C_1 B + \sqrt{2C_1C_2D}.$$

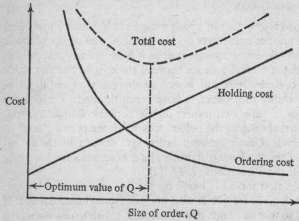

Fig. 3.22. Optimal stock ordering

This simple model can be elaborated in various ways, but the principle of minimizing over-all costs remains the same.

In most of the more interesting models, the level of demand for goods – and often of other factors, such as lead time – is not assumed to be constant but to be governed by a probability distribution, with the Poisson and Normal distributions the most popular. The distributions may, however, have to be obtained from observation; in such cases resort must usually be had to digital simulation. Indeed, stock control is essentially a queueing problem with a variable movement of goods into and out of stock.

GAME THEORY

The objective of *game theory* is to analyse how decisions should be made in competitive situations; in particular, in situations which can be modelled as a contest between a number of 'players' who act completely rationally in accordance with prescribed 'rules of the game'. An essential feature is that the outcome of a certain course of action by one player depends on the actions of some or all of the other players. At any stage all players are assumed to choose their courses of action simultaneously. As an example we may think of several companies engaged in competitive selling and price cutting. Another example might be that of a colliery reorganization, where the outcome would depend not only on the nature of the reorganization itself, but also on the attitude of the men and on the natural conditions prevailing in the colliery (for instance, the thickness of newly discovered coal seams).

Although the origins of game theory go back to the 1920s, the classic work on the subject, *The Theory of Games and Economic Behaviour*, did not appear until 1944. The subject has considerable mathematical interest and is closely related to the theory of linear programming. However, we shall not pursue it here because it has not yet proved a significant tool of management or planning, at any rate in civil life. Most 'real life' competitive situations – those that are studied in seminars on 'business games' – are too complex to be solved by the use of game theory in its present state of development. For further information, the reader is referred to the literature [7, 13].

REPLACEMENT AND RENEWAL

We have all had to face the question of when is the best time to replace an article that is wearing out. In the industrial context we may identify two main variants of this familiar situation. First, when should we replace an item whose maintenance costs increase with time: and second, when should we replace an item which may have little or no maintenance cost but is expensive to keep in service until it fails completely?

In practice problems of the first type are often solved by tabulating various functions of the costs involved. Let us first assume, for simplicity, that we wish to minimize the average annual cost of a piece of equipment whose scrap value is constant. If maintenance costs decrease or remain constant with time, the best policy is never to replace the equipment – a happy state of affairs not often encountered in practice! If, however, maintenance costs

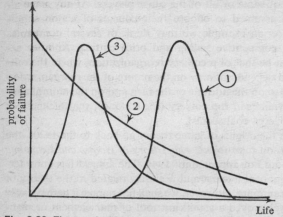

Fig. 3.23. Three typical life curves

increase with time, there is a general result [13] to the effect that the equipment should be replaced when the average cost to date (i.e., the total cost over n years divided by n) becomes equal to the current annual maintenance cost (i.e., the cost in the nth year). This simple replacement model can readily be adapted to allow for a variable scrap value. All one has to do is treat any change in the scrap value as an extra maintenance cost called depreciation. A more significant extension – and one that nowadays is essential in the interests of realism – is needed to take account of changes in the value of money. This topic may properly be deferred to later chapters.

To deal with the second type of replacement problem we need to know the probability density distribution governing the life of an item – what is termed the *life curve*. Fig. 3.23 shows three possible life curves. We shall not discuss the mathematical forms of these distributions [13], but shall only remark that curve (1) is typical of items which have one main kind of wear (e.g., car tyres); curve (2) applies when failure is likely to occur at random over time because of damage or misuse (e.g., windows, milk bottles, crockery); while curve (3) is characteristic of items that can wear out in various ways, or which contain many separate parts any one of which may fail (e.g., TV or radio sets). A similar approach can be applied to problems connected with the recruitment, wastage and promotion of staff. The mathematical treatment of replacement problems by means of life curves is known as *renewal theory*. It has been quite highly developed during recent years. The theory treats the life of an article – or an employee – as a random variable and the mathematics is largely derived from the theory of probability.

ECONOMIC FORECASTING

Forecasting the future demand for a product, let alone the future pattern of a national economy, is a hazardous undertaking in which, one suspects, guesswork (usually dignified by the term 'intuition') plays a major rôle. The more scientific methods of forecasting are based on some kind of statistical analysis of past behaviour. The data on such behaviour often exhibit some degree of regularity on which random fluctuations are superimposed. The problem is to isolate the underlying trend from the short-term 'wobbles'.

Various statistical techniques [3] for fitting curves to data are available for deciding what form of regularity to attribute to the data (linear growth, exponential growth or cyclical movement, for example). This is a 'once for all' procedure, however, and many situations demand a more flexible approach; they require *adaptive* forecasting in the sense that forecasts can easily be revised in the light of the latest information.

The two simplest techniques for doing this are the *moving average method* and what is called *exponential smoothing* [7]. A moving average is calculated simply by adding up the last n observations and dividing by n. (We are assuming that we are only forecasting the future value of a single entity.) When the next observation becomes available, the oldest one of the set of n is dropped and the new one takes its place. The question is: what should n be? If n is too large, we have excessive *stability* against random variation; if n is too small, our estimate is too *sensitive* to any changes in the trend. The best balance between stability and sensitivity depends on the context of the observations.

The basic idea on which exponential smoothing is based is this.

When we make a new observation, it is likely to be different from the current forecast. We must therefore construct our next forecast from the old one by taking account of the error we have just made. The correction should be some function of this error, so we have:

new forecast = old forecast + S (latest observation —

old forecast), (54)

where S is a smoothing constant between 0 and 1. Its value determines the balance between complete stability ($S = 0$) and complete sensitivity ($S = 1$). A value of S between 0·1 and 0·2 is found to be satisfactory in most cases.

Equation (54) can be written as

$$F_0 = F_1 + S(D_1 - F_1), \qquad (55)$$

where F_0 is the current forecast, F_1 is the forecast made one period ago for the current period and D_1 is the latest observation. But F_1 has been determined in the same way as F_0, and F_2 as F_1, and so on. We can therefore apply the same argument successively to previous forecasts, and so can express F_0 as

$$F_0 = S[D_1 + KD_2 + K^2D_3 + K^3D_4 \ldots], \qquad (56)$$

where $K = 1 - S$ and D_i is the observation made $(i - 1)$ periods ago. The new forecast is thus a weighted average of all previous observations, but with the weights decreasing by a constant factor, K, as the observations become more remote. It is this feature which gives the method its name. Note also that the sum of all the coefficients in (56) is unity.

Exponential smoothing is a simple and flexible adaptive forecasting technique and is widely used. The basic idea can be extended to deal with *time series* (i.e., a sequence of observations over a period of time) in which the trend is allowed to vary with time.

This completes our account of the newer methods of analysing operational problems and situations. It is perhaps appropriate that the final paragraphs should be about looking towards the future. One forecast we can make with confidence is that if a survey of this kind were to be made in, say, twenty-five years time, it would have a very different story to tell.

REFERENCES (More advanced books are starred)

1. HOLLINGDALE, S. H., and TOOTILL, G. C., *Electronic Computers*, Penguin Books, 2nd edn, 1975.

First published in 1965: revised in 1970 and 1975. Addressed to the general reader, the book explains how computers work, how problems are presented to them and what sort of jobs they can tackle.

2. MARTIN, J., and NORMAN, A. R. D., *The Computerized Society*, Penguin Books, 1973.

A comprehensive and readable factual survey and critical discussion of the individual and social implications of the computer and of complex computer-based systems. The authors' approach is indicated by the titles of the book's three main divisions: euphoria, alarm, protective action.

3. MORONEY, M. J., *Facts from Figures*, Penguin Books, 3rd edn, 1974.

First published in 1951; revised in 1953, 1956 and 1974. An excellent introduction for the general reader to the subject of statistics: its uses, its limitations, its mathematical basis in probability theory, and its working tools. The main statistical techniques are discussed and illustrated in some detail.

4. School Mathematics Project, Book 5, CUP, 1970.

This is the final O-level SMP text and contains a short chapter on linear programming.

5. VAJDA, S., *Readings in Mathematical Programming*, Pitman, 1962.

An excellent selection of worked examples, illustrating how mathematical programming techniques may be applied to a wide variety of planning problems. Most of the book is concerned with linear programming but several examples of integer and nonlinear programming are also discussed.

6. *VAJDA, S., *Mathematical Programming*, Addison-Wesley, Reading, Mass., 1961.

Although the book is designed as a university textbook, no more than an elementary knowledge of mathematics is assumed of the reader. The main emphasis is placed on expounding the theoretical foundations of the subject and the main algorithmic methods of solution.

7. MAKOWER, M. S., and WILLIAMSON, E., *Teach Yourself Operational Research*, English Universities Press, 1967.

One of the best of the 'Teach Yourself' series. The object of the book is to present the main ideas and techniques of operational research to the general reader. The authors adopt a problem-orientated approach.

8. BATTERSBY, A., *Mathematics in Management*, Penguin Books, revised edn, 1975.

First published in 1966. An elementary, highly readable introduction to the methods and applications of operational research. It is written primarily for managers and uses the minimum of mathematics. A valuable feature is the exercises, with solutions, that are included in each chapter.

9. BEALE, E. M. L., 'Mathematical Programming', *Bulletin IMA*, Vol. 6 (2), 1970.

A stimulating survey of the basic ideas and the main industrial uses of linear programming and its extensions. The article is based on a Symposium of the Institute of Mathematics and its Applications on 'Mathematics and Management'.

10. *FLETCHER, R. (editor), *Optimization*, Academic Press, New York, 1969.

The proceedings of an international conference organized by the IMA in 1968. Mainly for the specialist, but contains some review papers. Covers all the topics mentioned in the 'Nonlinear Optimization' section of this chapter.

11. *MURRAY, W. (editor), *Numerical Methods for Unconstrained Optimization*, Academic Press, New York, 1972.

Based on a joint NPL/IMA conference in 1971. The papers are largely expository in character and are addressed to engineers, physicists and economists rather than to mathematicians. Considerable emphasis is placed on the computational and numerical aspects of the methods discussed.

12. *ADBY, P. R., and DEMPSTER, M. A. H., *Introduction to Optimization Methods*, Chapman and Hall, 1974.

A clear and readable book designed as a university text. The basic optimization techniques are first discussed in some detail, with numerical examples. This is followed by a good survey of the more advanced methods for both unconstrained and constrained optimization that have been developed since about 1965.

13. *HOULDEN, B. T. (editor), *Some Techniques of Operational Research*, English Universities Press, 1962.

This short book, which has now achieved the status of a classic, was written by members of the O. R. Research Group of the National Coal

Board. It was based on a training course given to new entrants. The basic theory of the main techniques is admirably presented.

14. WELFORD, B. P., *Statistical Decision Theory, Bulletin IMA, 6* (2), 1970.

A most attractive introduction to the theoretical and the practical aspects of the subject.

15. LINDLEY, D. V., *Making Decisions*, Wiley, New York, 1971.

The book is aimed at business executives and politicians as well as scientists, and particularly to university students of *any* discipline. After a careful treatment of the fundamentals of the subject – and in particular, of how numerical measures may be attached to uncertain events and to the consequences of decisions – the author discusses the utility of money, the value of information, and the use of decision tables and decision trees.

16. *BELLMAN, R., and DREYFUS, S. E., *Applied Dynamic Programming*, OUP, 1962.

A comprehensive exposition of the theoretical basis of the subject and its wide range of application. The computational feasibility of the methods discussed receives particular attention throughout.

17. FLETCHER, A., and CLARKE, G., *Management and Mathematics* (2nd edn, revised and updated by C. W. Lowe), Business Publications Limited, 1972.

The aim of the book is to describe some of the mathematical techniques that are available to assist the manager (provided he has a good school mathematical background!). It succeeds in its purpose admirably. It is easy to read and its value is enhanced by a full discussion of a range of carefully selected 'real' examples.

18. *COX, D. R., and SMITH, W. L., *Queues*, Methuen, 1961.

An excellent account of the basic concepts and main mathematical techniques of queueing theory. The book seeks to provide a broad introduction to the subject, suitable for operational research workers and mathematical graduates.

19. *HAMMERSLEY, J. M., and HANDSCOMB, D. C., *Monte Carlo Methods*, Methuen, 1964.

Although written as a specialist textbook on the subject, it contains much of interest that will be comprehensible to the general reader with a good mathematical background.

20. HOLLINGDALE, S. H. (editor), *Digital Simulation in Operational Research*, English Universities Press, 1967.

The proceedings of a conference held in Hamburg under NATO

auspices. The papers range from descriptive accounts of specific applications to more specialist discussion of theory, techniques and computational topics. The level of the contributions is inevitably uneven, but there is much of general interest.

21. MITCHELL, G. H., 'Simulation', *Bulletin IMA*, Vol. 6 (3), 1969.

A very good short expository account of digital simulation, illuminated by the author's experience at the National Coal Board. The article is based on a contribution to the IMA Symposium mentioned in ref. 9.

22. MIRSKY, L., 'Pure and Applied Combinatorics', *Bulletin IMA, 5* (1), 1969.

A readable and stimulating introduction to a large subject.

23. *WELSH, D. J. A. (editor), *Combinatorial Mathematics and its Applications*, Academic Press, New York, 1971.

Proceedings of an IMA conference held at Oxford in 1969 – the first conference on combinatorics to be held in the UK. Most of the papers are addressed to specialists in the field and many unsolved problems are presented.

24. CHILLINGWORTH, D., 'Elementary Catastrophe Theory', *Bulletin IMA*, Vol. 11 (8), 1975.

An excellent introduction to a large, new and exciting subject. Contains useful references to recent work.

25. STEWART, I., *Concepts of Modern Mathematics*, Penguin Books, 1975.

An excellent popular exposition of difficult material; a delight to read. Contains a stimulating short account of catastrophe theory.

4 Networks

1. Introduction

A network is a simple concept exemplified in many familiar forms – the web of a spider, the road map of a city, the routes of an airline company, the connections in a telecommunications system. But until recently networks had not been used extensively by mathematicians. At school, probably the only network problem one encountered was the calculation of electrical resistances in series and in parallel; but a glance at a modern mathematics text [1] will show that times have changed!

What is new is principally two-fold: on the one hand an upsurge in interest in the intrinsic properties of networks, and on the other hand the recognition that important practical problems in industry, management and the social sciences can be formulated as network problems. It has been a fascinating example of the interplay between pure mathematicians developing general concepts of networks under the title 'graph theory', and applied mathematicians developing and solving network models under the general title 'operational research'. Newer theory and newer uses of networks have forged ahead together.

The starting point of the applied mathematician is invariably the formulation of a mathematical model of the problem he wishes, or is asked, or perhaps even paid to solve. For the problems which we shall discuss in this chapter, the formulation of a network model is sometimes immediate and transparent but in other cases requires careful probing and analysis. But the mere construction of the model can be very rewarding. To force an engineer in industry to state his problem with the clarity and definity needed for a network model can be the solution of the problem! The network models we describe usually involve some optimization procedure; something is to be maximized or minimized subject to certain constraints. What is the shortest, or quickest, or cheapest route from A to B on a given road network? What distribution of goods satisfying given supplies and demands minimizes the over-all cost or maximizes the total profit? What is the cheapest telephone switch which will guarantee that one subscriber's call to another will not be blocked? In constructing a building, how should the various jobs be scheduled so that the building can be erected as quickly as possible? How should a complex system be designed so it is most likely to continue to perform when certain of its elements fail?

It might be thought that the differential calculus would provide the mathematical apparatus for this optimization procedure – but for the simplest models we discuss this is not the case. A much simpler technique, an elementary form of mathematical programming, is what is required. For small network problems, pencil, paper, eraser, simple arithmetic and the occasional deep breath are sufficient, and the reader should be prepared for making this effort later in this chapter. For larger and most practical problems, a computer is essential. Indeed the development of the newer network techniques has gone hand in hand with the development of the computer. Network calculations are ideal for digital computation; they are simple, but there are lots of them. And the optimization procedures are usually best described by algorithms which can be easily translated into computer programs. The word *algorithm* is now popularly used for any method of solving a problem which is presented as a sequence of precisely stated successive steps. What this really means will be apparent from examples which appear later in this chapter.

Important in network problems is the effect of constraints, especially those such as road capacities or the limit on the number of available telephone lines which give rise to congestion. A proper microscopic analysis of network congestion involves the application of queueing theory; but for simplicity and brevity we shall avoid introducing this and instead take a more macroscopic viewpoint. This will prove adequate for the purposes of this chapter.

What is frustrating and confusing in network problems is the complete lack of a uniform terminology. 'What's in a name?' is Harary's [2] sarcastic quote from *Romeo and Juliet* as he complains about the 'personalised terminology' of graph theoreticians. Vertex, point, node, junction, are variously used for the same thing and a long glossary of pseudo-botanical names – tree, forest, blossom, branch, vine and so on – is hardly a pretty sight. We shall try to minimize the confusion by deliberately avoiding complete precision; the fastidious reader will easily detect this but can refer to Harary's formidable list of terms and definitions for consolation. The terminology most suited for our purposes will be introduced as required.

In order to describe the newer uses of network techniques, we begin in the next section with some examples of transportation networks. Not only are these familiar to the man-in-the-street, so to say, but he is well aware of transportation network problems and moreover often claims to be able to solve them. Some of these problems we shall formulate mathematically in the section 'Network Techniques', and by means of worked examples we shall introduce the reader to a newer network technique so that he can see for himself what this chapter is really about. In particular we shall give an algorithm for the solution of the cheapest path problem. In the section 'Telephone Networks' we turn to a completely different application and shall exploit their similarity with transportation networks to show how optimum telephone networks can be designed. Another field of application, project scheduling, will be considered in the following section and again we show that an optimum route algorithm has an important application. Finally (apart from a brief conclusion) we shall in 'Network Distribution' describe some network distribution problems and how the newer network techniques are being used to solve them.

2. Transportation Networks

UNDIRECTED NETWORK

As our first example of a transportation network, consider a map of a country giving the main arterial trunk roads. This is shown in Fig. 4.1 by an *undirected network* in which the numbered circles are *vertices*, representing major cities and road intersections, and the lines joining them are *edges*, representing the roads. The network is *planar*, i.e., it can be drawn on a sheet of paper without crossing of edges.

We could represent the structure of the network algebraically rather than geometrically by listing the vertices 1, 2, . . ., 24, and listing the edges by writing them as unordered pairs (1, 2), (1, 3), (1, 9), . . ., (23, 24). By an unordered pair we simply mean that (1, 2) is the same as (2, 1). In listing the pairs it is a useful convention, and helps to make sure that none is missed, to adopt dictionary order so that, for example, (1, 2) is listed before (1, 3) and (1, 14) is listed before (2, 3), and so on, up to (22, 24) and finally (23, 24).

To gain some small benefit from the inordinate school time now devoted to curly brackets, we could define the set of vertices

$$V = \{1, 2, . . ., 24\}$$

and the set of edges

$$E = \{(1, 2), (1, 3), . . ., (23, 24)\}.$$

The sets V and E describe just the structure of the network, ignoring any geographical significance of the edges as roads; together the sets define what is called a graph. But as we are usually concerned with more than just the structure, and indeed our chief concern will be with quantities such as lengths, costs and travel times associated with edges, we shall use the less defined term network.

TREE

As a second example of a transportation network, consider a road map which shows the best routes to and from a city for all major cities in a country. We have in mind the sort of map issued by automobile associations. Although we are not concerned at the moment with the criterion used in deciding which routes are the best, we give some idea from this quote: 'These routes are based on the easiest and quickest ways of travelling under average conditions; they are not necessarily the shortest.' Fig. 4.2 is an example of such a map for the routes to and from city 1 for the road network of Fig. 4.1. Given that the map shows only the best routes, without alternatives, the network is a tree. Perhaps Fig.

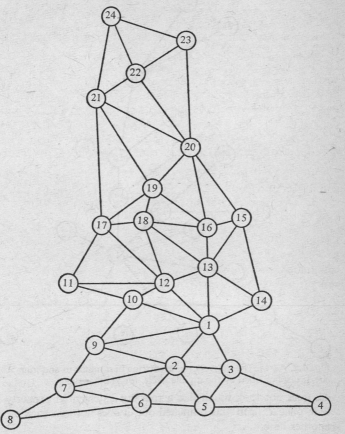

Fig. 4.1. An undirected network. The network represents the main trunk roads of a country. The numbered circles are the vertices of the network, the lines joining them the edges

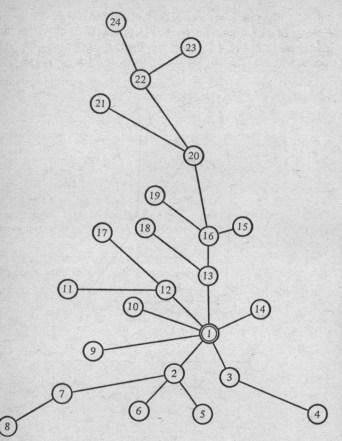

Fig. 4.2. A tree. The network represents the best routes to and from city 1, distinguished by the double circle, for all other cities

4.2 does not look much like a tree but its essential property is that, once a 'limb' has branched out, it does not join back on to another limb.

To define this tree structure more clearly we introduce the concept of a *path* from an initial to a final vertex as a sequence of distinct vertices and edges joining them; e.g., in Fig. 4.1,

9, (9, 10), 10, (10, 12), 12, (12, 18), 18
and   9, (9, 1), 1, (1, 13), 13, (13, 18), 18

are two paths from vertex 9 to vertex 18. It is important to note that the vertices are distinct so that in a path no vertex is visited more than once. A *cycle* is a path except that the initial and final vertices coincide; e.g., in Fig. 4.1,

2, (2, 3), 3, (3, 4), 4, (4, 5), 5, (5, 2), 2

is a cycle which could equally well be denoted by

3, (3, 4), 4, (4, 5), 5, (5, 2), 2, (2, 3), 3.

There are no cycles in Fig. 4.2.

Two vertices of a network are said to be *connected* if there is at least one path joining them. For a *connected network*, each vertex is connected to every other vertex. Figs. 4.1 and 4.2 both represent connected networks.

We can now formally define a *tree* as a connected network without cycles, but if you prefer informality just have a simple diagram such as Fig. 4.2 in mind!

COMPLETE NETWORK

As a third example of a transportation network consider Fig. 4.3, which represents the traffic movements between five adjacent cities. The numbers indicated on the edges of the network represent the traffic movements as, for example, the average number (in thousands) of week-day vehicular trips. Thus the total traffic between cities 1 and 3 in both directions is 5000 trips a day. In this network there is no geographical significance in the representation of the traffic movements as straight lines; these are not traffic routes and the crossings of the lines are irrelevant. Fig. 4.3 is an example of a *complete network*, each pair of vertices being joined by an edge. By contrast, Figs. 4.1 and 4.2 represent incomplete networks. It is a celebrated theorem of graph theory that a complete network with 5 or more vertices is non-planar; in other words if you start with 5 or more points on a sheet of paper and join each pair with a straight or curved line you will not be

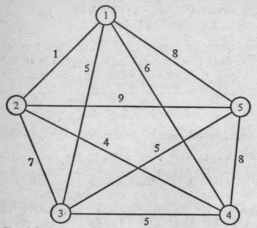

Fig. 4.3. A complete network. The network represents traffic movements between adjacent cities, e.g., between city 1 and city 3 there is an average of 5×10^3 vehicular trips each weekday

able to avoid having two or more lines crossed. What about 4 points?

DIRECTED AND MIXED NETWORKS

As a fourth example of a transportation network, consider a road map of the central district of a large city. It is safe to assume that the city has had traffic problems and it is safe to assume that the traffic engineers have tried to alleviate these problems with one-way streets. To indicate these on a network such as Fig. 4.4 we use lines with arrows and call them *directed edges*. Algebraically we represent a directed edge as an ordered pair of vertices so that (1, 2) is not the same thing as (2, 1). To the motorist it is the difference between driving the right way and the wrong way down a one-way street, a difference not to be ignored. Definitions of a *directed path*, such as 1, (1, 2), 2, (2, 3), 3, and a *directed cycle*, such as 8, (8, 6), 6, (6, 9), 9, (9, 10), 10, (10, 8), 8 are obvious extensions of those given for a path and a cycle; when we choose to ignore directions and arrows we can emphasize this by referring to *un-*

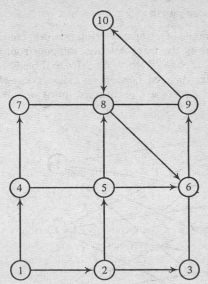

Fig. 4.4. A mixed network. The network represents the road map for the centre of a city with one-way streets indicated by directed edges with arrows and two-way streets by undirected edges

directed paths and *undirected cycles*. Any undirected path in a city network with one-way streets would be an allowable route for a pedestrian but not necessarily for a motorist.

A network with only undirected edges is called an *undirected network* (e.g., Figs. 4.1 and 4.2), a network with all its edges directed a *directed network* (e.g., Fig. 4.5) and a network with some of each a *mixed network* (e.g., Fig. 4.4).

The distinction between directed and undirected networks, edges, paths and cycles is important but can be very tiresome to maintain explicitly, and when it is clear from the text (and hopefully only when this is so) the distinguishing epithets, directed and undirected, will be dropped. And on other occasions we will not be frightened to sidestep the issue and use a common word such as route when, to be precise, we should talk about a directed or undirected path. 'What's in a name?'

BIPARTITE NETWORK

As a final example of a transportation network, consider Fig. 4.5, which represents the trans-Atlantic shipping routes from four ports of the east coast of North America to three European ports. This particular directed network is called a *complete bipartite network*. The vertices are partitioned into two subsets with the

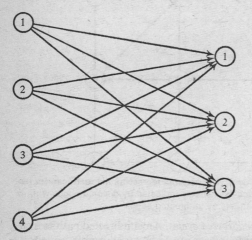

Fig. 4.5. A complete bipartite network. The network represents the shipping routes from four ports on the east coast of North America to three European ports

edges directed from each vertex of one subset to each vertex of the other subset. If some of the directed edges are omitted, the network is called an *incomplete bipartite network*.

The purpose of this section of this chapter has been to introduce various network concepts and terminology, by using examples of transportation networks. The same concepts and terminology are used in many different application areas. But before turning to these, we persist with transportation networks and use them in the next section to introduce some network techniques and show some of the newer uses for them.

3. Network Techniques

TRANSPORTATION STUDIES

Few large cities have escaped transportation problems, and few have escaped some sort of transportation study. Some of these studies have been vast in scale, conception and cost. Basic to these studies have been the analysis and evaluation of future road and public transit networks and it is just one aspect of this analysis – traffic assignment – which will be described here and used later to introduce a new and powerful network technique.

In formulating a long-range transportation plan for a city it is customary to divide the study area into *zones* which serve as major origins and destinations of traffic. Without reference to the details of the road network, the traffic between these zones is predicted for some future design year. These predictions could be represented by a complete network, such as Fig. 4.3, but for a large number of zones the diagram would be too complicated to interpret. In the next stage of the transportation study, various road networks are proposed and the predicted interzonal traffic assigned to them. The traffic engineers evaluate the networks to check whether the roads will carry the predicted traffic within the design limits. Finally a particular network is chosen as the best and recommended to the authority concerned, which as often as not rejects it!

The road networks used in these studies are very large, with as many as 16 000 vertices and 64 000 edges, and needless to say only a computer can handle them. How does the computer assign the predicted traffic to such a network? The simplest and most common method is to assign each interzonal parcel of traffic to the shortest path from its origin to its destination. It is easy to criticize this assumption. Time, convenience or ease of travel may be as good a criterion as distance. Such factors are certainly involved because motorists are unlikely to agree on the best route from A to B, and they are unlikely to choose the shortest. This objection can be partly met by associating with each edge (i, j) a cost $c(i, j)$ which is a mixture of distance, time and convenience. All we insist upon is that this cost is non-negative, finite and additive in the sense that the total cost in going from A to B is the sum of the costs of the edges in the path taken from A to B. We do not impose any constraint such as the triangle inequality (sum of the lengths of two sides greater than the length of the third). We can then talk about the cheapest paths from A to B, which admittedly transfers part of the difficulty of shortest paths to the determination of the edge costs $c(i, j)$. But leaving aside this issue, we wish to concentrate on the problem of the so-called all-or-nothing assignment in which, for given edge costs, the predicted traffic is assigned to the cheapest paths between origins and destinations.

Central to traffic assignment and central to transportation studies is the development of efficient network algorithms for determining cheapest paths on networks and many research papers and computer programs have been directed to this problem, with spectacular success. Computers can now generate cheapest paths extremely rapidly even for very large networks.

CHEAPEST PATH ALGORITHM

We shall describe a labelling algorithm, which is a very efficient algorithm for determining the cheapest paths from a given vertex of an undirected connected network to all other vertices, given the costs of the edges. With the proviso that in the case of a tie between equally cheap paths we choose but one, the cheapest paths together form a tree, such as Fig. 4.2, which gives not only the cheapest paths from a particular vertex, or origin, to all other vertices but also the cheapest paths to the origin from all other vertices. We shall by an example describe the algorithm precisely enough so that small networks can be solved with pencil and eraser, but this should not mislead the reader into thinking that the precise formulation and verification of network algorithms is straightforward (the over-confident reader will find appendix A of [3] a jolt).

Fig. 4.6(a) represents an undirected network with nine vertices (the omission of the number 2 is for later convenience) and ten edges with edge costs as indicated. We wish to determine the

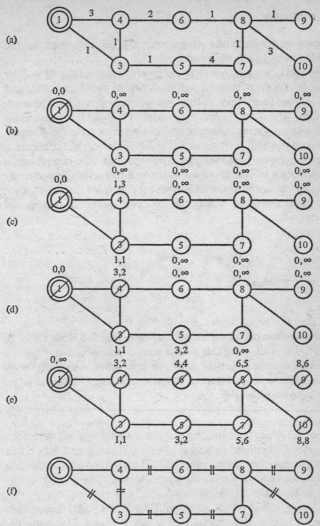

Fig. 4.6. Determination of cheapest paths: (a) the network with vertex **1** as origin and edge costs as indicated; (b) end of step 1 of the cheapest path algorithm; (c) end of step 2; (d) end of step 3; (e) end of Phase I; (f) end of step 1, Phase II

cheapest paths from the origin vertex 1 (distinguished by the double circle) to all other vertices.

The cost of edge (i, j) is denoted by $c(i, j)$, so that $c(4, 6) = 2$ for example. To each vertex i we shall attach two labels $P(i)$ and $K(i)$ whose temporary values are altered throughout the algorithm. The final permanent value $P^*(i)$ is the *predecessor vertex*, or the vertex immediately before vertex i, on the cheapest path, and the final permanent value $K^*(i)$ the cost of the cheapest path from the origin to vertex i. The algorithm is in two phases, a forward scan in which we fan out from the origin permanently labelling vertices in order of increasing cost from the origin, then a backward scan in which the cheapest paths themselves are determined.

Cheapest Path Algorithm

PHASE 1: *Forward Scan*

Step 1
(a) Label the origin vertex $P = 0$, $K = 0$.
(b) Declare the origin vertex permanently labelled, with $P^* = 0$, $K^* = 0$, indicating this with a stroke or slash.
(c) Label all other vertices with the temporary values $P(i) = 0$, and $K(i) = \infty$ (or some number much larger than the given edge costs). See Fig. 4.6(b).

Step 2
(a) For all edges (i, j) joining a slashed vertex i to an unslashed vertex j, calculate the sum $K(i) + c(i, j)$ and if and only if this is less than the current value of $K(j)$ make the value of the sum the new value of $K(j)$ and make i the new value of $P(j)$; otherwise leave the labels unaltered.
(b) The unslashed vertex, say j, with the K label with the smallest value is now slashed, its two labels $P(j)$ and $K(j)$ being declared permanent and denoted by $P^*(j)$ and $K^*(j)$. See Fig. 4.6(c), in which the vertex $i = 3$ has been permanently labelled.

Steps 3, 4, . . .
   Step 2 is repeated (part (*a*) being omitted if there are no edges joining a slashed vertex to an unslashed vertex) until all vertices are permanently labelled. See Fig. 4.6(*d, e*).

   At the end of Phase I, the value of $K^*(i)$ for each vertex i is the cost of a cheapest path from the origin vertex to vertex i.

PHASE II: *Backward Scan*

Step 1
For each vertex i (except the origin vertex) the edge $(P^*(i), i)$, i.e., the edge joining the predecessor vertex $P^*(i)$ of i to i, is marked with two short transverse lines, and declared a *critical* edge.

   The set of critical edges forms the cheapest path tree for the origin vertex. See Fig. 4.6(*f*).

Step 2
To determine the cheapest path from the origin to any vertex i we scan backwards from i to its predecessor vertex $P^*(i)$, then from this vertex to its predecessor and so on until the origin is reached.

   The following points on the algorithm should be noted:
   (*i*) It is harder to describe than to use! The reader should try pencil and eraser on the network given in Fig. 4.7.
   (*ii*) Ambiguity may arise in the algorithm when a tie occurs in Phase I Step 2(*b*). A tie can be resolved by any arbitrary choice or, for example, by always choosing the vertex with the smallest vertex number. Thus in Fig. 4.6(*d*) vertices 4 and 5 both have K labels equal to the minimum value 2; we choose to declare vertex 4 permanently labelled.
   (*iii*) At the end of Phase I, the K^* values give the costs of the cheapest paths; Phase II determines the paths, although for small networks they are by this time obvious.
   (*iv*) An automobile association map of best routes to and from a city makes this suggestion: 'You may find it easier to first trace the line of route backwards from your destination to your starting point.' The backward scan!

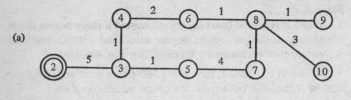

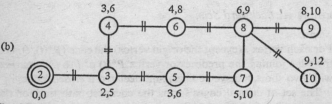

Fig. 4.7. Determination of cheapest paths: (*a*) the network with vertex 2 as origin and edge costs as indicated; (*b*) end of step 1, Phase II of the cheapest path algorithm

(*v*) It is worth-while emphasizing that the cheapest path algorithm solves an optimization problem with discrete rather than continuous variables and without the use of the differential calculus.

(*vi*) The cheapest path problem can be formulated in terms of dynamic programming (see Chapter 3) and this formulation is particularly convenient as a basis for the rigorous proof of the validity of the algorithm (see [3]).

NETWORK ASSIGNMENT

We now illustrate with a simple example how the paths determined by the cheapest path algorithm are used in the all-or-nothing traffic assignment to a road network, i.e., the assignment of *all* the traffic to the cheapest paths and *nothing* to any other paths.

Fig. 4.8(*a*), a complete bipartite network, represents the predicted traffic from two origins (vertices 1, 2) to two destinations (vertices 9, 10). The numbers attached to the edges are predicted traffic movements; e.g., 35 traffic units from origin 1 to destination 9. The unit may be, say, a hundred vehicle trips per hour; as it is a rate, the term *traffic flow* is common. The traffic is to be assigned to the road network shown in Fig. 4.8(*b*), with edge costs as indicated (cf. Figs. 4.6 and 4.7). We distinguish the origins and destinations by using double circles; the other vertices represented by single circles are simply intermediate points where traffic flow is conserved – what comes in there must come out!

In the all-or-nothing assignment we first determine the cheapest path from vertex 1 to vertex 9 (see Fig. 4.6(*f*)) and assign the traffic flow of 35 to it; similarly we assign the other three traffic flows and aggregate the traffic assigned to each edge. The assignment is shown in the second to last column of Table 4.1.

It is apparent that in this example the all-or-nothing assignment is unrealistic because in practice traffic would be more evenly distributed over the roads. But with many origins and destinations the all-or-nothing effect is not as dramatic as the result for this small example might suggest.

A more realistic assignment is obtained if some account is taken of the traffic congestion occurring with increasing traffic flow. This could be done by allowing the road cost to increase with the traffic flow, and this leads to a nonlinear optimization problem (see Chapter 3). Such a problem will be considered later; for the present we retain linearity by making the simpler assumption that

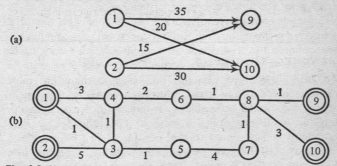

Fig. 4.8. Network assignment: (*a*) the traffic flows from origins 1, 2 to destinations 9, 10; (*b*) the road network to which the traffic is to be assigned, with road costs as indicated

TABLE 4.1   Network assignment for Fig. 4.8

	Cheapest path assignment				Total flows	Modified flows
Edge	1 to 9	1 to 10	2 to 9	2 to 10		
(1, 3)	35	20			55	5
(1, 4)					0	50
(2, 3)			15	30	45	45
(3, 4)	35	20	15	30	100	50
(3, 5)					0	0
(4, 6)	35	20	15	30	100	100
(5, 7)					0	0
(6, 8)	35	20	15	30	100	100
(7, 8)					0	0
(8, 9)	35		15		50	50
(8, 10)		20		30	50	50

a road may have a definite capacity which the traffic flow cannot exceed. This is indeed standard traffic engineering practice. A road may be planned as a four-lane highway with a capacity of 3000 vehicles per hour in each direction, experience having shown that, for the planned road widths and expected roadside activity, unacceptable breakdown of smoothly flowing traffic occurs when the flow exceeds this critical value.

The newer network techniques are readily applicable to such a *capacitated network*. Instead of the cheapest path assumption the basic optimization procedure used is to *minimize the total network cost*. The network model when formulated this way is an example of a *linear program* (see Chapter 3). Efficient computer programs are available for solving large linear programs so that assignment to large capacitated networks is not difficult. Solution by hand, even for small networks, is tedious. The following worked example, which is similar to that considered in Chapter 3 and which may with little cost be skipped by the weary reader, does illustrate the salient features of the minimum network cost technique.

In the example already considered (see Fig. 4.8) we now suppose that the road (3, 4) is one-way with a capacity of 50 units, all other roads being uncapacitated. This will force some traffic on to road (1, 4). The question is, how much? More notation is (unfortunately) needed and is listed in Table 4.2. All paths from the origins 1, 2 to the destinations 9, 10 have now to be listed (by showing the vertices visited). The associated path costs are

TABLE 4.2 Path data for Fig. 4.8

Path	Path cost	Path flow
1, 3, 4, 6, 8, 9	6	x_1
1, 3, 5, 7, 8, 9	8	x_2
1, 4, 6, 8, 9	7	x_3
1, 3, 4, 6, 8, 10	8	y_1
1, 3, 5, 7, 8, 10	10	y_2
1, 4, 6, 8, 10	9	y_3
2, 3, 4, 6, 8, 9	10	z_1
2, 3, 5, 7, 8, 9	12	z_2
2, 3, 4, 6, 8, 10	12	w_1
2, 3, 5 ,7, 8, 10	14	w_2

obtained by adding the edge costs in each path and the variables correspond to the path flows to be determined.

The network cost C which is to be minimized is

$$C = 6x_1 + 8x_2 + 7x_3 + 8y_1 + 10y_2 + 9y_3 + 10z_1 \\ + 12z_2 + 12w_1 + 14w_2. \quad (1)$$

The minimization is subject to various constraints. Thus, since the total traffic flow from vertex 1 to 9 is 35 units

$$x_1 + x_2 + x_3 = 35. \quad (2)$$

Similarly

$$y_1 + y_2 + y_3 = 20 \quad (3)$$

$$z_1 + z_2 = 15 \quad (4)$$

$$w_1 + w_2 = 30. \quad (5)$$

The capacity constraint on (3, 4) means that

$$x_1 + y_1 + z_1 + w_1 \leqslant 50 \quad (6)$$

where we have picked out those paths which use the edge $(3, 4)$. We replace this inequality by the equation

$$x_1 + y_1 + z_1 + w_1 + s = 50 \qquad (7)$$

by introducing the non-negative slack variable s.

Finally (and it is often easy to forget this) the variables x_1, x_2, ..., w_2 must also be non-negative; we cannot have negative flows (although cars can reverse!).

To summarize our network model, we wish to minimize C given by (1) subject to the equations (2), (3), (4), (5), (7) and all variables non-negative.

This is a standard form of the minimum network cost model, and it is a standard form of a linear program. It turns out to be a degenerate linear program, as already discussed in Chapter 3, and as given there the minimum value of C is 930 with the general optimum solution

$$x_1 = t, x_2 = 0, x_3 = 35 - t, y_1 = 5 - t, y_2 = 0, y_3 = 15 + t$$
$$z_1 = 15, z_2 = 0, w_1 = 30, w_2 = 0, s = 0 \qquad (8)$$

with $0 \leqslant t \leqslant 5$. Although this solution is not unique, the optimal assignment to the roads is unique and is shown in the last column of modified flows in Table 4.1. The capacity constraint has forced 50 units of traffic on to the previously unused road $(1, 4)$.

The important point, barely visible in this simple example, is that congestion can be taken into account by introducing capacity constraints; the network models take the form of linear programs which are easily solved by readily available computer programs.

Although cheapest path and minimum network cost models have been introduced here by appeal to transportation and traffic applications, they are widely applicable to many diverse fields, as other sections of this chapter will aim to show.

MAXIMUM FLOW

For a capacitated network in which the edges have definite capacities an important problem is the determination of the maximum flow possible from one vertex to another.

Let us, for the sake of economy, use Fig. 4.6(*a*) as an example and suppose that the numbers attached to the edges are not costs but are now interpreted as capacities. Suppose we wish to determine the maximum flow from the vertex 1 to the vertex 8 (we can ignore vertices 9 and 10 for this calculation). This determination is made by considering a separating *cut*, which is a set of edges of the network connecting one set of vertices including vertex 1 but not 8 to the remaining vertices. In the language of set theory we consider the set V of vertices partitioned into complementary sets X, $\bar{X}$ such that $1 \in X$ and $8 \in \bar{X}$ (the symbol '$\in$' means 'belongs to' or 'is a member of'); then the separating cut denoted by $(X, \bar{X})$ is defined to be

$$(X, \bar{X}) = \{(i, j); (i, j) \in E, i \in X, j \in \bar{X}\}.$$

The cut capacity is the sum of the capacities of the edges in the cut. The following examples should make this simple concept clear.

X	$\bar{X}$	cut	cut capacity
1	3, 4, 5, 6, 7, 8	(1, 3), (1, 4)	4
1, 3, 4	5, 6, 7, 8	(3, 5), (4, 6)	3
1, 3, 5, 7	4, 6, 8	(1, 4), (3, 4), (7, 8)	5
1, 3, 4, 5, 6, 7	8	(6, 8), (7, 8)	2

Since the flow from vertex 1 to vertex 8 must pass through the cuts which separate them, the maximum flow cannot exceed the capacity of any such cut. In fact the *max-flow min-cut theorem* states that the maximum flow from one vertex of a capacitated network to another vertex is equal to the minimum of the capacities of the cuts separating those vertices. For the present example the maximum flow from vertex 1 to vertex 8 is 2 units.

The max-flow min-cut theorem is an example of the so-called minimax theorems which form the backbone of the theory of network optimization. While something is being minimized, something else is being maximized. The theorems tend to be deceptively simple because they can be proved with simpler techniques than the traditional school mathematics methods.

Instead of the calculus, trigonometry, logarithms, coordinate geometry, complex numbers and so on, logical argument is mainly what is required; but conceptually this can become quite difficult. A constructive proof of the max-flow min-cut theorem, which shows how to attain the maximum flow in the network, follows an argument similar to that used above in explaining the cheapest path algorithm.

Perhaps the understanding of the max-flow min-cut theorem can be enhanced by appeal to some 'Z-car' terminology. Suppose the police wish to check all cars travelling from vertex 1 to vertex 8 for the road network of Fig. 4.6(a). They make two sensible assumptions: first, that the cars choose routes which avoid visiting a vertex more than once (in network terminology the cars follow paths); second, that they want to stop and check each car once, but only once. The problem: where should the police send their Z-cars to set up road blocks? The answer: to any set of roads which together form a separating cut. For example, the police could block the three roads (1, 4), (3, 4) and (7, 8), which do form a cut. Road blocks on (1, 4) and (7, 8), not a cut, would be insufficient because cars following the path 1, 3, 4, 6, 8 would get through unchecked. Road blocks on (1, 4), (3, 4), (6, 8) and (7, 8), again not a cut, would be unacceptable because cars checked on road (1, 4) would be checked again before getting to vertex 8.

To continue with the Z-car interpretation, now suppose that an edge capacity represents the maximum rate (in hundreds of cars per hour) at which the police can check cars on that road. For example, the capacity of 3 units of edge (1, 4) means that on road (1, 4) the police can check 300 cars per hour. For each acceptable set of road blocks, the police can determine the *total rate* at which they can check cars by simply adding the rates for each road block. With the roads (1, 4), (3, 4) and (7, 8) blocked, for example, the total rate at which cars can be checked is 500 cars per hour, corresponding to the cut capacity of 5 units. The problem: what is the maximum flow (cars per hour) possible from vertex 1 to vertex 8 and which roads should be blocked so that the police checkers are fully occupied? The max-flow min-cut theorem provides the answer. The maximum flow is 200 cars per hour, which is equal to the minimum total rate at which the police can operate effective road blocks. Road (6, 8) and either road (3, 5) or road (7, 8) should be blocked and the traffic directed so that half travels via vertex 6 and half via vertex 7. The police will not be idle.

The max-flow min-cut theorem has a simple interpretation for a city divided by a river, a very common situation. For an origin on one side and a destination on the other side of the river, the set of bridges and tunnels forms a separating cut which is usually of minimum cut capacity – and this capacity is the maximum flow possible.

A different application is to network reliability. In the design of a large complex system it is important to guarantee that it will function even though some of its elements fail by deterioration or ageing. One criterion for assessing network reliability is the minimum number of edges which must fail in order to disconnect the network. This can be determined by appeal to the max-flow min-cut theorem. Each edge is simply assigned unit capacity; then the maximum flow between two vertices is equal to the minimum number of edges which must fail so that all paths between the two vertices are broken. A typical problem might be the design of a network with, say, 1000 vertices so that it will not fail even if as many as any four edges fail. Efficient algorithms based on the max-flow min-cut theorem have been developed for solving such a problem.

4. Telephone Networks

TELEPHONY

The history of telephony [4] is a hundred-year story of spectacular technical achievement and social consequence, and in this story mathematics is a significant chapter not only for its contribution to telephone theory but also for the impelling feedback into mathematics of new mathematical techniques. The classic work of Erlang, for example, not only provided a practical and scientific basis for teletraffic theory, but helped to initiate applied probability as a new branch of mathematics [5]. Recent advances in telecommunications hardware are now demanding newer mathematical techniques and it is the application of some of these to telephone network problems which will be described in this section [6].

To provide each telephone subscriber with a separate independent circuit to every other subscriber would be quite impracticable and uneconomic and instead the connections are routed through exchanges where switching can be controlled, manually or automatically. A telephone network is therefore not a complete network but has more or less a tree structure as illustrated in Fig. 4.9. Although the terminology in different countries differs [7], it is common practice to subdivide the complete area to be served into a small number of *zones*, each with a *zone centre* exchange; each zone is further subdivided into group areas each served by a *group centre*; and within each group area there may be several *main* and finally *dependent* exchanges to which the subscribers are directly connected. A telephone call from one subscriber to a remote subscriber originates from the 'calling' subscriber to a dependent, then a main, a group centre and a zone centre exchange, then to the required zone centre exchange, through a group centre, main and dependent exchanges and finally to the 'called' subscriber.

The lines or circuits connecting subscribers to exchanges and exchanges to other exchanges are variously called *junction* or *trunk lines*, the latter term being reserved for long-distance lines. The inter-exchange lines may be 'both-way' or arranged as unidirectional 'outgoing' and 'incoming' lines.

The telephone network is best modelled as a hierarchy of networks, similar to the networks already described in transportation. The telephone trunk network (analogous to the main arterial road network) has the zone centres as vertices and the inter-zone junctions or trunk lines as edges (a junction not a vertex but an edge? – junction as in junction railway but not as in railway junction – 'What's in a name?'). For each zone, a more detailed network can be constructed with the group centre, main and dependent exchanges as vertices and the inter-exchange lines as edges. This hierarchical network description can be continued, right down to the representation of a single switch in an exchange as a network giving the connections between incoming and outgoing lines.

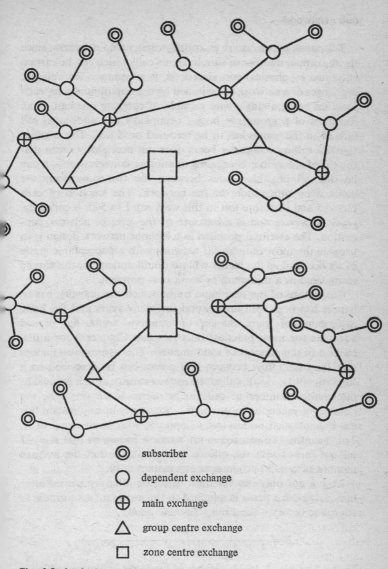

Fig. 4.9. A telephone network

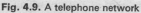

- ◎ subscriber
- ○ dependent exchange
- ⊕ main exchange
- △ group centre exchange
- □ zone centre exchange

Telephone networks are essentially capacitated networks, since the maximum number of simultaneous calls which can be carried is limited by physical considerations. It is common for adjacent exchanges of a network to be joined by a group of junctions, each junction being a single wire capable of carrying one call. If all junctions of a group are busy ('occupied') any additional call offered to the group has to be rerouted or is lost. The engaged signal a calling subscriber hears does not necessarily mean that the called subscriber is engaged in another conversation – it can be that all possible routes between the two subscribers are blocked by other users of the network. The fraction of calls blocked and therefore lost in this way, say 1 in 500, is called the *grade of service* and is a measure of the level of network congestion. The essential problem in telephone network design is to provide the most economical network with an acceptable grade of service, and as the reader will no doubt anticipate, this can be formulated as a minimum network cost problem.

The nature of the telephone traffic which the network has to handle has been well understood for many years and the basic assumption of 'pure chance' or 'random' traffic has proved adequate for most practical purposes (see Chapter 3 for a discussion of the concept of randomness). This assumption implies that the traffic flow between two points can be described by a single quantity, A say, called the *traffic intensity*, which is equal to the *average* number of calls in progress. More precisely, the number of junctions in use or the occupancy at any instant has the Poisson distribution (see Chapter 3) with mean equal to A. This quantity is measured in units called *erlangs* so that $A = 10$ erlangs means that the offered traffic is such that the average number of calls in progress at any instant is 10.

Erlang not only has this unit of traffic intensity named after him but also his name is attached to the most famous formula in teletraffic theory – the *Erlang Loss Formula*:

$$E = \frac{A^N/N!}{1 + A + (A^2/2!) + \ldots + (A^N/N!)}. \tag{9}$$

In this formula, N is the number of junctions in a group which is offered random traffic with mean A erlangs, and E is the proportion of time that all junctions are busy or the fraction of calls lost, i.e. the grade of service. Some readers will recognize the denominator $1 + A + (A^2/2!) + \ldots + (A^N/N!)$ as the start of the infinite series for exp A and often it is accurate enough to replace this denominator by exp A.

Without worrying about the derivation of the formula, let us illustrate its use in two simple cases:

N	A	A/N	E (approx.)
5	0·9	0·18	0·002
20	10	0·5	0·002

For the same grade of service $E = 0\cdot002$, i.e. a loss of 1 call in 500, a group of 5 junctions will carry 0·9 erlangs whereas a group of 20 junctions will carry 10 erlangs of traffic. The ratio A/N is sometimes called the *traffic efficiency*, equal to the number of erlangs per junction. In this sense, the larger group of junctions is much more efficient than the smaller group for the same grade of service. This is an important consequence of the randomness of telephone traffic. It is more efficient to combine small traffic flows on lines with small numbers of junctions into larger flows on larger groups of junctions.

Another important consequence of the randomness of traffic saves us from getting too deeply involved in probability theory. If two flows of independent random traffic are combined, the resultant flow is also random with an intensity which is simply equal to the sum of the individual intensities. In network calculations, the traffic intensities can therefore be combined as though the telephone traffic is a steady flow, like a fluid. Our previous work on flows on transportation networks can be directly translated to give a basis for the analysis of traffic on telephone networks. The range of application of the newer techniques we have described is considerably widened. The expert in transportation networks can, after learning a new jargon, soon become an expert in telephone networks.

In the following sections we shall consider the application of network techniques to problems of telephone congestion. First we shall consider optimum designs of switches which do not block calls; we then consider problems of routing traffic and finally the problem of designing an economical network within a given congestion limit. Throughout the discussion the striking similarity between telephone and transportation networks will emerge.

SWITCHING NETWORKS

The major service of a telephone exchange is the provision of a switching system which will connect incoming calls from calling subscribers to outlets which are eventually connected, through other exchanges, to the called subscribers.

The simplest 2×2 switch, with 2 inlet terminals, 2 outlet terminals, and 4 cross-points, is shown in Fig. 4.10(a) in which the 7 possible states of the switch are given. Fig. 4.10(b) is a conventional diagram used for this switch and Fig. 4.10(c) its representation as a complete bipartite network.

This switch is a trivial example of a *non-blocking network in the strict sense* – for each state of the network any idle inlet–outlet pair not in use can be connected without disturbing the calls already present, regardless of the way in which they have been connected. This is a desirable property of a switch, and to achieve this with minimal hardware is one of the important problems of switch design to which network analysis has been applied.

Fig. 4.11 represents a 3-stage switch using six 2×2 switches. It is not non-blocking in the strict sense but is *rearrangeable*. For the state of the switch with (1, 1) and (3, 3) connected as indicated by the cross-lines on the edges, it is seen that the idle pair (2, 4) cannot be connected so that a call (2, 4) is *blocked*; so also is (4, 2). But if the routing of the call (3, 3) is rearranged to use the top rather than the bottom of the two middle switches, the state is non-blocking and any of the idle pairs (2, 2), (2, 4), (4, 2) and (4, 4) can be connected. Fig. 4.12 is a 3-stage switch which is non-blocking in the strict sense.

For the general 3-stage switch shown in Fig. 4.13, the number of switch points (equal to the total number of edges within the

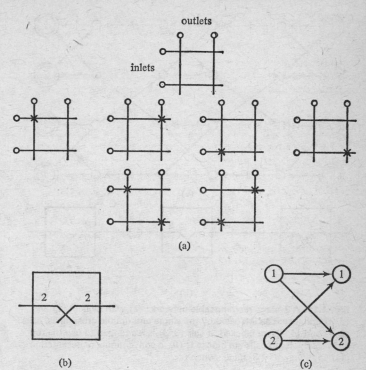

Fig. 4.10. A 2 × 2 switch: (*a*) the 7 possible states; × represents a closed cross-point; (*b*) a conventional representation of the switch; (*c*) representation as a complete bipartite network

switch network) is $2mnr + mr^2$. For this switch, the following theorems have been proved [6]:

(*a*) the switching network is non-blocking in the strict sense if and only if $m \geqslant 2n - 1$;

(*b*) the network is rearrangeable if and only if $m \geqslant n$.

To illustrate the significance of the first theorem, suppose that 100 inlets are to be connected to 100 outlets; for one 100 × 100 switch this would require 10 000 cross-points. With a 3-stage switch as in Fig. 4.13 and $n = r = 10$, $m = 19$, a network which

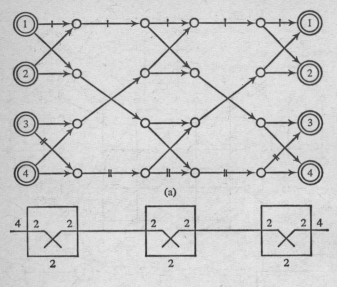

(a)

(b)

Fig. 4.11. A 3-stage rearrangeable network: (a) with calls (1, 1) and (3, 3) connected as indicated by the single and double cross-lines, calls (2, 4) and (4, 2) are blocked. If call (3, 3) is rearranged to use the top middle switch, no calls are blocked; (b) a conventional representation of the network as a 3-stage switch

is non-blocking in the strict sense is realized with only $2mnr + mr^2 = 5700$ cross-points, a considerable saving.

There has been considerable recent research into the design of non-blocking and rearrangeable networks, particularly for multi-stage networks which are of special importance in telephone applications [6].

ROUTING TELEPHONE CALLS

The concept of cheapest paths as described on p. 292 has important applications in the routing of calls in a telephone network. For long-distance calls, several different routes may be

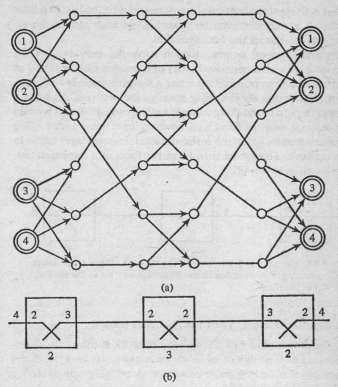

Fig. 4.12. A 3-stage network non-blocking in the strict sense: (*a*) there are two 2 × 3 switches in the first stage, three 2 × 2 switches in the second stage, two 3 × 2 switches in the third stage; (*b*) a conventional representation

available and the cheapest path algorithm can determine which is the best. It is usual to take for the cost associated with an edge the travel time or *propagation delay* which is additive over edges to give the cost of a path. The various systems which have been proposed [7] depend on the efficiency of the cheapest path algorithm. Their great advantage is that they can optimally route calls even when the offered traffic is subject to rapid variations.

What is the best route at one instant may not be the best at a later instant – it may not even be available. At each instant the algorithm determines the best available.

In more general communication networks *store-and-forward* service is provided whereby blocks of messages can be stored at vertices and then transmitted when a forward route is available. With the continually changing states of the network, choosing a best path right to the destination is not possible. Different routing procedures, such as fixed and random, have been tested using network models which minimize the total time messages spend in the network. The model then takes the form of a minimum network cost problem [8].

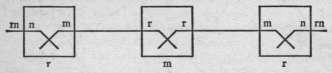

Fig. 4.13. The general 3-stage switching network. There are *rn* inlets and outlets, r $n \times m$ switches in the first stage, m $r \times r$ in the second, r $m \times n$ in the third

DIMENSIONING A TELEPHONE NETWORK

By the dimensioning of a telephone network is meant the determination of the number of junctions required to meet the traffic demands at a desired grade of service. A typical grade of service is a loss of 1 call in 500.

We have in mind a network for a single zone with about 50 exchanges and 200 lines connecting them. A common feature of such a network is a capability for *alternate routing*. For example, with three alternative routes, traffic is first offered to a *direct route*; if this is completely utilized, the traffic is then offered to an overflow or *second-choice route*, and if this is congested, to a *third-choice route*; and if this is congested the call is lost. The essential advantage of the alternate routing is that the second-choice and third-choice routes can be shared by traffic between different origins and destinations giving greater traffic on larger groups of junctions and hence higher efficiency. The overflow traffic will no longer be random as it will tend to come in concentrated bursts.

Until recently, dimensioning methods have been local in viewpoint. A particular edge of the network is investigated and the number of its junctions optimized, assuming the remaining background traffic to remain constant. Because edges are common to different routes, this local optimization is only approximate and a more global viewpoint is conceptually attractive. Such is provided by using a minimum network cost model: given the predicted traffic between exchanges, the numbers of direct and overflow junctions are determined so that the total network cost is minimized and the prescribed grade of service is maintained.

The minimum network cost model has already been encountered in network assignment (see p. 296). The present model is simpler in one respect and much more difficult in another. It is simpler because each parcel of telephone traffic is limited to three possible routes, whereas in traffic assignment the motorist has very many routes from which to choose. It is much more difficult because, for the telephone problem, the network cost is a non-linear function; instead of having a comparatively straightforward linear program we are forced to use a nonlinear program technique. The technique which has been used with success is a gradient method (see Chapter 3) which is a multi-dimensional extension of the simple method of finding the minimum point of a curve by successively following tangents until a horizontal tangent is reached. In practice a near optimal solution is acceptable and the iterations are terminated before too much computer time and money are used. In this method the variables – the numbers of junctions – are treated as continuous and have to be rounded off to integers for the final answer. To solve the problem as an integer program is beyond the capability of present methods.

The minimum network cost model, with a gradient method, has been successfully used for networks of varying size. How is the dimensioning tested? The common practice is to use simulation (see Chapter 3). Random traffic is generated and fed into the network stored in a computer. The number of lost calls is determined and after a sufficiently long run the simulation is stopped and the actual grade of service (i.e., proportion of lost calls) calculated. If this is within the prescribed limit, the dimensioning is satisfactory.

5. Project Networks

ACTIVITY DIAGRAMS

Much to a librarian's confusion, two books with the same title, *Network Analysis*, may be quite different, one concerned with electrical networks and the other with project scheduling. It is this second field of application to which we now turn.

From its beginnings in the 1950s, the analysis of project networks has now become widespread in many different engineering applications [9]. For practical systems the networks are too large for hand calculation, but again the computations are ideal for a computer to perform, and the development of computing techniques has been instrumental in helping to make network analysis an indispensable tool. The different terminology used by different users is again a source of confusion; we more or less follow that recommended by the British Standards Institution.

The project to be analysed, for example the building of a house, is first broken down into separate jobs or *activities*. The duration or time taken to complete each activity is estimated and the precedence relations indicating which activities must be com-

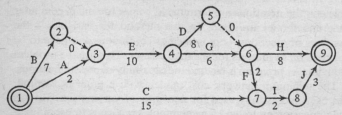

Fig. 4.14. Activity diagram for building a house. The activities *A, B, . . ., J*, their durations and the precedence relations between them are listed in Table 4.3

pleted before others start recorded (Table 4.3); to avoid possible arguments with builders and unionists the number of workers allocated to each activity has not been divulged.

An *activity diagram* or project network is now drawn by representing each activity by a directed edge, the vertices representing *events* or definite states in the progress of the project. For example, vertex 6 in Fig. 4.14 corresponds to the event for which

TABLE 4.3 Building a house

	Activity	Duration (days)
A	pour foundations	2
B	fabricate windows, doors	7
C	fabricate cupboards, furniture	15
D	install plumbing	8
E	erect walls	10
F	plaster walls	2
G	erect roof	6
H	landscape surroundings	8
I	install cupboards, furniture	2
J	paint inside, outside	3

precedence relations
D must follow E
E must follow A and B
F must follow D and G
G must follow E
H must follow G
I must follow C and F
J must follow I

the roof has been erected and the plumbing installed. The dotted edges in Fig. 4.14 are *dummy edges* which have no duration but are needed to maintain the correct precedence relations. The *start event* and *end event* are distinguished by double circles.

The activity diagram is an example of a directed network without directed cycles. A directed cycle would correspond to a logical inconsistency such as

B must follow A

C must follow B

A must follow C.

The mere drawing of an activity diagram for a complex project is an important aspect of project scheduling and assists, for example, in removing such logical inconsistencies.

The vertices of a directed network without directed cycles can always be numbered 1, 2, 3, ... so that, for each directed edge (i, j), $i < j$. A simple algorithm for doing this is:

Step 1. Number consecutively any vertices (chosen in any order) which do not have any incoming edges.

Step 2. Delete any numbered vertices and all edges emerging from them.

Step 3. Return to Step 1.

Stop when all vertices are numbered.

This algorithm has been used for the numbering of the activity diagram in Fig. 4.14. Computer programs are available for doing this numbering for large networks and they indicate when a directed cycle has erroneously occurred.

CRITICAL PATH ANALYSIS

The most important use of the activity diagram or project network is the determination of critical paths. A *critical path* is a directed path from the start to the end event which has the longest total duration. Any activity on a critical path is called a *critical activity* and the identification of these is important in the efficient scheduling of the project.

The following algorithm for determining the critical path is similar to the cheapest route algorithm described on pp. 292–6. The optimization procedure is maximizing rather than minimizing and the fact that the network is directed without directed cycles gives some simplification.

It will be supposed that the vertices have been numbered 1, 2, ..., n such that if (i, j) is a directed edge, $i < j$. Vertex 1 is the start event, vertex n the end event. The algorithm is in two phases, a forward scan in which we consecutively label the vertices, then a backward scan in which the critical path and critical activities are determined. The duration of the activity represented by edge (i, j) is denoted by $c(i, j)$ and to each vertex is attached two labels $P(i)$, $K(i)$. There is no need to use temporary values; those attached are immediately permanent (so that the asterisks used in the cheapest route algorithm are not necessary).

Critical Path Algorithm

PHASE I: *Forward Scan*

Step 1
Label the start event $P(1) = K(1) = 0$.
Steps $j = 2, 3, \ldots, n$
(a) For each edge (i, j) incident on j calculate the sum

$$K(i) + c(i, j).$$

(b) Choose from these sums that which is largest; let its value be $K(j)$ and let $P(j)$ be the value of i for which the sum is largest.

At the end of Phase I, the value of $K(i)$ for each vertex i is the total duration of the longest directed path from vertex 1 to vertex i.

PHASE II: *Backward Scan*

Step 1
The edge $(P(n), n)$, i.e., the edge joining the predecessor vertex $P(n)$ of the end event n to n is marked with two short transverse lines, and declared a critical edge.

Steps 2, 3, ...
Step 1 is repeated for the vertex $P(n)$ instead of n until finally the start event vertex 1 is reached.
The critical edges so marked together form the *critical path*.

In the terminology of critical path analysis, $K(i)$ is called the *earliest occurrence time of event i*, being equal to the earliest occurrence time of the start event plus the duration of the longest path from the start event to event i. The quantity $K(j) - K(i) - c(i, j)$ is called the *free float* and is the duration time by which the activity (i, j) may be delayed or extended without delaying the start of any succeeding activity. For each critical edge, the free float is zero, but the converse is not true. Thus the free float for the critical edge (2, 3) in Fig. 4.15 is zero but (7, 8) which has zero free float is not a critical edge.

As with the cheapest path algorithm, there may be a tie in (*b*) in Phase I; but this can be resolved by an arbitrary choice or, say, by choosing the smallest value of i.

Computer programs for critical path analysis are readily available and are extremely efficient. It is usual in practice to carry out frequent computer runs, updating the project network with new information as it becomes available.

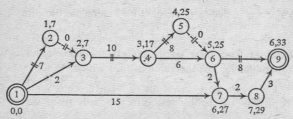

Fig. 4.15. Critical path analysis. The first label on each vertex is the predecessor vertex on the longest path from vertex 1 to the vertex; the second label is the total duration of this path. The critical activities are indicated by the edges marked with the two transverse lines

PERT

Of important historical interest in critical path analysis of project networks was the development of PERT (Program Evaluation and Review Technique) in the USA in the late 1950s. One of the features of PERT was the allowance made for the uncertainty in the durations specified for the activities of the project. As some durations can be estimated more precisely than others, it is reasonable to try to take this into account in the analysis. PERT does this by using three estimates of the duration of an activity – the best (a), the worst (b) and the most likely (m), which are combined to give a single expected time,

$$t_e = (a + b + 4m)/6,$$

which is used in the determination of the critical path. The three estimates can also be used to determine the probability of any event occurring at a particular time.

The PERT technique has initiated extensive mathematical research into the application of applied probability to project networks but the results do not seem to justify using the more complicated methods in engineering practice.

6. Network Distribution

HITCHCOCK PROBLEM

One of the early classic papers on optimizing the distribution over a network is that by F. L. Hitchcock on 'The Distribution of a Product from Several Sources to Numerous Localities' [10]. His problem, often called the transportation problem (see Chapter 3), can be formulated as follows.

Suppose that there are several factories producing the same commodity and supplying several warehouses. To be specific and introduce a helpful notation, let $i = 1, 2, \ldots, m$ denote the factories and $j = 1, 2, \ldots, n$ the warehouses. Factory i can supply a_i units of the commodity and warehouse j requires b_j units. The total supply

$$a_1 + a_2 + \ldots + a_m = \sum_{i=1}^{m} a_i \qquad (10a)$$

is supposed equal to the total demand

$$b_1 + b_2 + \ldots + b_n = \sum_{j=1}^{n} b_j. \qquad (10b)$$

The cost of transporting one unit of the commodity from factory i to warehouse j is denoted by c_{ij}.

The situation can be represented by a complete bipartite network as in Fig. 4.16 (a), which illustrates the case of three factories supplying four warehouses.

The problem posed by Hitchcock was to determine the best distribution of the commodity, satisfying the given supplies and demands. By best he chose to minimize the total cost of distribution, so that his problem is familiar to us as a minimum network cost problem.

Algebraically this can be formulated as a linear program, as on pp. 298–300, by letting x_{ij} denote the number of units to be sent from factory i to warehouse j. The minimum network cost problem written as a linear program is then

$$\text{Minimize } C = \sum_{=1}^{m} \sum_{j=1}^{n} c_{ij} x_{ij} \qquad (11)$$

$$\text{subject to } \sum_{i=1}^{n} x_{ij} = a_i, \ i = 1, 2, \ldots, m \qquad (12)$$

$$\sum_{=1}^{m} x_{ij} = b_j, \ j = 1, 2, \ldots, n \qquad (13)$$

$$x_{ij} \geqslant 0. \qquad (14)$$

In these expressions C is the total network cost or the total cost of distribution (a linear function). Equation (12) is interpreted as summing over all edges leaving vertex i to give the total supply at i. Similarly equation (13) implies summing over all edges incident on vertex j to give the total demand at j. The remaining inequalities are the non-negativity relations.

In a practical problem, say the distribution of motor cars, the supplies a_i and the demands b_j would be non-negative integers; and we would require the solutions x_{ij} to be non-negative integers. This additional requirement forces the linear program into an integer program which, in general, is a much nastier problem (see Chapter 3). However, in this particular case, we have the important result that when solved as a linear program we automatically obtain an integer solution.

Another important property of the solution to the Hitchcock

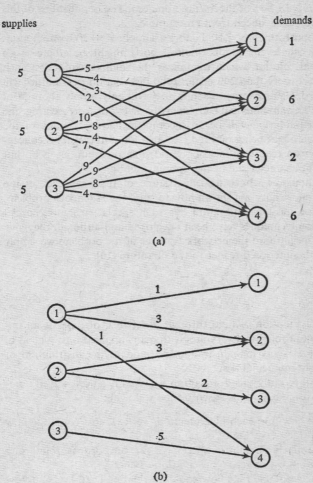

Fig. 4.16. Bipartite network for a Hitchcock problem: (*a*) the numbers on the edges are the unit costs; (*b*) the optimum distribution, the numbers on the edges giving the optimal number of units to be distributed

problem is this: there is an optimal solution such that not more than $m + n - 1$ of the x_{ij} are non-zero. The implications of this can be best deduced from an example.

Suppose that $m = 10$ factories supply $n = 200$ warehouses. Then, whatever the distribution costs might be, there is an optimal solution which minimizes the total network cost such that not more than 209 of the 2000 distribution possibilities will be used. On an average, each factory will only supply about 21 of the 200 warehouses. The bipartite network representing the optimal solution will be an incomplete network, many of the possible edges being omitted (see also Fig. 4.16(b)). This concentration on relatively few of the possible distributions is an important characteristic of the optimal solution.

Although we have described some of the properties of the solution we have not shown how to solve the Hitchcock problem. Because the linear program has such special structure, special solution techniques have been developed and although they give great insight into the network features of the problem we simply refer the interested reader to the literature [10].

HYDRAULIC NETWORKS

The distribution of fluid through a network of pipes is an important problem in hydraulics with many applications, such as in supply systems in industrial plants and in the distribution of water to municipalities.

Until recently this distribution problem has been solved iteratively using equations based on the following:

(a) the flow into each pipe junction (vertex) must equal the flow out (conservation of mass);
(b) the algebraic sum of successive pressure drops around any cycle must be zero (conservation of energy);
(c) an acceptable pipe flow equation relating head loss and flow for each pipe joint (edge) must be preserved.

Even with computer programs to help, the solution of the

numerous equations is cumbersome and has not been generally attempted for large practical networks.

A new approach to this distribution problem has been based on the theorem of least work or, in hydraulic terms, 'water seeks the path of least resistance'. The theorem has been stated as follows: the flow distribution in a pipe network is that flow which minimizes the total hydraulic energy loss, subject to the conservation of mass at each junction. Thus (a) above is retained, but a network minimization replaces (b) and (c).

So again we have a problem formulated as a network optimization subject to constraints. Since the expression for the hydraulic energy loss is nonlinear, the problem takes the form of a nonlinear program. It is similar to that discussed on p. 313 and a similar gradient method has been used for its solution (see Chapter 3). The technique is extremely flexible and can solve problems with mixed unknowns (e.g., pipe sizes, discharges and pressures), and it can handle unknowns defined by inequalities and network control problems. The network techniques now well developed in other applications seem assured of a promising future in hydraulics.

GAS PIPELINE NETWORKS

Network analysis has also been successfully used in problems concerning the distribution of natural gas in large pipeline networks [11]. The flow of gas through a pipe depends on the internal diameter, the length, the pressures at the end points, the grade, the temperature and the properties of the gas, and is formulated by some gas equation. There are various constraints to be satisfied, such as the limitation on the maximum pressure at any point of the system and the minimum pressure where the gas is delivered.

The essential problem is to choose the network and the pipe diameters so that the total cost of the network is a minimum. This is approached by first using tree networks which minimize the number of pipes that have to be used to connect the gas wells to the delivery point. The assignment of pipe diameters is made by determining critical paths, paths in the tree from the delivery point to the vertex where the calculated pressure is a maximum. Without giving any of the details, it is evident that the network techniques explained more fully above in other applications are just those which are useful in solving the problem of natural gas distribution.

7. Conclusion

In this chapter we have described some of the newer network techniques and the newer uses of them in a very wide range of problems. Other similar problems, such as power distribution, personnel assignment, sewerage disposal, airline routing and so on have not been mentioned but have also been successfully investigated by network analysis. It is a field of endeavour with considerable reward for the applied mathematician. Network mathematics in itself is elegant, a surprising variety of practical problems can be formulated as network models, and practical problems can be solved because the network techniques are well structured for computer calculations – in short it is very attractive mathematics which has proved very useful.

REFERENCES

1. School Mathematics Project, Book 3, CUP, 1970.

The SMP books have rightly won the reputation of being among the best of modern mathematics texts. Book 3 (for 13-year-olds) has an interesting chapter on networks which covers more than is needed for an understanding of this chapter.

2. HARARY, F., *Graph Theory*, Addison-Wesley, Reading, Mass., 1969.

For the pure mathematician – a definitive survey of the modern theory of graphs and their combinatorial properties by a leading research worker in this field.

3. POTTS, R. B., and OLIVER, R. M., *Flows in Transportation Networks*, Academic Press, New York, 1972.

This text covers some of the material of the present chapter from a more rigorous and more advanced point of view.

4. OVERMAN, M., *Understanding Telecommunications*, Lutterworth, 1974.

A popular, readable and modern account of telecommunications with a good chapter on understanding the telephone and how it works.

5. COLE, A. C., 'Aspects of Teletraffic Theory', *Bulletin IMA*, Vol. 11, pp. 85–92, 1975.

An excellent survey of the history and present state-of-the-art of teletraffic theory.

6. CATTERMOLE, K. W., 'Graph Theory and the Telecommunications Network', *Bulletin IMA*, Vol. 11, pp. 94–107, 1975.

Another excellent survey which covers in much more detail some of the topics in the present text.

7. HAMSHER, D. H. (editor), *Communication System Engineering Handbook*, McGraw-Hill, New York, 1967.

This handbook, using American terminology, has particularly good chapters on mathematical techniques, covering topics such as linear programming, network flow, minimum cost network, cheapest path algorithms, with discussion of their application to telecommunications.

8. KLEINROCK, L., *Communication Nets: Stochastic Message Flow and Delay*, McGraw-Hill, New York, 1964.

A mathematical approach to the problem of optimizing communication networks based on a minimum network cost model.

9. BATTERSBY, A., *Mathematics in Management*, Penguin Books, 1975.

This recent revision of a deservedly popular book has an excellent chapter with interesting and realistic examples of network analysis.

10. DANTZIG, G. B., *Linear Programming and Extensions*, Princeton University Press, 1963.

An outstanding text on linear programming with a very extensive treatment of the Hitchcock and other network models considered in this chapter.

11. FRANK, H., and FRISCH, I. T., 'Network Analysis', *Scientific American*, Vol. 223, pp. 94–103, July 1970.

This interesting article in a readily accessible journal gives a readable account of maximum network flow, network reliability and vulnerability, and of the successful application of network techniques to the design of a pipeline network for the natural gas fields in the Gulf of Mexico. The claims for the savings of millions of dollars are staggering justification for the use of network analysis. It pays!

5 Finance

1. Introduction

If mathematics is regarded as basically the science of relationships between quantities, in general measurable in numerical terms, then the origins of the material presented in this chapter lie in the remote past when man was developing an ordered social structure.

Before the invention of money, value relationships such as terms on which corn might be exchanged for cattle would have been found to be mainly dependent on quantity, a shortage of one commodity, resulting perhaps from crop failure from adverse climatic conditions, forcing up its value in terms of other commodities not so affected. Although not recognized explicitly, the fact that values were basically dependent on supply and demand would have been implicit in the social organization. Thus one of the main structural features of modern economic theory, the idea of an equilibrium point where the forces of supply and demand balance, has a long history.

Another requirement for the evolution of man was the recognition of the need to husband resources against uncertainties of the future. Here again there was a supply/demand relationship but modified by the intrusion of a time factor. The development of cultivated crops and domesticated animals for food reduced substantially the uncertainty of relying upon natural sources but increased the need for allocation of resources between current consumption and future need. Lessons of bitter experience would have developed empirical rules concerning the relationships between, say, the amount of a crop that should be reserved for future needs and that used for current consumption. Thus another important principle, the concept of saving, also has a long history – in fact, without it man could not have survived.

But underlying both of the principles is the uncertainty of knowing what the future has in store. Early man survived when he devised methods of reducing the impact of these uncertainties to a sufficiently low level. Once having achieved this, the way was clear for major evolutionary growth but, because there was no obvious limitation on resources, the lessons of uncertainty were largely forgotten and the world came to be regarded in terms of simple direct relationships between relevant quantities. In the natural sciences this endeavour to express phenomena in concise mathematical terms reached its peak about the turn of this century, but various inconsistencies in the descriptive models led to the introduction of uncertainty into the models, and the remarkable developments of this century have flowed from this change.

Although there were some early stirrings away from the deterministic view towards the introduction of uncertainty principles, development has tended to lag behind applications in the natural sciences. Only in relatively recent years has the movement gained momentum and in this area many of the newer uses of mathematics can be found.

It is, however, necessary to comment that financial activities are closely related to social structure. They are concerned with relationships between people in their everyday pursuits and are thus affected by political, social, legal and other influences. Observations may be made of features such as the consumption of a particular food, the incidence of sickness, the prices of shares or the amount of life assurance sold, and endeavours made to find qualitative and quantitative relationships between such features and others considered to be relevant. In general all relationships, even though expressed in concise mathematical terms, are at best only approximations since they arise from a great many inter-related factors, many of which cannot be reduced to mathematical or numerical forms.

The situation is different from that existing in the natural sciences. In physics, for example, experiments will be designed to measure a relationship between two quantities when all other relevant factors are controlled, so that it becomes reasonable to seek a mathematical justification for the observed relationship. In the social area such controlled experiments are very rarely possible, and the interpretation of observations is strongly dependent on the skill of the analyst in recognizing the influence of unmeasured or unidentified factors. Nevertheless, there is an essential requirement, as in the natural sciences, that mathematical models should not be used unless and until confidence in them has been gained by some form of critical testing.

One general further observation is relevant. A large amount of work in finance is done by professional advisers on behalf of their clients, normally business enterprises engaged in competitive activity. The members of the professions have undergone an intensive course of specialist studies based on the distillation of knowledge accumulated over many years. Thus there exists a fund of methodology which has been well tested by experience and provides standards against which competence can be measured. Innovations will, therefore, be slow to be recognized as new methods will not have the same authority until they have been adequately tested. It is likely that newer techniques are being tested and developed without general dissemination because they are not yet part of the accepted professional background, or because they are private to clients, and publication could be considered detrimental to the clients' interests. The extent of innovation can then only be surmised, but it is likely that much more is being done than would be apparent from published papers.

An interesting aside to this point arises where there is an academic development associated with the professional activity. The relative freedom of academic study provides scope for publication of research without the inbuilt conservatism of the professions, so that one constraint on development is removed. Thus the growth in university departments concerned with accountancy has provided an environment for cross-fertilization with other academic disciplines, in particular the developments in mathematics and allied subjects.

One concept frequently encountered in the construction of models is that of optimization. This is clearly linked to the idea of equilibrium, which was introduced on p. 326 as the relative levels at which prices for two different commodities would settle down, having regard to the pressures of supply and demand. The equilibrium point could be looked at as the optimum position in the light of the specified conditions, only two of which were stated explicitly. If other conditions are introduced explicitly, such as hours of work and labour costs, then these would influence the relative price level and the equilibrium would settle down at a different point. Thus the equilibrium could then be described as optimal in terms of the specified conditions.

The introduction of constraints is perhaps the condition which brings much realism into the mathematical formulation of business activities, whether in the private or the public sector of an economy. A simple example is the requirement that companies prepare accounts which include a statement (balance sheet) in which the totals of assets and liabilities valued on prescribed rules are separately shown. The financial position is apparent from the amount that must be included to equate the two quantities. This balancing is not to be confused with the sixteenth-century invention of double-entry accounting, which is a very effective method of securing an automatic control of book-keeping entries. A more potent instance is perhaps the need to ensure that prices charged for goods manufactured are sufficient to meet all the associated essential items of costs connected.

It may also be noted that the application of mathematics has been significantly helped during the past thirty years by developments in the techniques largely introduced during the Second World War for operational research purposes. The advent of computers has had an impact in various directions: firstly in the facility for the handling of the very large quantities of accounting and allied data which are an essential adjunct of business; and secondly in the heavy computations that arise from the solution of the mathematical models required to provide an adequate representation of business processes.

Finally a word must be said about the mathematical background. By far the largest proportion of the data arising from the financial sector is of a statistical nature, and quantities measured usually involve a great many variables which it is not practicable to measure precisely. This means that the ideas of mathematical statistics underlie a great deal of recent work, with a strong bias towards the use of probability theory as an effective way of making the models more realistic by the introduction of random variation.

In considering the more detailed aspects of newer uses of mathematics, it becomes immediately apparent that there is considerable overlap. The actuary, for example, in his concern with the financial management of life assurance operations will have a direct interest in the problems arising from the administrative needs of companies employing perhaps many thousands of employees, problems of marketing and of data processing, in respect of all of which mathematical techniques will have some part to play. However, he is also concerned with the investment of substantial funds which are in principle largely held for the benefit of policyholders. His investment strategy must have regard to the nature of the liabilities, largely the values of the policies, so that he will be interested in principles of portfolio management (see p. 377) as well as the assessment of individual securities.

Bankers have a similar division of interest between the management of large multibranch operations and the problems of investment of the funds held on behalf of customers.

Accountants are ubiquitous in business and finance with a diversity of interests which spread from their statutory duties as auditors to advisory functions on company finance, investment, taxation and, in recent years, in management consultancy.

While the Stock Exchange as such is essentially a trading forum, it does exercise an important discipline on the financial markets through its own rules for the conduct of its member firms and its requirements for companies who wish to have a market where their own securities can be traded. From the point of view of this chapter the important area is the activity of brokers and consultants in providing analyses of companies appropriate to the needs of institutions and others with responsibilities for the investment of funds.

All of these activities are pursued against an economic background, and economists, or perhaps econometricians, will be found advising these various institutions in the pursuit of their own, or their clients', requirements.

Within this network of inter-related topics, several are selected for more detailed description. However, in seeking to provide a more readable presentation, it should not be overlooked that the allocation of topics is to some extent arbitrary. Also no specific reference has been made to the statutory functions of accountants and actuaries, the former in regard to companies and the latter in regard to assurance companies and pension funds.

2. Accountancy

GENERAL

Finance involves two main streams of activity, which may be broadly described as reports for external users and reports for internal (management) users. The former arises since most forms of human activity involve stewardship and accountability to some external authority: the directors of a firm have a responsibility to their shareholders, the management of a club to its members and so on. This means that some measures have to be devised to provide succinct descriptions of the various activities involved. Most such measures are reduced to numerical terms, so that it is a relatively short step to use mathematical techniques as a means of interpretation and of discovering and using relationships between the various quantities involved, which is the second of the two main streams.

In most, if not all, circumstances there is a compression of substantial quantities of information so that properly relevant aspects can be distilled from largely irrelevant detail. The information also consists of two basic components: (a) measurable facts such as quantities of goods manufactured during a particular time period; and (b) estimated items such as the expected prices and the time when goods will be sold. The former is basically a counting process, although statistical techniques based on mathematical theory can be used to reduce the labour and cost of the counting process; the latter involves judgement. Thus the information is a mixture of objective and subjective components and this fact must be kept in mind when interpreting trends or other features. The presentation problem turns on the skill in estimating the proper weight to be given to the subjective components; mathematical techniques can be used to analyse past experience to identify significant factors which are efficient for projection purposes. Market research techniques are an example of the way in which these principles are applied in practice.

It is also of value to note that inherent in any operation are various time lags. There is an inevitable lag between identifying the existence of a demand for a commodity and the marketing and sales effort to meet a changed demand. In conditions of stable prices and wage costs, such time lags will not give rise to difficulties, but if, for example, the rate of inflation should be changing then serious errors could be made if the lag effects were ignored. As a corollary to this the traditional interval for reporting on enterprises is 1 year. This evolved when changes in the value of money were small and the rate of inflation no more than a few per cent. In these conditions yearly intervals were adequate for most presentational and management purposes. However, in the conditions of the UK, particularly in recent years, the rate of change is such that so much can happen between yearly reports that internal investigations must be made at much shorter intervals if management control is to be effective in anticipating changes; this gives rise, of course, to the need for much sharper tools for analysis, and the publication of quarterly reports is now commonplace. Complications are introduced by the fact that with yearly studies the seasonal patterns to which most businesses are subject can be conveniently averaged out, but with shorter intervals allowance must be made for such variations. In particular, the effect of the very high rates of inflation has highlighted the problems of presenting accounting information.

Complementing the detailed analysis of the various activities entering into a business operation is the management of the business as a whole, and in recent years considerable development has taken place on studies of the whole organization. In formulating the mathematical models to describe the inter-relationship between the component features, practical realism is introduced by means of equations which define constraints on the variables dictated by internal or external operational limitations. These techniques of linear, integer and dynamic programming are referred to in Chapter 3.

To illustrate some of these points simple examples covering sampling methods, time lags, linear programming and probability follow, which have already been mentioned in Chapter 3.

SAMPLING

Sampling methods in accounting were referred to above as a means of simplifying counting operations and can be divided into three techniques: (a) *estimation sampling*, by which the characteristics of a large group of items are determined within specific limits by the detailed examination of a selected sample; (b) *acceptance sampling*, by which the quality of a batch of items is measured by the estimated proportion of defective items in a selected sample; and (c) *discovery sampling*, in which the sample is so designed that it will contain, with a pre-assigned probability, one particular instance of a specified class.

Suppose that a manager is considering the position of the amounts due from 2000 clients and that he wishes to form an opinion of the proportion of the total amount due which has been outstanding for, say, 6 months. It is assumed that because certain records were accidentally destroyed, no summary of the time distribution of the individual balances exists, but that the total amount is available. He could, of course, at some considerable cost, arrange for each account to be analysed and the required information obtained by summing the results, but the situation hardly justifies this expense. He knows from experience that the accounts are fairly homogeneous and decides to take a 1 in 20 sample, which he selected by taking every twentieth account starting from a randomly selected number in the first 20. These cases are then analysed in some detail and, contemplating the possibility of further questions, the clerk concerned divides the overdue amounts in intervals of 3 months.

The following figures are obtained from the sample:

Time (months)	0–3	3–6	6–9	9–12	over 12	Total
Number of accounts	49	34	12	5	0	100

The sample shows that 17% of the cases have balances older than 6 months, so that an answer to the question posed is provided if we assume that the individual balances do not show a wide dispersion about their average. However, the sample is a random selection of 100 from the total of 2000, and other samples would be expected to show different results because of the effect of random sampling.

To complete the answer it is necessary to find some measure of the way in which the values from the sample will be spread around the true value, a problem well known in statistical theory. For this particular example a sufficiently accurate result can be obtained by use of the Normal probability function set out in Fig. 5.1. The area of the curve above the x-axis represents the total number of cases, the values for each case being measured from the over-all average in terms of the standard deviation (denoted by σ). This is found by taking the square root of the sum of the squares of the individual measurements from the average divided by the number of measurements. The mathematical form of this curve ($ke^{-x^2/2}$) is known and hence the proportion of the total number of cases which lie between any two values can be found. The natural scale gives the theoretical values and the adjusted scale is derived from the data.

Thus the Normal curve tells us that about 95% of cases will lie between 1·96 standard deviations either side of the mean. In this example the mean is 0·17 and the standard deviation can be taken as $\sqrt{0.17 \times 0.83 \div 100} = 0.0375$ so that it can be said that 95% of samples will give values between 0·0965 and 0·2435.* In other

---

*This is arrived at by using the formula: Estimated standard deviation $= \sqrt{p(1-p)/n}$ where $p =$ the proportion of occurrences of an event in question, $n =$ the number in the sample. This is also known as the standard error of the mean.

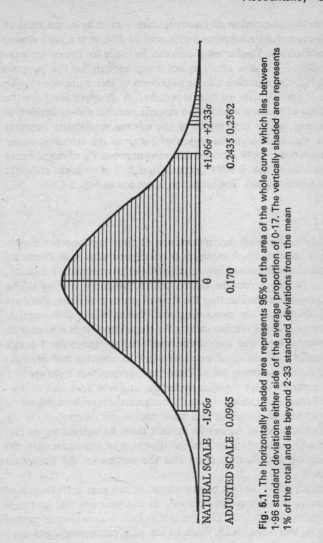

Fig. 5.1. The horizontally shaded area represents 95% of the area of the whole curve which lies between 1·96 standard deviations either side of the average proportion of 0·17. The vertically shaded area represents 1% of the total and lies beyond 2·33 standard deviations from the mean

NATURAL SCALE	0.0965	0.170	0.2435	0.2562
ADJUSTED SCALE	−1.96σ	0	+1.96σ	+2.33σ

words the proportion of balances over 6 months in the total of 2000 cases will lie between 9·65% and 24·35% with a high degree of probability. Similar estimates can be made for the other over-due periods. If this result is not close enough for the purpose required, the precision can be improved by increasing the sample size. It may be that all that is required is an upper limit for the purpose of making a provision against possible non-collection of balances; in this case tables of the normal probability function show that an upward deviation of 2·3 times the standard devia-tion will occur from random sampling in about 1% of cases. From this it follows that it is 99% certain that 25% of balances are more than 6 months old. The situation is set out in Fig. 5.1.

TIME LAGS

For an example of the importance of time lags another simple case is considered. A business is concerned with manufacturing articles from imported raw materials; the process takes 6 months, and a further 6 months elapse before the goods are sold. The conditions are ideal in that the flow of goods is regular, there are no hold-ups in the processing and the sales are also regular. Prices and wages remain steady. In these conditions it is easy to see that at any time the business will require space for 1 year's throughput, made up of one half work in progress and one half shelf stocks awaiting sales. Thus the business will have spent 1 year's purchase for the raw materials, and will have had to pro-vide space for storing 1 year's materials either as finished or partly finished goods. It will have had to obtain money (in the form of capital or loans) to finance these items and the selling price will have had to cover the purchase of raw materials, the processing and storage costs and the service on the borrowed money.

Suppose that the cost of raw materials in a year is 1000 units of currency, that processing charges in the year are 500 and that storage costs are 250. The net outgo before considering capital costs is, therefore, 1625, since in the year processing costs on 1 year's throughput will have been spent, but one quarter of this will have been recouped from sales of goods completed during the first half of the year. If we assume that the service of capital is $12\frac{1}{2}\%$ per year then the total capital required is

$$1625 \div (1 - 0 \cdot 125) = 1857$$

i.e., 1625 plus $12\frac{1}{2}\%$ on 1857.

The manufactured goods must be sold for $1750 + 232 = 1982$ units to balance the books.

Suppose now that the foreign country from whom the raw materials are bought decides to increase the price on 1 January by 50%. If all other conditions remain unchanged, the outgo becomes 2125, the capital required 2429 ($= 2125 \div 0 \cdot 875$), and the required sales $2250 + 304 = 2554$. However, if for some reason it is not possible to increase the sale price for 6 months so that the receipts from sales during the year amount to 2268 ($991 + 1277$) only, then the business will show a loss of 286 ($= 2554 - 2268$). It might be argued that the loss should be 214 only, i.e., the difference between the outgo of $2254 + 232$ and the income of 2268, but this would be fallacious as the company would have had to borrow this difference from some source, or carry it forward to be recovered in the future.

In the following year conditions are assumed to remain stable on the basis of the revised figures and, provided the capital has been provided, the income and outgo will again be in balance. However, the company has had to face two problems: (a) the increase in capital from 1857 to 2429; and (b) making good the loss of 286 which it incurred because of the lag before selling prices could be adjusted. In fixing the revised selling price based on sales of 2554, it is assumed that the additional capital could be provided at a cost of $12\frac{1}{2}\%$. So far as the loss is concerned the management have to decide how this is to be dealt with. If profits in earlier years have been retained they could be drawn upon or there may be other activities which are profitable, so providing an offset. In the long run, of course, the loss must be reflected in selling prices.

This example is, of course, a very simplified version of real life situations, but the essential principles remain. The development of more realistic models requires a separation between quantities and prices, assumptions regarding growth and the introduction of irregularities in the pattern of deliveries of raw material, processing times and the incidence of sales. The complex equations can be solved by modern techniques.

When the time of payment is involved, one of the earliest techniques was to introduce a rate of discount into the calculations, to reflect the fact that, as money earns interest, the present value of a sum due some time in the future is less than the value of a similar sum due immediately. This discounting technique has for many years been an integral feature of the valuation of assurance and annuity funds by actuaries, where the contracts are of a very long-term nature. However, when applied to the circumstances of an individual situation some care is needed if unsatisfactory results are to be avoided. For example, it may be decided that a suitable criterion for reaching decisions is the excess of the discounted value of income over that of outgo taken over, say, the next 10 years. If, however, there was an inherent time lag so that income was received later than the related outgo the net income during the early years might be negative, a feature which would not be apparent from the difference between the totals of the discounted values. A numerical illustration is given on p. 352.

CONSTRAINTS

In many practical problems constraints of various kinds may be involved and the linear programming technique, set out in Chapter 3, provides a convenient method of incorporating them into the description models. A further simple example may be of interest.

We consider a company with facilities for manufacturing three kinds of articles, A, B and C, each of which involves a fabrication process from raw materials, assumed to be in unlimited supply, and a finishing process. The available weekly installed machine capacity is assumed to be 60 hours for the fabrication process and

80 hours for the finishing process, and the hourly requirements for a unit of each kind of article are:

	Fabrication	Finishing
A	5	4
B	3	8
C	7	6

It is also known that the selling prices of the articles are £20, £28 and £36, respectively, these prices having been fixed to show the same percentage profit margin.

It is assumed that the company's objective is to obtain the largest profit, which will be secured in this case by maximizing the total sales, with regard to the limitations on productive capacity. This equivalence flows from the simplifying assumption that the selling prices have been fixed to show the same percentage profit. In practice the model should be extended to include fixed and variable costs and to bring in assumptions regarding labour utilization which it is normally worth-while to maximize. Although these extensions are needed for practical application the aim here is to set out in the simplest terms the principles of linear programming which have been described in some detail in Chapter 3.

If we denote by x_1, x_2 and x_3 the numbers of articles A, B and C, respectively, the conditions can be succinctly described by the expressions:

$$\left. \begin{array}{l} 5x_1 + 3x_2 + 7x_3 < 60 \\ 4x_1 + 8x_2 + 6x_3 < 80 \end{array} \right\} \text{Capacity limitations}$$

and a maximum for $20x_1 + 28x_2 + 36x_3$ Profit Index (objective function).

The two inequalities provide for a situation in which the maximum profit may be achieved without utilizing all the available machine capacity. We can formulate them as equations by introducing the two new variables y_1 and y_2 which represent the unused capacity, i.e.,

$$5x_1 + 3x_2 + 7x_3 + y_1 = 60 \tag{1}$$

$$4x_1 + 8x_2 + 6x_3 + y_2 = 80. \tag{2}$$

If the optimum solution (in terms of the 'profit' criterion) is such that y_1 or y_2 is not zero, it follows that there is idle capacity, and since idle capacity means that capital resources are not being fully utilized it implies that the model should be modified to reduce the profit function by a factor to represent the relevant cost.

The formal solution of this type of problem is by a process of successive approximation. First a set of values x_1, x_2 and x_3 is found which satisfies the inequalities, remembering that the quantities are essentially positive. Then, by considering the effect of small adjustments in the x's on the profit function, the optimum solution can be obtained.

So, considering equations (1) and (2) it is readily seen that if x_1, x_2 and x_3 are each taken as unity $y_1 = 45$ and $y_2 = 62$. Thus these values are a feasible solution, but the large values of unused capacity show that the solution is not very practical as the profit index is only 84. Another feasible solution is found by making $x_1 = x_2 = x_3 = 4$, in which case $y_1 = 0$ and $y_2 = 8$, and the profit index is £336.

Starting from this second case we note that if x_2 is increased by a small quantity δ the total fabrication time will be increased by 3δ so that, to keep within the total of 60 hours, x_1 and/or x_3 must be reduced. But as the profit from x_3 is larger than from x_1 we consider reducing x_1 by $3\delta/5$. The effect on equation (2) is to decrease the hours from x_1 by $4 \times 3\delta/5$ and increase the hours from x_2 by 8δ, i.e., a net increase of $5 \cdot 6\delta$ hours. If we then increase x_2 by $8 \div 5 \cdot 6 = 1 \cdot 429$ we have

$$x_1 = 3 \cdot 143, \qquad x_2 = 5 \cdot 429, \qquad x_3 = 4 \cdot 000$$

and
$$5x_1 + 3x_2 + 7x_3 = 60$$
$$4x_1 + 8x_2 + 6x_3 = 80$$

with
$$20x_1 + 28x_2 + 36x_3 = £359.$$

Capacity is now fully utilized but the profit may not be maximized. It is noted that the profit from a unit of x_3 is much greater than from x_1 so we consider the effect of eliminating production of x_1 entirely, and concentrating on x_2 and x_3. If we reduce x_1 by

3·143 units, we can increase x_3 by 2·245 units using the 60 hours' fabrication time, but find that this would require 80·902 hours' finishing time. We then adjust again and find the result:

$$x_1 = 0, \ x_2 = 5·263, \ x_3 = 6·316$$

with a profit of £375, which is a maximum.

Thus the optimum solution is reached by ceasing production of item A. Suppose, however, that marketing considerations require that a minimum of 2 units of A should be produced. Since x_1 produces the lowest contribution to profit the solution will only require the minimum, so we can set up the equations as

$$3x_2 + 7x_3 = 50$$
$$8x_2 + 6x_3 = 72$$

using the maximum resources. In this case, with only two unknowns, an exact answer can be found as $x_1 = 2·000$, $x_2 = 5·369$ and $x_3 = 4·842$ with a profit figure of £365. The reduction of £10 in the profit arises from a gain of £40 on type A, of £3 on B and a drop of £53 on type C, the pattern of these being dictated by the time constraints on the fabrication and finishing processes.

This example illustrates the nature of the problems towards the solution of which considerable effort has been devoted in recent years. It is a natural development from the principles of cost analysis which could have been used as a basis for further illustrations. Equally the example could have been generalized by specific treatment of the capital aspects or of the various refinements that can be introduced; particularly developments in statistical techniques. It will not have escaped notice that the values of x_1, x_2 and x_3 found involve fractions, which are incompatible with the production of unit articles. This blemish can be removed by a technique known as integer programming (see Chapter 3), in which the variables are constrained to be whole numbers. In practice the numbers will usually be large, so that the effect of the fractions may be unimportant and the above solution will be adequate.

PROBABILITY

Probability can be introduced in various ways, but the ideas involved can be illustrated by considering the situation of a company wishing to make plans for future production facilities and having to make some assumptions about the level of future sales. From its records the company will have information about the past and may have found that significant fluctuations are related to factors external to the company. It could make a projection for the future based on the average of the past figures and thus make the implicit assumption that the external factors operate as in the past. It could, however, analyse the effect of changes in the external factors, associating a probability to each of the alternatives, thus leading to a set of projections that would cluster round the average value. Knowledge of this spread would enable a more realistic approach to be made to its future production problems.

As a simple example, consider an ice cream merchant contemplating a short-period campaign extending over 6 days. Records show that the chance of a fine day is 0·4 and a 'non-fine' day is 0·6. On a fine day 300 units could be sold, but only 150 on a 'non-fine' day. He can only make a fixed number of units each day, and units not sold have no value. Production costs per day are £100,

TABLE 5.1

x	Cost/day	Unit sales Fine	Non-fine	Expected sales per day	Expected profit per day
200–300	x	x	150	1·30 (0·4x +90)	117 − 0·48x
150–200	100 + 0·5x	x	150	1·30 (0·4x +90)	0·02x + 17
0–150	100 + 0·5x	x	x	1·3x	0·8x − 100

plus £0·5 for each unit, plus a further £0·5 for each unit in excess of 200. Each unit sold yields £1·3.

If we denote the daily production of units by x we can set down the scheme shown in Table 5.1. The expected sales are based on the expected number of fine and non-fine days and it is readily seen that the expected profit is a maximum of £21 per day when $x = 200$. The situation is set out in Fig. 5.2.

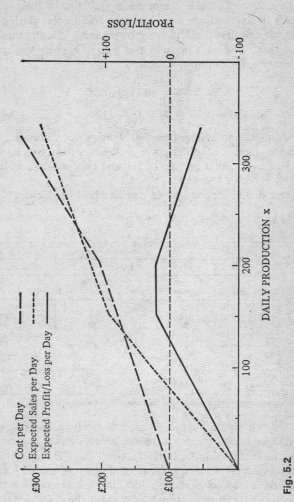

Fig. 5.2

If now the position is examined in the light of the various numbers of fine and non-fine days that could occur during the period, we can first find the expected pattern by noting that the chance that r fine days occur out of 6 days is given by the expression (see, for example, Chapter 2)

$$\frac{6!}{r!(6-r)!}(0{\cdot}4)^r(0{\cdot}6)^{6-r}$$

('r!' means '$r \times (r-1) \times (r-2) \times \ldots \times 2 \times 1$')

on the simplifying assumption that there is no relationship between the chances on different days. We can then associate the chance of a particular pattern of fine and non-fine days with the resulting profit or loss as in Table 5.2.

The maximum expected profit is given with a production of 200

TABLE 5.2

No. of fine days	Chance	Daily production					
		250 units		200 units		150 units	
		Result	Expected profit	Result	Expected profit	Result	Expected profit
6	0·004	450	2	360	1	120	—
5	0·037	320	12	295	11	120	5
4	0·138	190	26	230	32	120	17
3	0·276	60	17	165	46	120	33
2	0·311	−70	−22	100	31	120	37
1	0·187	−200	−37	35	6	120	22
0	0·047	−330	−16	−30	−1	120	6
Total expected profit for 6 days			−18		+126		+120

The total expected profits agree with the formula in the preceding paragraph.

units per day, but should there be no fine day (which will happen, on average, 1 in 20 times) there will be an over-all loss on the operation. In this example, the loss is small, but it should be noted that a production of 150 units a day will generate a profit regardless of the outcome of the weather, in many respects a more satisfactory proposition. It may be possible, however, for insurance to be obtained to compensate for the adverse result of 6 non-fine days, so that the choice of the number of units to be produced would then depend on the cost of such protection. If this proposition were being considered as a basis for financial help, clearly the security is greatly improved if the unprofitable days could be eliminated.

3. Actuarial Work

GENERAL

Among the group of activities covered by this chapter the work of actuaries has long been recognized as involving a significant mathematical content, the basic reason being the long-term nature of contracts of life assurance and annuities. Much of commerce and industry is built on short-term contracts or agreements between firms and customers, so that external events which may influence business relations can be quickly reflected by pricing changes. On the other hand, the long-term contracts of life assurance operations are renewable at the option of the customers so that pricing changes on existing contracts cannot be used to counter-balance external changes which are adverse to the company and favourable to the policyholder.

LIFE TABLE

The history of insurance operations can be traced back some thousands of years, but the application of mathematical, as distinct from accounting, methods started with the concept of the life table introduced in 1662 by John Graunt, FRS. In its simplest form the life table sets out the survival pattern of a group of lives of the same initial starting age. It is conventional to adopt a convenient number for the number of lives at age 0, to calculate the expected number of deaths during the next year of age, and thus to find the expected survivors at age 1. The process is then repeated until there are no survivors.

The broad pattern of the variation in the rate of mortality according to age is a rapid decrease from age 0 to a minimum around age 10 and, thereafter, a steady rise until the extreme age of rather more than 100 years. However, the level of mortality depends on factors other than age; for example, important variations are shown between geographical regions, partly related to social and economic factors. In particular the experience of persons taking out insurance policies is more favourable than that of the total population. The Institute of Actuaries and Faculty of Actuaries (in Scotland) have for many years compiled data for insured lives and prepared mortality tables for insurance (and pensions) purposes. The latest of such tables relate to the mortality for the period 1967–70, and the rates for ages 60–65 have been used to illustrate the basic calculations for premiums and the principles underlying actuarial mathematics.

The essential features of a life table are set out in Table 5.3. The rate of mortality is the probability that a person aged x will die before reaching $x + 1$. Thus if we start with 100 000 persons aged 60 then, on the basis of the table, 1471 will die before age 61. The remainder, 98 529 will be the survivors at age 61 and so on. From the above table the total numbers of deaths from age 60 to 64 is 8799, so that the probability that a person aged 60 dies before reaching age 65 is simply $8799/100\,000 = 0.08799$, and the complement of this, i.e., 0·91201, is the probability of survival to age 65. This is, of course, equal to 91 201/100 000 or l_{65}/l_{60}.

TABLE 5.3

Age	Rate of mortality (q_x)	Number living (l_x)	Number of deaths (d_x)
.	.	.	.
.	.	.	.
.	.	.	.
60	0·01471	100 000	1471
61	0·01632	98 529	1608
62	0·01808	96 921	1752
63	0·02001	95 169	1904
64	0·02213	93 265	2064
65	.	91 201	.
.	.	.	.
.	.	.	.
.	.	.	.

PREMIUMS

Suppose we now wish to calculate the annual premium for an assurance on a person aged 60 to pay £1000 on death before age 65 and £1000 on survival to that age (this is commonly described as an endowment assurance policy). It is clear from the life table that the major part of the premiums will be required to provide the survival benefit and thus the company will be in a position to earn interest on the premiums. If it is assumed that money earns interest at $i\%$ per annum then the present value of a unit due 1 year hence will be $v = 1/(1 + i)$, 2 years hence $v^2 = 1/(1 + i)^2$, etc. It is then a simple matter to calculate the present value of the expected premiums by combining the probability that the person will be alive to pay the premiums with the appropriate present values, i.e., to discount the expected payments. Similarly, the claims outgo can be discounted. By equating these two expectations the required premium can be found.

The calculations, based on an interest rate $i = 4\%$, are shown in Table 5.4. It is assumed that premiums are payable at the

TABLE 5.4

t	$v^t = 1/(1 + i)^t$	$v^t l_{60 + t}$	$v^{t + 1} d_{60 + t}$
0	1·00000	100 000	1414
1	0·96154	94 740	1487
2	0·92456	89 609	1558
3	0·88900	84 605	1628
4	0·85480	79 723	1696
5	0·82193	74 961	

beginning of the year, and that claims are paid at the end of the year. Thus the value of expected premiums of P per annum at the outset of the policy in respect of 100 000 persons is

$$\{l_{60} + v l_{61} + \ldots + v^4 l_{64}\} \times P = 448\ 677P$$

and the value of the expected claims

$$\{v d_{60} + v^2 d_{61} + \ldots + v^5 d_{64} + v^5 l_{65}\} \times 1000 = 82\ 744 \times 1000.$$

Hence $P = £184·42$.

This premium can also be separated into £17·35 (the annual premium to provide for the death claims), and £167·07 (the annual premium to provide for the claims on the survivors to age 65).

This, however, illustrates only the principle of calculation, as in practice it is necessary to include in the calculations management expenses to allow for their heavier incidence in the early years of the policy, the incidence of taxation, safety margins and, for with-profit policies, an allowance for future bonuses. The appropriate allowances for these items are based on analysis of the operations of the company, on opinions regarding the expected mortality experience and the interest earnings, all of which may be relevant over the period of the policies, some of which can persist for 80 years or more.

RESERVES

Now consider the way in which the finances of the policies described on p. 347 develop. Assuming that the experience follows the life table, we can calculate the flow of premiums and claims in each year and add in the interest earnings, remembering that premiums are assumed to be received at the beginning of each year and the claims paid at the end (Table 5.5).

TABLE 5.5

	Year				
	1	2	3	4	5
Premiums	18 442	18 170	17 874	17 551	17 200
Claims (death)	1471	1608	1752	1904	2064
Claims (maturity)	–	–	–	–	91 201
Net	16 971	16 562	16 122	15 647	− 76 065
Interest	738	1436	2143	2860	3586
Net flow	17 709	17 998	18 265	18 507	− 72 479
Cumulative flow	17 709	35 707	53 972	72 479	0

(Figures in 000's)

The cumulative flow represents the reserves held by the company to meet future claims and just amounts to the sum required to meet the maturity claim at the end of 5 years. The reserves at the end of each year can be expressed mathematically in terms of the life table functions, and actuarial techniques have been developed to facilitate these calculations in respect of the many types of policy on the books at any instant.

If we now consider the policy as made up of the two components, i.e., death and maturity, we can separate the reserve into two parts and obtain the figures shown in Table 5.6. The proportion

TABLE 5.6

	Year				
	1	2	3	4	5
Reserve for death claims	334	518	535	369	0
Reserve for maturity claims	17 375	35 189	53 437	72 110	91 201
TOTAL	17 709	35 707	53 972	72 479	91 201

of the reserves required for the death claims is quite small, by far the larger proportion being needed for the maturity claims. The traditional operations of life assurance companies are based on careful investment of the policy reserves to produce at least the interest earnings implied by the premium calculations and also to provide the cash at maturity required to meet the claims. Policies are expressed in money terms so that in principle investments have to be sold for money to pay the claims. If the interest earnings fall short of the premium basis, or if the investment shows a loss, then the proceeds would fall short of the amounts guaranteed under policies and the shortfall would have to be met from other resources within the company. In fact over many years the earnings have exceeded the premium basis, so that more than enough has been available at maturity, and companies have distributed this to policyholders in the form of bonus payments.

In practice, of course, the position is not so clear cut, since the total business consists of many policies of different types, and investments are not earmarked against individual policies.

EQUITY-LINKED CONTRACTS

About fifteen years ago there was a movement to modify the traditional operations to give policyholders contracts under which the benefits, instead of being expressed in money terms, were linked to the value of the securities in which the reserves were invested. The requirements for death claims are relatively small. In the above example about 10% of the premiums are required for this purpose, the balance being used for investment towards the maturity benefits. Effectively, this means that the fluctuations in investment proceeds are passed direct to the policyholders, so that if there is a pronounced fall near maturity, their policy proceeds may be well below the money value of a corresponding traditional policy. Equally, if there happens to be a boom, their proceeds will be greater. However, in order to make the contracts more attractive, some companies offered a guarantee that the maturity proceeds would not fall below the nominal amount of the contract (say the total premiums paid), and one of the interesting questions then arising was the amount that should be charged for this guarantee. Since the future rate of return on investments, including any capital gains or losses, is unknown the only way to provide meaningful answers is to establish from past records the pattern of variation of the out-turn of investments, and from this to estimate the chances of the different outcomes, similar to the method used on p. 344. The required premium can then be found.

DISCOUNTED CASH FLOW

The calculations on p. 348, referred to as 'emerging costs' in actuarial terminology, provide some insight into the special features of actuarial funding, either for life assurance or pension funds. The pattern for the contract illustrated is of an excess of income in the early years, followed by a net outgo in the later years. When the over-all position of a growing company is considered, the usual situation is that the current outgo is more than covered by current income, so that should the prices of investments held fall, there is no urgency to sell, and thus market values have only a minor significance in a balance sheet drawn up in traditional form. The reality is the flow of income and outgo so that active insurance funds with traditional business can survive fluctuations in investment conditions, since they have some scope for timing their investment activity. In fact, it can be shown that in certain circumstances, if there is a sufficient supply of fixed interest securities redeemable at various dates in the future, a portfolio of investments can be selected which just matches the emerging costs, and the operation is then immunized against market fluctuations.

One of the basic uses of a discounted model in life assurance and annuity funds is to provide a standard against which to compare the actual experience, and thus to provide warnings of developing trends so that corrective measures can be applied before the financial position of the fund becomes affected. This differs from the use of discounted cash flow methods in appraisal of business projects where the period is relatively short and the object is to use rates of interest which are realistic in the short term and in relation to the financial activities of the company concerned. We can, however, use the actuarial emerging costs to illustrate these latter principles. From the figures on p. 347 the net flow assumed to occur at the beginning of each year can be calculated and discounted to the starting time. The commercial situation is that there is an outward flow of expenditure on projects followed by the inward flow of income from the projects. The operation is thus the mirror image of the insurance situation and the same basic calculations can be used provided it is remembered that a positive net flow in the commercial case corresponds to a negative net flow in the insurance case and *vice versa*. Calculations at three rates of interest are shown in Table 5.7. If these figures represented outgo in years 1 to 5 and income in year 6, then at 4% interest the two items are just in balance. Column 1 corresponds to an interest rate of 0%, and the position is that there will be a positive return. However, at interest rates higher than 4% the balance becomes negative, i.e., the value of the expected outgo is greater than the expected income. It will be

TABLE 5.7

		Discounted value		
Year	Net flow	4%	6%	8%
1	18 442	18 442	18 442	18 442
2	16 699	16 057	15 754	15 462
3	16 266	15 040	14 477	13 945
4	15 799	14 045	13 265	12 542
5	15 296	13 074	12 116	11 243
6	− 93 265	− 76 658	− 69 693	− 63 474
Total	− 10 763	0	+ 4 361	+ 8 160

apparent that the use of interest in the calculations, generally referred to as discounted cash flow, could be of significant importance in appraising the relative merits of different projects where the patterns of income and outgo differ.

RUIN THEORY

Early in the 1900s Scandinavian actuaries spent a considerable amount of time in developing an approach to insurance company financial management, based on treating the operation as a collective. They considered the flow of premiums and claims and, building on the fact that the number and amounts of claims are random events, built models which gave numerical values for the reserves that should be held to cover the random fluctuations. These models were quite general as regards the type of insurance but, so far as conditions in the UK life business were concerned, found little application as they covered only a small part of the actual fluctuations. However, from about 1950 onwards it has become increasingly apparent that this collective risk approach is very suitable for non-life insurance and development has been rapid, with actuaries becoming increasingly involved as the techniques became more familiar. The problems of the linked contracts can be more conveniently approached from a collective point of view, with the proviso that it becomes necessary to build the models round the concept of random variation in the value of assets in addition to the random variation in the liabilities.

If we consider the operation of an insurance company over an accounting period, say 1 year, the company is assumed to start the year with resources, i.e., capital and free reserves, equal to U_0. During the year it will receive a flow of premiums in respect of policies exposed to risk of claims during the year. If we denote the premiums for a year as P and assume the flow is uniform, the accumulated resources at time t will be $U_0 + Pt$, reaching $U_0 + P$ at the end of the year. During the year claims will arise on the policies and if the total of claims notified (and paid) up to time t is denoted by $x(t)$, the net resources at time t will be $U_t = U_0 + Pt - x(t)$ and $U_1 = U_0 + P - x(1)$ at the end of the year. If the net resources at any time become negative, i.e., if $x(t) > U_0 + Pt$, the company would be unable to pay its debts and is described as ruined. One problem of risk theory is to calculate the probability that ruin occurs during the year, or the complementary probability that the company is not ruined. Fig. 5.3 sets out the position.

The solution of the problem requires advanced mathematics, but if a simplified case is considered in which ruin occurring during the year is ignored, i.e., it is assumed that the claims are all paid at the end of the year, whenever they occur, it is possible to illustrate the principles involved. It will be appreciated that the ruin probability found for this simplified model will be less than the 'true' value and may, therefore, be unsatisfactory for practical use.

In this simpler model, the resources at the end of the year before payment of the claims will be $U_1 = U_0 + P$. If the claims $x(1)$ amount to less than this, then $U_1 = U_0 + P - x(1)$ will be positive. Ruin occurs if the claims exceed $U_0 + P$. Thus the problem can be solved if it is possible to calculate the probability that the total claims $x(1)$ do not exceed a given value Z, where Z can take all values from 0 to the maximum that would be paid if all policies gave rise to total claims during the year. It should be noted that the policies being considered are indemnity policies,

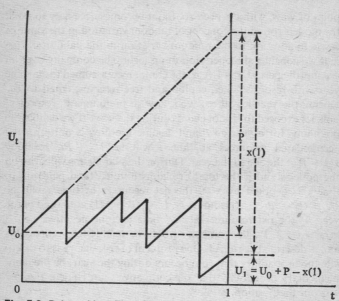

Fig. 5.3. Ruin problem. The initial resources U_0 are supplemented by the flow of premiums and depleted at random intervals by the claims

i.e., the amount of a claim is limited to the damage suffered by the policyholder or the sum insured, whichever is the smaller. This implies that the amount of any particular claim that may be incurred in the future cannot be forecast, but, fortunately, statistical analysis has shown that the distribution pattern of claims by amounts of payment is sensibly stable in time, provided corrections are made to eliminate the effect of changes in the value of money and other inflationary influences.

Suppose now that n claims have arisen during the year. If it is assumed that each claim can be regarded as having arisen from a common distribution, statistical theory tells us that the expected total of these claims will be $n \times m$, where m is the average value from the common distribution and the variance will be $n\mu_2$ where μ_2 is the variance of the common distribution (see for example Lewis and Fox, p. 67, or Moroney, Chapter 5). We do not, of course, know how many claims will arise during a year but we can use the so called Poisson distribution (see for example Lewis and Fox, p. 167, or Moroney, Chapter 8) to determine the probability that exactly n claims will arise during the year as

$$p_n = \frac{(\lambda N)^n e^{-\lambda N}}{n!}$$

where λN is the expected (i.e., average) number of claims that will arise during the year and e^x is the exponential function.

If we now combine these probabilities with the expected means and variances for $0, 1 \ldots n$ claims we can find the means and variance of the total expected claims in the year.

Thus, expected value of

$$
\begin{aligned}
x(1) &= \Sigma \frac{(\lambda N)^n e^{-\lambda N}}{n!} \, n \,.\, m \\
&= \frac{(\lambda N) e^{-\lambda N}}{1!} 1 \,.\, m + \frac{(\lambda N)^2 e^{-\lambda N}}{2!} 2 \,.\, m + \ldots \\
&= \lambda N m e^{-\lambda N} \left\{ 1 + \frac{\lambda N}{1!} + \frac{(\lambda N)^2}{2!} + . \right\} \\
&= \lambda N m e^{-\lambda N} e^{\lambda N} \\
&= \lambda N m
\end{aligned}
$$

and variance of $x(1)$

$$
\begin{aligned}
&= \Sigma \frac{(\lambda N)^n e^{-\lambda N}}{n!} \left\{ nm^2 + n(n-1)m^2 \right\} - (\lambda N)^2 m^2 \\
&= \lambda N m^2 .
\end{aligned}
$$

We now have the mean and variance of the distribution of total claims and, provided λN is fairly large and the shape of the individual claim distribution is such that m^2 is of similar magnitude to m, can use tables of the Normal probability function to determine the probability that the total claims expected in a year do not exceed any given value, and can thus determine the required ruin probabilities.

Should the ruin probability, i.e., the risk of insolvency, be considered too large, this simple model shows that various corrective measures can be taken. Consideration of Fig. 5.3 shows that, if the total resources at the end of the year can be increased, then the ruin probability will be reduced. Thus, U_0 can be increased, i.e., more capital provided, or P may be increased, i.e., premiums raised. Alternatively arrangements could be made with a re-insurance company which would have the effect of reducing the impact of the larger claims and thus reducing the value of m^2 leading to a reduction in the ruin probability.

The wide extent of mathematical application by actuaries is only partly apparent from published sources and it is, therefore, difficult to speak with authority about new techniques being applied. Certainly techniques required in the management of the administrative side of the insurance operations are in daily use and the need to invest the large funds held on behalf of policy-holders means that the methods of appraising the performances of companies of all kinds are utilized, quite apart from the special aspects of the financial management of the life assurance and pensions funds and, to a growing extent, the non-life operations.

4. Banking

The broad heading banking covers a wide variety of activities, all of which are ultimately expressed in numerical terms and thus, in principle, provide scope for the application of mathematical techniques.

It is, perhaps, convenient to make a separation into those activities which involve the internal organization of the banks and those which arise from the operations of banks in their functions either as central bankers, or in the provision of financial resources for business and private individuals.

In developed communities banks will employ many thousands of staff distributed over a widely dispersed branch organization. The management of these organizations will give rise to many problems, for the solution of which mathematical techniques have been developed over recent years. For example, the proper planning of staff requirements can be materially helped by the models developed for graded populations. A major problem is one of communication between branches and the central office, and some help in the determination of the best method, in the sense of balancing costs and customer service, of dealing with the many thousands of daily transactions may be found from models developed in the communications field.

However, these internal problems are essentially similar to those arising from business operations generally, so it is not necessary to elaborate on them and the subsequent emphasis will be on the financial rather than administrative aspects.

The functions of central banks in regulating the flow of funds and the availability of credit in the community can be looked upon as decision-making processes which involve judgement elements, such as overseas political developments which are hardly amenable to the application of mathematics. The detailed way in which finance is made available for the conduct of commercial and industrial businesses, as well as the financial requirements of the personal sector, depend on the structure of the banking industry and thus vary from country to country. However, some of the common denominators are (a) the need for some appraisal methods for determining the form and extent of loan and/or credit facilities of individual enterprises; (b) similar criteria for private individuals; (c) the provision of advice regarding investment in private enterprise, for purposes such as pension scheme finance; and (d) the maintenance of a proper balance between deposits and advances to avoid liquidity or solvency problems. A few comments will be made on each of these.

Reduced to the essentials the bank accepts money from its depositors, for which it pays interest, and lends money to its customers for their business activities. The amount it charges for such advances has to cover its expenses and interest in respect of the deposits. Thus in regard to (a) above it will have a good idea of the cost of providing finance, but each case presented will have to be examined to see whether the enterprise is likely to be able to meet the service on the loan and also to repay the loan at an agreed date.

This means that an examination must be made of the whole operation of the enterprise to determine whether it is being operated on a basis which ensures that the income from its activities is sufficient to meet all the outgo, including in the outgo the costs in respect of the capital it has borrowed to build its factories or to finance work in progress. The analysis must, therefore, be made on a flow basis, as the very simple example on p. 337 shows, and the techniques developed in recent years have been directed towards this end.

The situation in regard to the determination of credit facilities for individual customers has parallels with the problem outlined later (p. 362). The quality of an account depends on a number of factors related to the personal circumstances of a customer, some of which can be 'measured', e.g., marital status or stability of employment, but some of which cannot. By analysis of the customer's credit experience in relation to the measurable factors a pattern can be established and the problem of personal appraisal, which must always be involved, simplified.

Item (c), advice on investment matters, arises in various ways, and, of course, is not restricted to the banking field since actuaries, accountants, stockbrokers and others offer such advice. The common denominator is the analysis of particular investments in the light of the personal circumstances of the customer or, in the case of funds, the selection of a portfolio of investments suited to the requirements of the particular fund. The former offers little scope for formal mathematics, but there has been considerable activity in the latter field which is dealt with later (p. 375).

As mentioned earlier, the essential operations of a bank consist in borrowing money from customers on a short-term basis and using this money in investment in industrial and commercial concerns involved in various aspects of trade. Since customers may wish to use their money from time to time, the financial management involves a nice balance between the relative movement in the customers' requirements and the investments in which the monies are placed. There will be a normal pattern in the net over-all movement of the deposits so that a pattern can be established for the amount of money which should be readily available to meet the daily fluctuations. There will, however, be longer-term and seasonable variations in the demands by customers for their deposits and thus a need for readily available cash to match these demands. The way in which the monies are invested must have regard to these fluctuations so that, for example, repayments of loans do match the expected requirements.

Since uninvested money will earn no profit, it is an essential part of the operation that the liquidity, i.e., the free cash, is maintained at as low a level as practicable, it being remembered that, should an unanticipated demand develop, it will be necessary to realize investments which may then result in a loss. While it is theoretically possible to develop mathematical models to describe the situation, many of the variables involved relate to future conditions which are subject to political and other considerations and thus not susceptible to precise numerical evaluation. However, as in the operations of insurance companies which also have a problem in matching the pattern and flow of assets and liabilities and where models built on probability patterns are being developed to provide quantitative guidance in management, so scope for similar models exists in the banking field.

5. Econometrics

GENERAL

Econometrics is the statistical and mathematical analysis of economic relationships, so in a certain sense the other activities discussed in this chapter could be regarded as included within its boundaries. To illustrate some of the principles, two examples from non-life insurance have been selected, although these particular developments have been largely associated with extensions of traditional actuarial methods.

Certainly the economist or econometrician advising on the day-to-day problems of business is close to the consequences of the results of decisions which may be based to a greater or lesser degree on his analysis. He is thus made aware of situations where the results diverge from the expectations upon which the decisions were made. Since business decisions have, in principle, to be expressed in numerical terms and are determined by a selection from alternative courses of action judged against some criterion, lack of success could result from bad judgement regarding the relative importance of different factors, or from their inter-relationship or perhaps the complete omission of significant factors.

There are basically three aspects involved in any scientific activity, namely: (a) the formulation of a hypothesis regarding the relationships between observed data, such as the rate of growth of the national income and the level of employment; (b) the collection of statistics relative to the hypothesis and the expression of the hypothesis in concise or mathematical terms; and (c) modification or improvement of the hypothesis.

However, even though a particular mathematical form may be found to represent the numerical relationship between observed quantities, this may be no more than a concise description of the particular statistical data and use of the form outside the range of the data may lead to erroneous conclusions. The point can be stated explicitly in that a statistical relationship between quantities is a limited association and is not in any sense a causal relationship, which can only be established by an objective analysis of the way in which the relationship between the particular quantities arises.

The caution in the preceding paragraph is essential as so many economic data are basically a complex of many variable factors, few of which can be measured independently of each other. A simple example will fix the ideas. Suppose we are told that the sales of a small firm have been steady at 200 for each of the past 3 years and that profits have been 15%, $17\frac{1}{2}$% and 20%. A statistical forecast for the next year would be $22\frac{1}{2}$%. However, the situation is that the business consists of types of goods with different rates of profit as shown in Table 5.8. Since the increase in the rate of profit has arisen for a reduction of good A balanced by an increase of good B, the profit for the next year, assuming the total sales are 200, is more likely to be 20% than $22\frac{1}{2}$%.

Many economic models are built up on elaborate statistical analyses (regression analyses) in which the observed values of a quantity y are related in a linear or more complicated manner to

TABLE 5.8

| | Year | | | | | |
| | 1 | | 2 | | 3 | |
	Sales	Profit	Sales	Profit	Sales	Profit
Good A	100	10	50	5	0	0
Good B	100	20	150	30	200	40
Total	200	30	200	35	200	40
Rate		15%		$17\frac{1}{2}\%$		20%

many other observed quantities x_r; in symbols, equations of the form

$$y = a_1x_1 + a_2x_2 + \ldots + \delta$$

where δ is a residual which sweeps in all the differences between the y's and the set of x's, the coefficients a_r being found from observations. If there are unsuspected complications like that outlined in the preceding paragraph, then to use this expression as a basis for projection could be equally misleading. Since statistical models may be made extremely sophisticated and involve extensive computer-based calculations based on many variables the analyst must always be alert regarding the potential dangers.

MULTIFACTOR ANALYSIS

A practical example that brings in a number of mathematical applications is provided by the premium structure of motor insurance. The basic problem arises from the complex manner in which the claim cost depends on many variable factors, a number of which cannot be conveniently measured. Thus it is apparent that an important factor is that of annual mileage, but there are considerable practical difficulties in using this as a rating factor. The practical solution is to find a number of readily observed factors, such as the power of the car measured in terms of horse power per ton, age (of policyholder and car), geographical location, use, each of which has some relationship to the true risk which can be measured by the actual claims cost. Suppose, however, that five such factors were used, each of which was measured at five levels. This would give rise to $5^5 = 3125$ separate rating categories, and if only 100 000 cars were insured the average number of cars in each group would be 32. Since the chance of a claim is of the order of 10% per annum, the experience in each rating group would be subject to very large fluctuations from random sampling errors, and it would be very difficult to assess the true relationship between the experience of the various groups. Furthermore, administration with this large number of separate groups would be complicated and expensive.

To overcome these complications, rating structures based on finding the relative importance of each factor on its own, i.e., eliminating the relationship between the factors, have been developed. The rating formula can then be built on an additive (or a multiplicative) model in which points are allotted to each factor, and the aggregate of the points for a given case gives the relative risk level. In this way, the number of separate classifications in the above instance would be reduced from 3125 to 25, a very significant saving in work.

The principles involved are illustrated by the following hypothetical example involving two risk factors. It is assumed that the investigation is in respect of 10 000 policies which have been classified at three levels for each of the two risk factors. Each policy is assumed to be at risk for a year and the numbers of claims arising during the year of observation have been recorded, but only the analysis of the exposure according to the separate risk factors is available, i.e., the two-way classification is not available for the claims.

It is also assumed that the average cost per claim is the same for all cases so that the analysis may be made in terms of the claim rate only, i.e., the ratio of the number of claims to the number exposed to the risk. The data are shown in Table 5.9.

An additive model is required, i.e., one in which the claim rate Q_{rs} for level r in factor A and level s in factor B is represented by the sum of three components, namely, a constant m, independent of the risk factors; an amount a_r solely dependent on factor

TABLE 5.9

| | | Exposed to risk (car years) | | | | Number | Claim |
	Level	1	2	3	Total	of claims	rate
Risk	1	400	300	1600	2300	547	0·23783
factor	2	200	400	1980	2580	415	0·16085
A	3	100	600	4420	5120	538	0·10508
Total		700	1300	8000	10 000		
Number of claims		184	225	1091		1500	
Claim rate		0·26286	0·17308	0·13638			0·15000

A; and an amount b_s solely dependent on factor B. Symbolically

$$Q_{rs} = m + a_r + b_s \ (r, s = 1, 2, 3).$$

The problem is to determine the 'best' values of a_r and b_s using as a criterion the claims expected from the exposed to risk in the various subgroups.

The marginal totals of actual claims at the various risk levels provide six pieces of information, but since the totals for each risk factor are the same the effective number is only five. The model involves seven unknowns, three for each factor and a constant. We can however reduce the number of unknowns by making m equal to the over-all claim rate (0·15). It may also be noted that if the term a_r be increased by a constant amount δ, and b_s be decreased by a similar quantity, the values of Q_{rs} will not be altered. Hence if we assume that a_1, say, is zero we are left with five unknowns and can obtain a solution. Another criterion will then be needed to fix a value for δ and a suitable basis is to require that the total expected claims according to each risk factor is zero – it will be seen that by making the constant factor m equal to the over-all claim rate, the total expected claims from the a and b factors will be zero.

Thus the first equation between the factors is

$$2300m + 2300a_1 + 400b_1 + 300b_2 + 1600b_3 = 547.$$

Putting $m = 0.15$ and $a_1 = 0$ gives

$$400b_1 + 300b_2 + 1600b_3 = 202.$$

The following equations are derived similarly:

$$2580a_2 + 200b_1 + 400b_2 + 1980b_3 = 28$$
$$ 5120a_3 + 100b_1 + 600b_2 + 4420b_3 = -230$$
$$200a_2 + 100a_3 + 700b_1 = 79$$
$$400a_2 + 600a_3 + 1300b_2 = 30$$

The solution of this set of linear equations is straightforward and leads to the values:

$$a_2 = -0.07, \; a_3 = -0.12; \; b_1 = 0.15, \; b_2 = 0.10, \; b_3 = 0.07.$$

To find the value of δ, the expected claims according to factor A are $2580 \times -0.07 + 5120 \times -0.12 = -795.0$, and according to factor B, $700 \times 0.15 + 1300 \times 0.10 + 8000 \times 0.07 = 795.0$. The total exposed to risk is $10\,000$ so that we adjust a_r by $+0.08$ and b_s by -0.08 leading to the final result

$$m = 0.15; \; a_1 = 0.08, \; a_2 = 0.01, \; a_3 = -0.04; \; b_1 = 0.07,$$
$$b_2 = 0.02, \; b_3 = -0.01.$$

The corresponding values from the marginal distributions are

$$m = 0.15; \; a_1 = 0.088, \; a_2 = 0.011, \; a_3 = -0.045; \; b_1 = 0.113,$$
$$b_2 = 0.023, \; b_3 = -0.014,$$

the value of b_1 in particular being markedly different.

When many factors with different numbers of levels are involved the differences between the relativities calculated as above and the figures derived from the marginal totals may become very large arising from inter-relationships between the factors. Unless these are eliminated as above it is likely that a rating structure would be dangerous in practical conditions of intensive competition. For these more complicated situations the solution of the equations requires a computer, but the principles are the same.

One company has made calculations for eight risk factors with the following numbers of levels for each factor.

	Number of levels
Age of policyholder	15
Rating district	6
Age of car	8
No-claim discount category	7
Car rating group	5
Use	3
Standard/Non-standard	2
Voluntary excess	4
Total	50

A system of points is allotted to each factor and the total number of points for a given case is used to fix the level of premium to be charged. This reduces the premium calculations to a series of simple additions and references to a single simple scale. Since the claims from motor insurance are highly linked to the rate of inflation and it has been found that the relativities are stable, premium revision reduces to the adjustment of the single scale, a very important saving in costs.

Reference has been made (p. 358) to an application in banking in which a closely allied method is used. Another interesting variation is in regard to the derivation of criteria for testing the financial situation of US non-life insurance companies. Records for several companies which ran into financial difficulties were available and also for a sample of companies which had not failed. Values of a large number of different criteria were calculated for the companies in each of the two groups. For each criterion the over-all average and the averages for companies in the 'distressed' group and for the solvent companies were calculated.

In respect of each criterion companies were classified as 'distressed' or solvent depending on whether their value fell the same side of the over-all mean as the average for the two subgroups. The number of criteria were then reduced by retaining only those which gave a minimum of about 70% correct classifications. These criteria were then further reduced by eliminating those showing high interdependence. Finally this last group of five criteria was used to develop a single function made up of a weighted linear relationship, i.e.,

$$Z = v_1x_1 + v_2x_2 + \ldots + v_5x_5.$$

In this expression the weights v_r are derived from a model which maximizes the differences between two group means and which minimizes the likelihood of mis-classification. The function Z is then a compression of all the information for a given company and its value, relative to a pre-determined scale, provides a criterion for indicating whether a given company falls into a distressed or solvent category.

EQUILIBRIUM

Essentially the classical mathematical models postulate relationships of a deterministic nature between observed characteristics. Stochastic models recognize that this idealized description is seldom, if ever, realized and that the more appropriate model is to allow a spread of alternative outcomes with pre-assigned probability levels. The results of such models will resemble those derived from deterministic considerations in so far as the average behaviour is concerned, but will reflect the basic uncertainties by leading to a range of answers, of varying probability, which describe the real knowledge of the outcome. In some circumstances the average behaviour may be sufficient for the purpose in mind but in other situations the alternative outcomes, even though relatively improbable, may be unacceptable and thus affect any decision processes that may be involved.

A further advantage of such models is in the study of equilibrium processes, which underlies a great deal of economic study. The introduction of stochastic variation into the prescription can lead to solutions which, while behaving on the average in a reasonable manner, show fluctuating behaviour ranging from minor ripples to explosive deviations.

An example that will be familiar to most readers is the no-claim discount scheme in use in motor insurance. The premium rate for an individual risk may be determined as described earlier from a number of factors which are capable of precise definition. However, it is likely that not all of the systematic variation will be identified by these factors and the no-claim discount schemes provide for premium adjustment in the light of actual claims experience of the individual case. The analysis of such schemes provides an example of a probability model which shows clearly how equilibrium positions can arise in real-life situations, but also serves to illustrate the dangers of interpreting such positions as 'laws of nature'.

There are many no-claim discount schemes in operation, each of which differs in minor ways from the others, and the following scheme has been developed to show the basic principles involved. It is assumed that there are 5 premium 'states' denoted 1, 2, 3, 4, 5 for which the relative premiums are 1·0000, 0·9000, 0·8100, 0·7290 and 0·6561, respectively. Cases in Class 1 are those which had a claim in the preceding year; cases in Class 2 have had one claim-free year, i.e., their last claim was made in the year before last. Cases in Class 5 have made no claims for 4 or more years. The premium is revised at the end of each year and cases with no claims in the year are moved up one class (to a higher number). Cases with claims in the year move down 3 steps if they are in Class 5, down 2 steps if in Classes 3 and 4 and 1 step if in Class 2. Non-claiming cases in Class 5 remain there and claiming cases in Class 1 remain there. The movement rules are illustrated in Fig. 5.4.

New cases enter the scheme in Class 3 and the problem of interest is to determine the relative numbers in the separate classes at various future dates. It is assumed that the probability $q(= 1 - p)$ of a claim occurring in a year is independent of the class and remains constant in time. The situation may be expressed concisely in matrix notation (see Chapter 2) but for the present purpose a direct calculation has advantages. We denote by N_r^t the proportion of cases in Class r at time t, where t is measured in years.

Consider now the number N_3^t, the proportionate number of cases in Class 3 at the beginning of year t, who will pay a relative

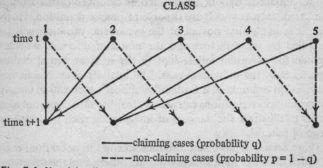

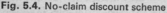

Fig. 5.4. No-claim discount scheme

premium of 0·81. During the year a proportion of these cases will incur claims and will be transferred to Class 1 and will be called upon to pay a relative premium of 1·000 in year $t + 1$. Those cases who do not claim will be transferred to Class 4 and will pay a reduced premium of 0·729 in year $t + 1$.

By considering all the groups in this way the net movement in each class can be expressed in terms of the proportions at time t and the claim probabilities as follows:

$$N_1^{t+1} - N_1^t = - pN_1^t + qN_2^t + qN_3^t$$
$$N_2^{t+1} - N_2^t = pN_1^t - N_2^t + qN_4^t + qN_5^t$$
$$N_3^{t+1} - N_3^t = pN_2^t - N_3^t$$
$$N_4^{t+1} - N_4^t = pN_3^t - N_4^t$$
$$N_5^{t+1} - N_5^t = pN_4^t - qN_5^t.$$

If the group is assumed to be closed, i.e., no new entrants or exits, the over-all net flow must be zero. The total proportions at time $t + 1$ is $\sum_{r=1}^{5} N_r^{t+1}$ and must equal 1; similarly at time t is $\sum_{r=1}^{5} N_r^t = 1$, so that the total sum of the left-hand terms is zero. Similarly the sum of all the terms on the right-hand side will be found to be zero.

When the scheme has been operating for a long time an equilibrium situation will be reached when the net flow in each group is zero, i.e., $N_r^{t+1} = N_r^t$. We can thus find the equilibrium values by putting the left-hand side in each equation equal to zero and then have 5 equations in 5 unknowns, which can be solved by successive elimination (remembering that $\sum_{r=1}^{5} N_r^t = 1$) giving:

value when $q = 0.3$, $p = 0.7$

$$N_1^T = q^2(1 + p)/(1 - p^2q)$$
0·1794

$$N_2^T = qp/(1 - p^2q)$$
0·2462

$$N_3^T = qp^2/(1 - p^2q)$$
0·1723

$$N_4^T = qp^3/(1 - p^2q)$$
0·1206

$$N_5^T = p^4/(1 - p^2q)$$
0·2815.

The numerical values when $q = 0.3$ and $p = 0.7$, i.e., 3 persons in 10 claim per annum, are given in the last column and show the proportions in each class in the stationary situation. If we now bring in the relative premiums charged for each class we can find the over-all average premium as 0·8132. This is not far removed from the premium for Class 3, so that having decided that new cases enter the scheme in Class 3, the over-all annual premium income will be largely unaffected by the no-claim discount scheme, a feature which could be of value in a number of different ways. We therefore examine how the scheme develops in time, assuming new cases enter into Class 3. We could solve the difference equations in the preceding paragraph mathematically but the resulting expressions are rather cumbersome and numerical methods are quite convenient, either by matrix methods (see Chapter 2) or by following the scheme through stage by stage. The results by this latter method are set out in Table 5.10.

As will be seen the proportions in each class show some oscillations before the distribution settles down to its stationary condition. This damped oscillation feature is one reason why the mathematical solution is cumbersome. The weighted average premium does not show any very large variation with development year t so that this scale has a reasonable degree of stability.

It is not without interest to examine the situation where it is assumed that there is an annual intake of one case per annum in Class 3. The proportions in the various classes in successive years

TABLE 5.10*

t	Premium class					Weighted premium
	1	2	3	4	5	
0			1.0000			0·8100
1	0·3000			0.7000		0·8103
2	0·0900	0·4200			0·4900	0·7895
3	0·1530	0·2100	0·2940		0·3430	0·8052
4	0·1971	0·2100	0·1470	0·2058	0·2401	0·8127
5	0·1662	0·2717	0·1470	0·1029	0·3122	0·8096
6	0·1755	0·2409	0·1902	0·1029	0·2905	0·8120
7	0·1821	0·2408	0·1686	0·1331	0·2754	0·8131
8	0·1774	0·2500	0·1686	0·1180	0·2860	0·8126
9	0·1788	0·2454	0·1750	0·1180	0·2828	0·8130
10	0·1797	0·2454	0·1718	0·1225	0·2806	0·8131
11	0·1790	0·2467	0·1718	0·1203	0·2822	0·8130
Ultimate	0·1794	0·2462	0·1723	0·1206	0·2815	0·8132
Class						
premium	1·0000	0·9000	0·8100	0·7290	0·6561	

* Originally presented at the 18th Congress of Actuaries, 1968 (Vol. II, Subject 4)

can easily be found by successive addition of the values in the preceding table. Thus after 12 years the respective proportions are 1·9788, 2·5809, 2·6340, 1·7235 and 3·0828, adding up to 12. Standardized to unity these become 0·1649, 0·2151, 0·2195, 0·1436 and 0·2569. These are significantly different from the theoretical equilibrium values.

The example has been chosen to show, among other things, how stable distributions can arise in practical situations. For example, if statistics were kept of the proportions of cases in each class, the figures would show a stability from year to year – the particular pattern depending, of course, on rate of growth of the business, since the over-all total would be comprised of cases in various development levels. If calculations or decisions were based solely on the equilibrium (stable) position, which is generally the situation for economic data, it is possible to arrive at wrong interpretations or forecasts since the equilibrium arises from the interplay of a number of factors not prescribed by the model. Thus the equilibrium distribution is the same for entrants in each class but the development towards the equilibrium depends on which class they enter the scheme. Entrants in Class 1 for example would start by paying a premium of 1 and the weighted average would decrease towards the equilibrium premium of 0·8132. It follows that the average premium will be dependent on the rate of growth. If this feature were not allowed for in the analysis of statistical data, important errors in premium calculations could be made.

The no-claim discount described is an idealized presentation in that the expected claim rate (q) has been assumed to apply to all classes and at all times. The modifications required to reflect practical situations are not dealt with here, but reference is made to one feature known as 'hunger for bonus'. A person who is in Class 5, and who has an accident, will consider whether his estimated loss of discount will be more or less than the cost of the claim. If the cost of the claim is significantly below the amount of the discount, he will lose, and particularly if his renewal date is near, so he may well decide to meet the cost himself and stay in Class 5. To introduce 'hunger for bonus' into the model requires information about the distribution of claims by size, and the specification of behaviour rules by which decisions to claim or not to claim are reached. Such more complicated models can be tackled by the techniques of dynamic programming, which have been developed over recent years, and referred to in Chapter 3.

TIME SERIES

Many questions arising in economics involve the analysis of observations of quantities measured at successive points of time and a considerable amount of work has been devoted to the analysis of time series of this type, the object being to try to devise models that show the relationship between the quantities at successive intervals for the purpose of predicting future values.

One of the simplest of such models is the auto-regressive process in which the value of a quantity at time t is assumed to be a proportion of its value at time $t - 1$ plus a random component. In symbols the simplest model is

$$f(t) = af(t - 1) + \varepsilon_t$$

where $f(t)$ is the value at time t, a the proportion and ε_t the random component. By repeated substitution we can derive

$$f(t) = a^t f(0) + (\varepsilon_t + a\varepsilon_{t-1} + \ldots + a^{t-1}\varepsilon_1).$$

If a is less than 1 and the system has continued for a long time the term $a^t f(0)$ becomes vanishingly small and the value of $f(t)$ a function of the random elements ε. If the random fluctuations ε can be described by a statistical distribution about a mean m, then the values of $f(t)$ constitute a series of values fluctuating around the average value $m/(1 - a)$. It is necessary to note that implicit in this formulation are the assumptions of a long period during which the process has operated and the constancy of the form of the random elements ε, so that in practical economic situations this idealized solution is likely to be a rarity, and care must be exercised in interpreting a series of observations without adequate regard to these inherent conditions.

Since most econometric problems involve time lag relations between various quantities, which have to be expressed in a common unit such as money, considerable complications arise if the value of money is itself changing. The simple example on p. 336 illustrates the situation, but with more complex situations, such as the inter-relationships between prices and incomes, such changes can have major consequences on the equilibrium situations. For example a simple model for the behaviour of price levels might be a second-order differential or difference equation (i.e., an equation which includes relations between the rate of change of quantities and the rates of change of the rates of change as well as the values of the quantities) reflecting the situation that the level is determined by the interplay of forces of supply and demand. The solution of such an equation takes different forms according to the relationship between the various terms in the equation. Thus it may be that the solution is periodic, a common occurrence in economic data. Implicit in the model may be relationships between quantities separated by a time lag of 2 years; a significant change in the value of money over this period would change the relationships and hence the form of the solution, which may move from the periodic to a rapidly increasing or decreasing form.

Thus a high rate of inflation could disturb equilibrium situations and analysis of the way in which the effect will emerge will be closely dependent on the time lags on which the equilibria were based. It is perhaps not surprising that prediction in this field is notoriously difficult.

6. Stock Exchange

GENERAL

The Stock Exchange is basically a market place where shares representing ownership of corporate enterprises or title documents in respect of corporations or loans made to governments can be traded. It has other important functions arising from its trading activities so that all that has been said before about the scope for applying mathematical methods in regard to industrial and commercial enterprises is applicable.

However, the area that has shown a high degree of attraction for mathematical analysts in recent years has been the study of the behaviour of share prices for the purpose of devising rules for the investment of funds such as pension funds, charitable and other trusts. There has also been a parallel activity in the study of the movement of prices of redeemable securities, such as government bonds, which form an essential part of the financial scene.

MOVEMENT OF SHARE PRICES

Consideration will first be given to investment in ordinary shares which represent the ownership of corporate entities and thus the equity in the organization. The price of a share in a particular company is influenced by many factors, such as the demand relative to other shares, the expectations of dividends, the quality of management and fiscal aspects such as taxation. The weight attached to these different factors will vary from time to time so that the price, which is essentially determined as an equilibrium point between a willing buyer and a willing seller of the share, will fluctuate. If it were possible to find some causal relationship for the movements then, by a judicious programme of buying and selling, the value of the portfolio of shares could be steadily increased.

If the market is efficient then of course, in the absence of any real increase in the over-all value of the market, the advantage would be short-lived as counter forces would develop to maintain the equilibrium. The structure of the market, i.e., the position of intermediaries such as stockbrokers, who are remunerated on a (tightly controlled) commission basis related to the value of the shares, and of the financial analysts who help to form and crystallize opinions on the weight to be attached to the various factors influencing the price, helps to generate such counter forces because temporary fluctuations in prices are quickly recognized and buyers and sellers found at the new equilibrium.

This very simple analysis shows that short-term changes in share prices are the result of a great many influences operating in different directions, a situation well known in physical sciences. The resultant pattern is well described by a statistical model of a cocked hat shape (see Fig. 5.5) where there are many observations clustered about an average value with a tailing-off as the values increase either side of this average. If observations of actual price movements are collected it becomes possible to express this cocked hat pattern in mathematical terms, and it is then possible to develop quantitative methods for use in devising or testing rules for directing investment of funds.

Given a set of such observations there are many alternative mathematical expressions which can be devised to express the pattern. However, the description of a set of data by a mathematical expression is no more than a succinct way of describing the pattern of the particular set and, because of the many possible alternatives, it is essential to see that the mathematical model used can be derived from some rational assumptions in regard to the underlying causes which generate the particular pattern.

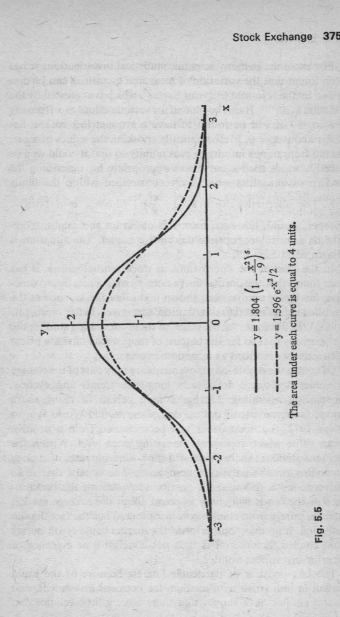

$$y = 1.804 \left(1 - \frac{x^2}{9}\right)^5$$

$$y = 1.596 \, e^{-x^2/2}$$

The area under each curve is equal to 4 units.

Fig. 5.5

For example, in many scientific statistical investigations it has been found that the variation of measured quantities can be described by the 'Normal Curve of Errors', which is expressed by the function $k_1 e^{-x^2/2}$. If this be plotted for various values of x from say -3 to $+3$ it will be found to have a symmetrical cocked hat shape about $x = 0$. Mathematically speaking the values of x can extend from minus infinity to plus infinity so that it could well be asked whether such a curve was appropriate for describing the pattern of quantities which were constrained within the limits ± 3. The alternative form $k_2\left(1 - \dfrac{x^2}{9}\right)^n$, where n is a positive number, would, however, meet this objection and might, therefore be a more appropriate descriptive model. The situation is illustrated in Fig. 5.5.

In the case of the fluctuations in stock market prices, it has been found that the normal curve does provide a first approximation, but closer analysis has shown that other forms, such as the so called stable Paretian distribution, may give a better representation. Whatever may be the result of these continuing studies, the fact remains that so far the pattern of movements of share prices is effectively described as a random process.

Of course the day-to-day movements are only part of the change in values and there are clearly long-term trends and cyclical movements operating. During a long period of rising share prices, such as existed from the end of the Second World War to about 1971, the fluctuations will be measured from a positive mean value which represents the rising price level. Within the market individual share prices will show differing rates of change depending on the particular companies. This means that if an active policy is followed for portfolio investment, the performance of the funds will show a spread about the average market increase purely from chance considerations. Thus the fact that an individual fund may 'outperform' the market could result purely from chance factors and is thus not conclusive as evidence of superior investment ability.

The last point is of particular interest because of the rapid growth in unit trusts as a medium for personal investment, and several studies have shown that there is very little relationship between the relative performance of individual funds in successive intervals of time, evidence of the random character of the movement of prices.

The nature of the changes in the long-term movement in prices of securities has been and is still the subject of many investigations. This type of analysis, i.e., changes over a period of time, has been developed for economic series of many types. The mechanism is well known since in free market conditions the price is the resultant of forces of supply and demand. If demand exceeds supply means will be developed for the latter to be increased, and an equilibrium price will be reached when supply equals demand. However, precision in such matters is difficult to achieve so that the supply tends to exceed demand. Either demand picks up or the supply is cut back, so that the price level shows marked cyclical variations, the period of the cycle varying with the characteristics of the relevant market and production situations. All of this can be succinctly described by systems of second-order differential equations.

The time series analysis is directed to finding out if there are regularly repeating features of the variations which can be described in cyclical form, i.e., the search is for waves and for their period and amplitude. Now it is well known that wave motion can be measured by the number of waves passing a given point during a specified time unit. The commonest example is perhaps radio waves, which are expressed either as a wavelength of so many metres or as a frequency of so many kilocycles. A great deal of effort has been devoted to the analysis of share prices in terms of frequency, the so-called spectral analysis, as an alternative procedure to the classical forms of harmonic analysis, but no clear-cut patterns have emerged which would help to frame investment policies.

PORTFOLIO MANAGEMENT

An important development in investment activity was the Markowitz theory of portfolio selection published in 1952. The portfolio investment problem is the technique of selection of particular investments from a given population to meet some predetermined criteria, i.e., a problem in constrained optimization, some varieties of which have already been discussed.

In formulating the model, Markowitz considered two characteristics of each security, namely riskiness and expected returns. Riskiness is measured by the variance of rate of return on a stock, i.e., a measure of the way in which the return tends to be concentrated round its average value. The rate of return may be determined in various ways but the essential point is that it measures the change in value over a period, such as 1 year, of an investment including any dividends paid.

It is postulated that a characteristic of investors is that they are risk-averse, i.e., they prefer less risky to more risky investments and that a suitable criterion to be used in portfolio selection is that they should seek to maximize the expected return at a given risk level, or alternatively to minimize expected risk at a given expected return. Since there is a free choice over a wide selection of securities there are many possible alternative investment patterns to be considered in order to satisfy the criteria so that a numerical solution is practicable only if a computer is available.

A considerable economy in computer usage can be achieved by reformulating the basic model, relating the return on each individual security to the return on the market index. This model, due to Sharpe, may be summarized in the equation:

$$R_i = \alpha_i + \beta_i I + u_i$$

where R_i is the return from the ith stock, α_i is the expected return from the stock in a steady market, β_i is a coefficient which measures the sensitivity of the price of the stock to movements in the market, I is the market index, and the final term u_i sweeps in all other variations. To use the model, values of α_i, β_i and the standard deviation of u_i have to be found for each stock in the market. The riskiness is now measured by the β_i rather than by the standard deviation of the rate of return. The two are equivalent when the portfolio selected is efficient, which means that no other selection from the market has a greater return for a given level of risk. It will be appreciated that the mathematical apparatus involved in these problems has to be supplemented by a good working knowledge of modern statistical methods.

As a final comment reference is made to another facet of stock exchange investment concerned with redeemable bonds rather than equity stocks which have formed the major consideration of the preceding comments. The matching of the term of assets to the term of liabilities is an essential part of the management of financial institutions, whether they be the very long pattern associated with pension funds, the rather shorter term for life assurance portfolios or the much shorter need of banks. The readily available and marketable supply of government securities which mature at different future dates, and carry different rates of interest, which are important with regard to the differing basis of taxation of the various funds, provides a separate area of study. If the yield on each security is related to the time the security has to run to maturity, a more or less smooth curve will be found which is termed the yield curve. The precise shape of the curve will depend on market conditions and expectations and, much in the same way as investors will seek equity stocks that appear to be cheap or expensive at a particular time, so variations in the yield curve will present opportunities for switching investments from one security to another at a modest marginal gain. Since very large amounts of stock may be involved the yield variation has only to be fractionally over the (low) costs of intermediaries for transactions to be worth-while.

Various attempts have been made to express the yield curve by empirically derived mathematical expressions but, in the absence of a formal model which leads up to such expressions, the value of such expressions as an instrument for application outside the range of the actual data is questionable. Research is still continuing.

ACKNOWLEDGEMENTS

While the views expressed in the foregoing are personal, I would like to express my appreciation to Professors R. Brealey, B. V. Carsberg and C. B. Winston, and to Messrs J. Durnin, P. Wood and A. Young for advice on a number of matters.

BIBLIOGRAPHY

Much of the recent development in the field covered by this chapter is contained in the publications of the specialist professional and academic bodies concerned. However, a number of books provide detailed treatment of many of the topics mentioned and the following list comprises a representative selection. Many of the references presume a working knowledge of statistics and useful introductions at a modest mathematical level are

MORONEY, M. J., *Facts from Figures*, Penguin Books, 1975.
LEWIS, T. W., and FOX, R. A., *Managing with Statistics*, Oliver and Boyd, 1969.

Accountancy

BOUTELL, W. S., *Contemporary Auditing*, Dickenson Publishing Company, Encinco, California, 1970.
   A collection of papers by various authors on different aspects of auditing procedures, including a section on applications of sampling.
Corporate Planning, Firm, A computer model for financial planning, Research Committee Paper No. 5, Institute of Chartered Accountants.
   A detailed treatment of an application of linear programming.
SOLOMONS, D., *Cost Analysis*, Studies in Cost Analysis, Sweet and Maxwell, 1968.
   A comprehensive selection of papers covering many of the practical methods used by accountants.
SALKIN, G., and KORNBLUTH, J., *Linear Programming in Financial Planning*, Accounting Age Books, 1973.
   A detailed description of the practical application of modern programming techniques.
MACE, R., *Management Information and the Computer*, Accounting Age Books, 1974.
   Describes applications of computers and the use of information theory.
Mathematics in Management, Symposium arranged by the Institute of Mathematics and its Applications, London, June 1973.
   A succinct presentation with a minimum of mathematics of a number of topics from a management point of view.
Report of the Inflation Accounting Committee, HMSO, Cmnd 6225, September 1975.

Non-mathematical but essential background material in conditions of high rates of inflation.

Actuarial Work

There are few popular books on actuarial work but the following will be found of value in appreciating the nature of actuarial problems.

OGBORN, M. E., *Equitable Assurances*, Allen and Unwin, 1962.

A very readable account of the early history of the development of scientific life assurance.

COX, P. R., and STORR-BEST, R. H., *Surplus in British Life Assurance*, CUP, 1962.

Describes the evolution of the financial management of life assurance over the past 200 years.

Mathematics in Actuarial Work, Joint Symposium organized by the Institute of Mathematics and its Applications and the Institute of Actuaries, October 1971.

A number of papers concerned with some of the current applications of actuarial techniques.

Claims Provisions for Non-Life Insurance Business, Symposium organized by the Institute of Mathematics and its Applications to discuss a problem of important current interest, May 1974.

BEARD, R. E., 'Some Evolutionary Threads in Actuarial Science', *Bulletin IMA*, Vol 10 (9/10), 1974.

A largely non-mathematical description of the development of some principles of actuarial work.

Banking

COHEN, K. J., and HAMMER, F. S. (editors), *Analytical Methods in Banking*, Irwin, 1966.

Non mathematical, but covers a wide range of the practical aspects.

EILON, S. F., and FOWKES, T. R. (editors), *Applications of Management Science in Banking and Finance*, Gower Press, 1972.

A selection of papers covering a number of topics including some references to applications in banking.

Economics

The field of economics is so wide that a selection of books is difficult. The following are recent collections of readings which are representative of some of the topics.

SEN, A. (editor), *Growth Economics*, Penguin Books, 1970.

ARCHIBALD, G. C. (editor), *The Theory of the Firm*, Penguin Books, 1971.

ATKINSON, A. B. (editor), *Wealth Income and Inequality*, Penguin Books, 1973.

More advanced books dealing with uncertainty and stochastic variation are:

BORCH, K. H., *The Economics of Uncertainty*, Princeton University Press, 1968.

TINTNER, G., and SENGUPTA, J. K., *Stochastic Economics*, Academic Press, New York, 1972.

BORCH, K., and MOSSIN, J. (editors), *Risk and Uncertainty*, Proceedings of a conference held by the International Economics Association, 1966, Macmillan, 1968.

Reports Nos. 1 and 2 of the Royal Commission on the Distribution of Income and Wealth, Cmnd 6171, 6172, July 1975, give a comprehensive background survey.

Stock Exchange

BREALEY, R. A., *An Introduction to Risk and Return from Common Stocks*, MIT Press, 1969.

A very readable non-mathematical account of the stock market and its variations.

WEAVER, D., *Investment Analysis*, Longman Group, 1971.

Deals with company account analysis, stock markets, portfolio selection and allied topics.

TAYLOR, B. (editor), *Investment Analysis and Portfolio Management*, Elek, 1970.

A comprehensive collection of articles from British publications.

GRANGER, C. W. J., and MORGENSTERN, O., *Predictability of Stock Market Prices*, Heath Lexington Books, Lexington, Mass., 1970.

A moderately advanced text dealing with, *inter alia*, spectral analysis as applied to market prices.

Mathematics in the Stock Exchange, Symposium arranged by the Institute of Mathematics and its Applications, October 1972.

A number of papers concerned with mathematical and statistical aspects of stock market variations.

6 Planning

Planning is the continuous process of thought and analysis that precedes action. We need to plan whenever we have to deal with a set of decisions which interact on each other. Such decisions form an elementary *system* in the sense that they possess three properties:

(a) each decision affects at least one other decision;
(b) each decision is affected by at least one other decision;
(c) within the system there is no independent subset.

There are three words commonly used in discussing planning at the highest level – strategic, corporate and long-term. It will be useful to comment on these words and their implications. All decisions we take have effects which extend over different periods of time and also these effects vary in the extent to which they can be reversed. Tactical decisions are those with shorter-term effects and are easier to reverse. Strategic decisions are those with longer-term effects and are more difficult to reverse. Hence strategic planning and long-term planning can be taken as interchangeable.

Similarly tactical plans deal with parts of an organization, while strategic planning deals with the whole. Hence strategic planning and corporate planning are also interchangeable. We will therefore use the terms strategic, long-term and corporate synonymously.

It is within this concept of strategic planning that we shall discuss the mathematical developments that are currently occurring.

1. The Boundary Problem

This book has discussed the use of mathematics in creating models of decision situations. A model is a simplification of a real situation and hence is a many–one transformation in which the richness and complexity of reality is compressed to within the compass of our capacity to control and analyse. In this process of simplification and model-building we inevitably extend the span of control of a manager. Consequently we introduce another form of complexity in that the objectives to which the manager is working will be increased and their dimensions changed. (This means, of course, that while models should improve the quality of decision-making they make the task of the manager himself more difficult.) In passing, we can observe that regarding a typical organization, such as that shown schematically in Fig. 6.1, as the level at which the problem occurs moves up, so does the sensitivity of the decisions taken, and the heterogeneity of the variables also increases. The final characteristic is that decisions at the upper levels tend to have a longer gestation time before results are visible and a longer period over which the consequences of these decisions will have an effect.

This brings us to the vexed question of the planning horizon. In considering the consequences of decisions we must take into account not only 'altitude' in the organization, with the consequent straddle of affected area, but also the time period into the future over which we must estimate the consequences of the decision. Chapter 5 referred to the effect of time and the ways in which it can be taken into account, but in considering the planning horizon we must first understand the reversibility of the decisions and the time period over which these decisions, like it or not, will bear fruit. We must be prepared to forecast the environment over that time period and even, sometimes, resist the temptation to forecast further ahead than we really need.

2. The Nature of the Organization and its Environment

All organizations consist of people linked together to achieve sets of objectives while working within constraints. These constraints

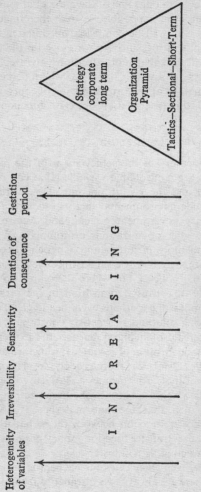

Fig. 6.1. Effect of 'altitude' on class of decisions

can be legal, social and/or economic; often they can be expressed in mathematical form but some of them are qualitative and it will then be difficult, even impossible, to treat them mathematically. The objectives may well be comprised of different, and incommensurable, units of measure. As we move up the organization pyramid we will find that more decisions involve these incommensurables. As we saw in Chapter 3, where a single objective can be cast in mathematical form (implying of course commensurable variables), this is termed an objective function and can be expressed

$$E = f(x_1, x_2, \ldots x_m; y_1, y_2, \ldots y_n)$$

where E is a measure of the achievement of the particular objective, the x's are controllable variables that can be set by the decision-maker and the y's are uncontrollable variables. These latter will include environmental factors (such as raw material prices), arbitrary government policies and Acts of God, and policies set by other organizations or competitors. It is in this sense that the methods of forecasting (Chapter 5) become of importance. Where there is only a single objective it is often possible to derive optimal policies, but where there are two or more incommensurable objectives optimization in the present state of knowledge is impossible (perhaps even meaningless) and the task is that of deriving solutions that satisfy constraints.

The characteristic of living systems is the existence of three basic functions: acquisition of resources; transformation of these resources; and emission: or, more crudely, IN, THROUGH, OUT. This is as true of the earthworm as it is of, say, Unilever Ltd.

The elements which are fed into the organization are, in their simplest form, men, machines and materials. It is usual to use money as a transfer function between them, but it must be remembered that as a transfer function money has no value in itself but only when acting as a transfer agent, i.e., when it is being spent. However, even when only one element is being considered (say, money) there are always two extraneous factors to be added – time and probability. It is of interest to note that the money and time conjunction was dealt with in the discounted cash flow methods of Chapter 5 and the money and probability conjunction in the utility statements of Chapter 3.

3. Linear Deterministic Models

The simplest form of decision is one into which neither time nor probability enters. This means that the decision taken has no effect on any other decision and also that its consequences are known exactly in advance (i.e., deterministic). Within this class, the simplest form of all is that of a single linear objective function of continuous variables:

$$E = \sum a_i x_i$$

where x_i themselves are subject to linear constraints of the form

$$\sum_i b_{ij} x_i \leqslant B_j$$
$$\sum_i c_{ij} x_i = C_j.$$

These are the linear programming problems of Chapter 3. They become more complex when the x_i are non-continuous – for example, if they can only take whole number values, or if the functions are nonlinear. But essentially this class of problem is well researched and is probably unlikely to be as rewarding of future study as other areas.

4. Manpower Planning

Manpower planning has become of increasing importance with the introduction of laws of redundancy and of industrial relations and with the increasing cost of labour in relation to capital.

Government, management and unions alike have come to recognize the need to relate manpower and personnel policies (recruiting, promotion, retirement, training and development, etc.) to each other and to the business strategy of the enterprise. Manpower planning is defined as the set of concepts and methods that enables one to do this, to ensure the availability of people with the right training and experience, to minimize the disruptive and debilitating effect of redundancy and to ensure the availability of satisfying and productive employment. A range of techniques is available to help the personnel director to understand the qualitative implications of possible policies. Most progress has been made in modelling the processes of turnover and wastage on the one hand and the flows of people within the organization on the other. Both assist in estimating the possible future supply of people.

We may consider a crude wastage rate for a group of people as

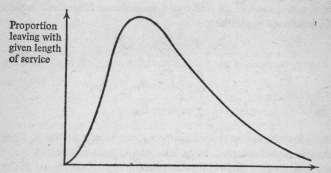

Proportion leaving with given length of service

Length of service

Fig. 6.2

the number of people leaving the group divided by the average number in the group. A common pattern often relates this crude wastage to the length of service as in Fig. 6.2. Such a relationship shows that there is a tendency for some people to leave soon after taking a job, the leaving rate building up to a peak followed by a decline which represents the long-service employees. Hence it can be seen that a feature of groups of people which are expanding is that there will be an immediate (if transitory) increase in turnover. On the other hand it can be observed that if a group wishes to contract by natural wastage (with recruitment being stopped) there will be a tendency for the wastage rate to decline. If this is not realized then forecasts of time to run down manpower will be too optimistic.

There are, of course, many other ways of measuring wastage.

The whole problem of measurement is one that is often skated around, for measurement is never neutral. When we decide what should be measured to describe a particular phenomenon, we imply that that measure is relevant. (For example, we can have a view of the productivity of a hospital by measuring the rate at which patients in different diagnostic categories are discharged. In this way hospital A with an average stay of 2 weeks for a specific complaint might be felt to be more efficient than hospital B where the average stay was 3 weeks. But if all patients in hospital A die within a month of discharge we would question whether the measure of average stay is a good one. For average stay ignores the condition of the patient on discharge and hence is not a relevant measure of hospital performance.)

All measures of labour wastage are attempts to describe the above diagram in terms of a single number, and it is difficult to do this sort of thing. Measures which have been used are the average of the distribution (which is the expected life as employed by actuaries) but this has the difficulty of dealing with distributions that have a long tail. Other measures are the half-life (i.e., the time for half the people to leave) and the proportion surviving for some specified period.

There are obviously two ways of analysing this type of problem. Firstly, since length of service is a continuous variable, as are the ages of those involved, one can build models in the form of mathematical functions. For example, consider a group of men beginning their employment at time $t = 0$. Let the frequency distribution of their length of service until they leave be $f(t)$, then if on leaving a man is replaced by another man with a similar $f(t)$ distribution, the renewal equation is

$$L(T) = f(T) + \int_0^T f(t)L(T - t)dt$$

where $L(T)$ is the rate of leaving at time T.* Renewal theory can

---

*In words, this equation is saying the following. At any time T, those leaving are the sum of two sorts of people. There are those who have been in the system from the very beginning and they comprise the term $f(T)$, for this represents the proportion of those who joined at the start (when $t = 0$)

be used to forecast recruitment needs, provided the $f(t)$ used is amenable.

On the other hand one can look at a group of people by, as it were, taking a series of snapshots at intervals of time. The data available would enable the analyst to impose the appropriate number of drop-outs from the group in each time-interval (from data on death, resignation, promotion) and also entry to the group in the form of recruits of selected ages. If the interval was 1 year, then the individuals in the group are updated from snapshot to snapshot by adding a year to their ages and lengths of service. This approach is akin to that of the meteorological studies in Chapter 1: for weather (like men's ages) is a continuous variable, but can profitably be studied in snapshot form.

One of the areas of application of manpower studies is education planning. In broad terms this is concerned with the adequate provision of teachers, buildings and equipment for education of students. We shall illustrate the manpower aspect of this by looking in general terms at the estimation of student numbers, using this to forecast the requirements for teachers, and then more specifically we shall consider a problem studied at the University of Adelaide in estimating numbers of PhD students in the university.

In estimating the needs for teachers the first step is to forecast numbers of students to be taught. At a national level all that may be necessary are gross numbers of children of different ages. But an individual institution will need to estimate the number of students taking particular subjects at different levels.

From the predicted student numbers, the required number of

---

and who leave *now* (at time $t = T$). There are also those who have replaced someone who left. Looking backwards from *now* ($t = T$), these 'substitutes' will have been in the system various periods of time, and if they came in at time t, they will *now* have been in for time $(T - t)$, with a leaving rate *now* of $L(T - t)$. These substitutes replaced those who left at time t, and the rate of leaving then was (by definition) $f(t)$. Hence the sum total of the rates of all substitutes leaving now is the sum of products $f(t) L(T - t)$ taken over all possible values of t, that is from the beginning to *now*, i.e., from $t = 0$ to $t = T$. This is the second (integration) term.

teachers is calculated on the basis of an assumed student:teacher ratio (e.g., twenty students to one teacher).

The process steps forward year by year, as this is the fundamental time step. After all, most of us have a birthday each year and yearly move into successive age groups. Most students stay in a class for a year and then progress to the next. Educational institutions run in yearly cycles. So it should not be surprising to find that the mathematical models used in education planning are discrete models with a fundamental time step of 1 year (with the notation of Chapter 1, $Dt = 1$ year).

But there are two different ways of using this 1-year time step. One way is to imagine taking yearly snapshots of the education system being studied and then to use these snapshots to predict what is likely to happen in future years. For example, a headmaster may record for 5 successive years the number of third-year mathematics pupils in his school as 37, 39, 45, 32 and 40, and from the trend in these figures try to predict the number of such students in future years. However, it is more fundamental, and usually more successful, to concentrate on a single *cohort* of students entering the institution in a particular year and then follow this cohort through the institution year by year; the number of students in different subjects and classes is deduced from the cohorts. This student cohort concept is explicit in the American convention of calling the class of students entering college in, say, 1980 the 'class of 1984' – that is, the cohort of students expected to graduate in 1984.

As an example of a mathematical approach to estimating student numbers for education planning, we shall consider an analysis of the PhD course at the University of Adelaide, South Australia (this example is based on the PhD research of L. H. Campbell). The PhD students undertake research for about 3 or 4 years, but no longer than 7 years; some drop out without graduating. Students enrolled for a Masters degree may transfer at the end of their first or second year of research to the PhD course which is then backdated to their starting year. These eventual PhD students have therefore to be included in the analysis.

By following the cohorts of enrolled PhD students, and Masters students destined to transfer to the PhD course, we can identify the following *transition* proportions:

$p(i, t)$ = proportion of students enrolled in the PhD course at time t who began $(i - 1)$ years previously and who are still enrolled at time $t + 1$, $i = 1, 2, \ldots, 6$;

$s(t)$ = proportion of students enrolled in a Masters course at time t and are still enrolled at time $t + 1$ in the Masters course but will eventually transfer to the PhD course.

For example, $p(6, 1970) = 0\cdot25$ means that a quarter of the PhD students who began the course in 1965 and who are enrolled in 1970 will still be enrolled in 1971; three-quarters will either have graduated or have dropped out during the period 1970–71. And $s(1970) = 0\cdot5$ means that a half of the Masters students enrolled in 1970 who will eventually transfer to the PhD course will still be enrolled as Masters students in 1971; the other half will have transferred during 1970–71.

We now define:

$M(i, t)$ = number of PhD students at time t who began their research $(i - 1)$ years previously, $i = 1, \ldots, 7$;

$m(i, t)$ = number of Masters students at time t who will eventually transfer to the PhD course and who started $(i - 1)$ years previously, $i = 1, 2$.

Note that the ranges of values of i correspond to the 7-year limit for PhD students and the 2-year limit for Masters students wishing to transfer.

The progress of the student cohorts year by year can now be formulated in terms of the following difference equations:

$$M(2, t + 1) = M(1, t)p(1, t) + m(1, t)[1 - s(t)],$$
$$M(3, t + 1) = M(2, t)p(2, t) + m(2, t)$$
$$M(i, t + 1) = M(i - 1, t)p(i - 1, t), \quad i = 4, 5, 6, 7,$$
$$M(2, t + 1) = m(1, t)s(t).$$

Initially we let $M(1, t + 1) = E(t + 1)$ = number of new PhD students admitted in the period t, $t + 1$, and similarly $m(1, t + 1) = e(t + 1)$ = number of eventual PhD students admitted to the Masters course in the period t, $t + 1$. A geometrical representation of the situation is given in the flow chart of Fig. 6.3.

It is now not difficult to deduce the total number of students in any year by adding our cohorts. But before doing this we introduce a simplification which became evident after an analysis of historical data. The values for the transition proportions $p(i, t)$ and $s(t)$ were found to vary little from one year to the next; so that in our mathematical notation we just drop the variable t and write $p(i)$ and s. The total number $N(t)$ of PhD students enrolled at time t can then be found from the difference equations, giving the lengthy expression

$$N(t) = E(t) + p(1)E(t - 1) + p(1)p(2)E(t - 2) + ..$$
$$.. + p(1)p(2)p(3)p(4)p(5)p(6)E(t - 6)$$
$$+ (1 - s)e(t - 1) + [s + (1 - s)p(2)]e(t - 2)$$
$$+ [s + (1 - s)p(2)]p(3)e(t - 3) + ..$$
$$.. + [s + (1 - s)p(2)]p(3)p(4)p(5)p(6)e(t - 6).$$

How can such an expression be a help in the strategic planning of education? One benefit of the approach has been the identification of the importance of the transition proportions. Although they have been taken to be the same values each year, it is easy to allow for variation by introducing upper and lower limits and deducing a range of values for the predicted number of PhD students in future years. The approach also makes it easy to test the effect of a sudden slump in enrolments.

But most important has been a reverse use of the difference equations. Instead of deducing the total numbers N from given enrolments E and e, the University of Adelaide has actually been faced with the problem of deducing E and e for given N. The Universities Commission, as a result of its national planning, has decided on target values for the total number of PhD students for each year of the next triennial planning period and it is left to the university to control the admission of post-graduate students to meet this target. If the targets are exceeded, the university gets no compensatory funds for the excess students, but if the targets are not met, promised funds are likely to be reduced. In this situation

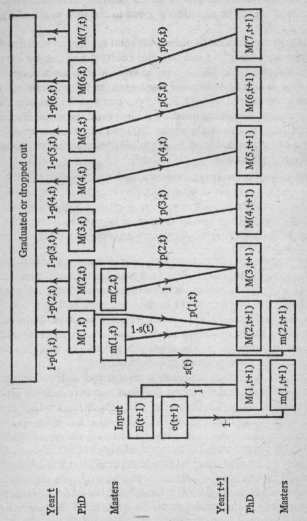

Fig. 6.3. The change in the student PhD population from year t to year $t + 1$

the mathematical approach has proved very beneficial. Although it is easy in principle to choose enrolment figures to match the target totals, in practice these tend to oscillate – a large intake one year, a small intake the next, and so on. To solve the difference equations so that the targets are achieved as smoothly as possible is an optimization problem – and it turns out to be an integer program (see Chapter 3). Fortunately this has a special structure which has enabled an efficient computer algorithm for determining the optimum solution to be devised.

Similar approaches are not uncommon in much larger organizations. The North Thames Gas studies * have included research into numbers and grades of gas distribution manual workers.

The grade structure for distribution fitters is shown in Fig. 6.4. Higher grades require a high degree of technical skill and vacancies at these levels have to be replaced by promotion. Consequently, entry to the above system tends to be by recruitment at the base and there will obviously be the possibility of lags if the system is not optimally staffed.

The number in a grade at a particular time period (for example at the end of a year) will be the algebraic sum of the number in the grade at the beginning of the year together with promotions into and out of the grade and external recruitments and losses.

As has been seen, it is possible to structure this problem as a Markov process (as in the education planning problem). It is also possible to look at the problem in terms of a renewal process.

Markov models assume that promotions and recruitment numbers leaving each grade are known and use this as a basis for prediction. In addition the North Thames Gas model also assumes that there is a fixed promotion rate out of each grade and a fixed wastage rate from each grade.

If $N(I, T)$ is the number employed in grade I at time T,

$\qquad R(I, T)$ is the number externally recruited into grade T during time $(T, T + 1)$,

$\qquad W(I)$ is wastage rate from grade I and $P(I)$ is the promotion

* 'Manpower Planning Models in North Thames Gas', S. R. Hartley, Operational Research Department, presented at the O R Society Conference, Loughborough, September 1975.

rate out of grade I (note that both $W(I)$ and $P(I)$ are constant over all time periods),

then
$$N(I, T+1) = N(I, T) \times \{1 - W(I)\}\{1 - P(I)\}$$
$$+ \sum_{I^*} \{N(I^*, T) \times \{1 - W(I^*)\} \times P(I^*)\}$$
$$+ R(I, T)$$

The I^* notation covers those in grades below I who were promoted into grade I during the year (T to $T+1$). As an example, referring to Fig. 6.4,

$$N(2, T+1) = N(2, T) \times \{1 - W(2)\}\{1 - P(2)\}$$
$$+ N(3, T) \times \{1 - W(3)\}P(3)$$
$$+ N(6, T) \times \{1 - W(6)\}P(6)$$
$$+ R(2, T)$$

and $N(5, T+1) = N(5, T) \times \{1 - W(5)\}\{1 - P(5)\} + R(5, T)$.

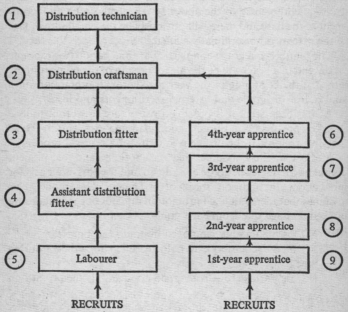

Fig. 6.4

The arithmetic involved is straightforward but tedious, and a computer is obviously a labour- and time-saving aid. In addition it is important to have precise definitions of N, R and W. The model provides planners with annual predictions for each grade of the numbers promoted out of it, the numbers leaving it and the number in it.

The renewal model has the same assumption of the constancy of wastage rates but assumes that where it is not possible to fill vacancies in a grade by promotion then the shortfall will be met by recruitment from outside. The object of the model is to forecast the numbers of promotions and recruitments necessary each year to maintain the number in the grades.

The required promotion rate out of a (lower) grade during a year is the ratio of the shortfall in the grade above to the number eligible for promotion from the lower grade. In practice in the model this rate can be overridden by imposing upper and lower bounds on it (notice here the difference from a Markov type model where this rate is assumed constant from year to year).

For each grade I let
$Q(I, T)$ be the calculated outward promotion rate during $(T, T + 1)$,
$E(I, T)$ the required number at time T,
$RR(I, T)$ external recruitment during $(T, T + 1)$,
$UP(I)$ and $LP(I)$ upper and lower bounds on Q,
$W(I)$ wastage rate.

Then shortfall in grade 1 at $T + 1 = E(1, T + 1) - N(1, T) \times \{1 - W(I)\}$.

Similar equations can be derived to determine relevant features of the manpower system. For example it can be shown that the number eligible for promotion from grade $2 = N(2, T) \times \{1 - W(2)\}$

and so $Q(2, T) = \dfrac{E(I, T + 1) - N(I, T) \times \{1 - W(I)\}}{N(2, T) \times \{1 - W(2)\}}$.

$Q(2, T)$ must lie between $UP(2)$ and $LP(2)$. If it is greater, then $Q(2, T)$ is set at $UP(2)$; and if it is less then it is set at $LP(2)$.

The shortfall in grade 2 at time $(T + 1)$

$$= E(2, T + 1) - N(2, T) \times \{1 - W(2)\}\{1 - Q(2, T)\}.$$

The numbers eligible for promotion $= N(3, T) \times \{1 - W(3)\}$ from grade 3, and $N(6, T) \times \{1 - W(6)\}$ from grade 6.

An initial value θ has to be set such that $Q(6, T) = \theta \times Q(3, T)$. Then we find that

$$Q(3, T) = \frac{E(2, T + 1) - N(2, T) \times \{1 - W(2)\}\{1 - Q(2, T)\}}{N(3, T) \times \{1 - W(3)\} + \theta N(6, T) \times \{1 - W(6)\}}$$

(provided this is between $UP(3)$ and $LP(3)$),
otherwise it is set at $UP(3)$ or $LP(3)$, respectively.
In similar ways the other numbers can be calculated to give

$$\begin{aligned} N(I, T + 1) = {} & N(I, T) \times \{1 - W(I)\}\{1 - Q(I, T)\} \\ & + \sum_{I^*} \{N(I^*, T) \times \{1 - W(I^*)\}Q(I^*, T)\} \\ & + RR(I, T). \end{aligned}$$

The calculations for the renewal model require as input for each grade:

(1) initial numbers each year;
(2) required numbers each year;
(3) upper and lower bounds on outward promotion rate;
(4) wastage rate;

and the model yields for each grade each year:

(1) predicted numbers;
(2) calculated outward promotion rates;
(3) predicted outward promotion numbers;
(4) predicted numbers leaving (other than by promotion);
(5) predicted recruitment.

The two models are used in a complementary fashion. The Markov model can be used experimentally. For example it can test the effect on future numbers employed in each grade of different recruiting and promotion policies and hence the optimal policies can be estimated.

On the other hand if we start with the ideal manpower distribution, the renewal model can be used to determine the promotions and recruitments necessary. In practice the Gas Board uses both methods to derive its 5-year plans.

5. The Continuous Approach

As an example of the continuous approach there is the concept of a salary progression curve. An analysis of individual salary histories, corrected for inflation, in one organization led to an empirical fit of the form

$$Y_t = \log \frac{S_t}{I_t} = a + bx_t + cx_t^2,$$

where $\quad S_t$ = salary in year t

I_t = Index applicable to year t

x_t = age at year t

and a, b and c are parameters for the individual.

Values of a, b and c were determined for groups of individuals. It was found that both b and a, and c and a, were correlated and could be related linearly. This, of course, reduced the parameters to only one and the above equation could be rewritten

$$Y_t = (\alpha + \gamma x_t + cx_t^2) + \lambda(\beta + \delta x_t + \zeta x_t^2),$$

where λ was a parameter for each individual lying between 0 and 100. λ then became the 'apparent potential' for an individual. The model was used to estimate 1967 salaries for a sample of people using their salaries between 1956 and 1958. Their salaries in 1967 were predicted very successfully both as a group and as individuals. This method could then be used to forecast the grades that staff would achieve, simply by transforming salaries into grades, and these relationships could be used in a simulation approach.

Manpower problems can also be structured in linear programming form. Let us divide time into series of slices and in each time period, denoted by t, there will also be a number of types of job, called activity, and denoted by i.

For each time period t, and each activity i, let

$X_A(i, t)$ = recruits
$X_R(i, t)$ = redundancies
$X_p(i, j, t)$ = promotions from i to j
$X_T(i, j, t)$ = transfers from i to j
$X_o(i, t)$ = overmanning
$X_u(i, t)$ = undermanning
$b(i, t)$ = manning required
$C_A(i, t)$ =
$C_T(i, t)$ = $\Big\}$ unit costs of recruiting, transfers, etc.
$r(i, t)$ = retirements (including death)
$X(i, t)$ = initial manning.

The total number of people required for activity i in time t is the sum of a number of plusses and minuses, i.e.,

PLUS Initial manning
Recruits
Promotions into activity
Transfers into activity (not involving promotion)
Undermanning (i.e., short staffing)

MINUS Redundancies
Promotions out to other activities
Transfers out to other activities (not involving promotions)
Overmanning (i.e., excess staff)
Retirements (including death).

In algebra this becomes

$$b(i, t) = X(i, t) + X_A(i, t) - X_R(i, t) + \sum_{k \neq i} X_p(k, i, t)$$
$$- \sum_{k \neq i} X_p(i, k, t) + \sum_{k \neq i} X_T(k, i, t) - \sum_{k \neq i} X_T(i, k, t)$$
$$- X_o(i, t) + X_u(i, t) - r(i, t)$$

('$\neq$' means 'is not equal to'). Such equations can be written for each year to form a matrix. Policy constraints may also be introduced, such as

$$X_o(i, t) \leqslant o(i, t)$$
$$X_u(i, t) \leqslant u(i, t)$$
$$X_A(i, t) \leqslant K \cdot X(i, t).$$

If the cost of recruiting, redundancy, overmanning, undermanning, etc. can be estimated then the total cost of the system is given by an expression of the form $\sum CX$ and the problem becomes one of minimizing $\sum CX$ subject to all $X \geqslant 0$.

6. Dynamic Programming and Planning

The description of the planning process at the beginning of this chapter and the method of dynamic programming outlined in Chapter 3 have an obvious affinity for each other. In particular, dynamic programming operates by moving through a series of stages from the here-and-now to some planning horizon. It does this by evaluating, at each stage, what the next step should be (as the hymn writer puts it 'I do not wish to see the distant scene, one step enough for me'). Although the forms of objective function may seem limited, the further reading suggested will show a variety of functions that can be maximized in this way. Clearly the network problems of Chapter 4 are a special case of dynamic programming.

Take the case of a chocolate manufacturer who is buying cocoa as a raw material. The price will vary from day to day and he will wish to have a secure supply of the commodity. If he thinks prices are rising then it is prudent to buy more at today's price so as to see him through a period of higher prices. Conversely if his forecast is that prices will fall, then it is sensible to stay out of the market (providing he has sufficient stocks to do so) until the trough in prices is reached.

If he knew in advance what prices were going to be then he could buy at an absolute minimum. For example, suppose we know that over the next 2 years the cost per ton would be as shown in Fig. 6.5, then the optimal amounts to buy at each buying decision would be those to cover consumption up to the next buying decision.

At the starting point, A, enough would be bought, taking account of stock already on hand, to cover consumption up to B. From B to C, a period of constantly declining prices, we would live hand to mouth; and so on. The point is that if we were clever enough to know the future, we would not need the mathematics. If we do not know the future, the forecast would be in the form of a set of probabilities of future prices at different times.

The price, x, at the end of each month will not be known beforehand and will, of course, be a variable quantity. In the simplest

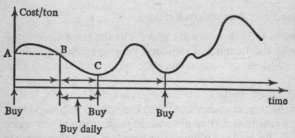

Fig. 6.5

case, suppose this price is, for the next 2 years, forecast to be going to vary about the same average level and with the same degree of uncertainty month by month. Then we are going to have 24 opportunities to buy and we have the constraint that if we consume one unit of stock each month, then we must always, after we have bought at a month's end, have at least one unit, and at most n units, in stock where n is the number of months remaining before we get to the end of the 2 years.

The first decision is how to react to the first price in month 1. If it is high we may wish to buy only one unit (the minimum). If it is low we will wish to buy more; how much more depends on how low the price is. It is this relationship between price and optimal stock on hand, so as to buy at minimum long-run price, which dynamic programming establishes, by a development of the methods outlined in Chapter 3.

An extension of this decision rule for month 1 would tell us, in the light of the price yielded each month, how much we should then buy at that price in order to have a minimum total price over the 24 months. In reality the month by month price will not vary about the same average and, as we look further ahead, our uncertainty will increase. At any particular time we will have what we can term a set of distributions for all future months. A characteristic of these will be that as we move further ahead the distributions become wider, reflecting this increasing uncertainty about the future. Nevertheless the dynamic programming method can still be applied to decide how much we should buy at today's price. Such methods are not uncommon in purchasing commodities which have varying prices. The critical factor, of course, is the validity of our forecasts. No amount of advanced mathematics will turn a bad forecast into a good one.

7. Competitive Models

In those aspects of organizational activity where the consequences of a decision depend partly on decisions being taken by our competitors, the forms of approach divide sharply into two categories. In the first category the effect of our competitors' response (or even the response itself) is so delayed that it will not change the consequence of our original decision. As will be seen, in this type of problem the allocation procedures (linear and dynamic programming) can then be employed. The second category is that in which a competitor responds fast enough for our decision to interact with his response.

An actual case study might illustrate the application of mathematics in the case of competitive tendering. In this study an oil company asked assistance in formulating bids to acquire exploration and exploitation leases. Periodically the states of Louisiana and Texas put on the market a number of pieces of land that were potentially oil-bearing (perhaps twenty pieces at a time). The geologists of all the companies surveyed these tracts and then submitted sealed bids. On a given day all bids were declared and the highest bidder on each piece of land was awarded the rights. The problem as posed by the company was: 'given the total amount of money that we have to invest in bids, how should we

divide this out so as to maximize our return?' This led to the first
question – what is meant by return? In this case it was decided
that since the company's geologists assigned an index (x) on a
scale of increasing attractiveness (0 to 100) to each piece of land,
a strategy would be derived to maximize the total of x's received
in a sale for a given total amount bid.

It was determined, from analysis of past sales, that for any
particular piece of land, the average of the bids (A) made by the
competitor companies was related to the x value assigned by our
company. In addition the number of rival companies tendering

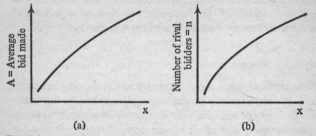

(a)   (b)

Fig. 6.6

(n) was also similarly related. Neither relation was exact, of
course, but the relation was strong enough to be usable (see Fig.
6.6). Hence in any future sale of a set of tracts, with differing x
values, the number of bids and their average could be estimated
for each tract. This was not, of course, sufficient to estimate the
probability that a particular bid on a particular tract would be
successful. However, an analysis of all the bids made showed that
if the jth bid on tract i, B_{ij}, is related to the average A_i of the bids
on tract i, the resulting probability distribution of $\dfrac{B_{ij}}{A_i}$ was 'well
behaved' (see Fig. 6.7), the scale running from (approximately)
0.2 to 4.0, the distribution being centred, of course, at 1.0.

It was now possible to forecast the probability that a bid B on a
tract with given x would be successful, for the relationships above
will estimate A and n.

A bid B will beat any other single bid whose ratio to the given

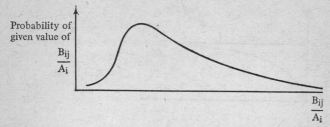

Fig. 6.7

value of A is less than $\dfrac{B}{A}$, i.e., for any given B the proportion of
Fig. 6.7 which lies to the left of the corresponding $\dfrac{B}{A}$. Let this
proportion be $p(B/A)$. However, since our bid B must beat an
estimated n rival bids in order to be successful, the probability of
the bid winning is $\left\{ p\left(\dfrac{B}{A}\right) \right\}^{n}$. This can be plotted (Fig. 6.8). The
expectation of return $\left[\left\{ p\left(\dfrac{B}{A}\right) \right\}^{n} \cdot x \right]$ tends to the particular x
(Fig. 6.9). The only remaining problem is how to allocate a total
amount T of bids between alternative tracts so that the total
expectation of x is maximized.

Let the relationships in Fig. 6.9 for n tracts each with different
x values be $f_1(b_1)$, $f_2(b_2)$, ... $f_m(b_m)$. Then the problem is
maximize $\sum f_i(b_i)$ subject to $\sum b_i = T$.

Consider the case of just two tracts, with expected returns

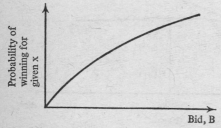

Fig. 6.8

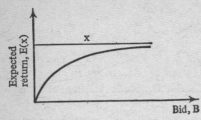

Fig. 6.9

$f_1(b_1)$ and $f_2(b_2)$. We then have to select b_1 and b_2 such that $f_1(b_1) + f_2(b_2)$ is as great as possible while, at the same time, $b_1 + b_2 = T$, the fixed amount of total bid (Fig. 6.10). It is clear that the allocation as shown cannot be the best. For since the tangent at A_2 is steeper than that at A_1, the increase in $f_2(b_2)$ caused by moving b_2 slightly to the right is greater than the decrease in $f_1(b_1)$ caused by moving b_1 the same distance to the left. That is, if the slopes of the tangent are different, the bids can be improved and conversely in any optimal allocation of bids between any number of tracts the slopes of the tangents are all parallel. We conclude that the optimal allocation with two tracts only has

$$f_1'(b_1) = f_2'(b_2) \quad \text{where} \quad b_1 + b_2 = T.$$

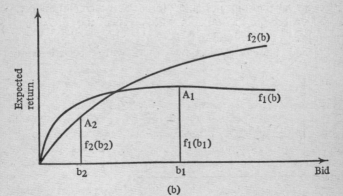

(b)

Fig. 6.10

Returning to the original problem, we have at the optimum

$$f_1{}'(b_1) = f_2{}'(b_2) = \ldots = f_n{}'(b_n) \quad \text{where} \quad \sum b_i = T.$$

We therefore form the graphs of $f_i{}'$ (b). (We note in passing that a characteristic of the curves in Fig. 6.9 is that the steeper the slope at the origin, the lower is the corresponding value of x where the curve levels out horizontally. Consequently in Fig. 6.11, f' curves corresponding to low x values, cut the horizontal line at any given height λ at higher bids (b).)

Take a trial value of λ in Fig. 6.11 and for each curve of $f_i{}'(b)$

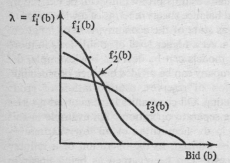

Fig. 6.11

drop the perpendicular to obtain the corresponding b_i. Form $\sum b_i$. If this is more than T, then the next trial should be at a greater λ and *vice versa*. Continue in this way until the λ is obtained such that all $f_i{}'(b_i) = \lambda$

and

$$\sum b_i = T.$$

This set of b_i gives the bids to maximize $\sum f_i(b_i)$.

This method has proved very successful in practice. We note that it can be adapted to deal with the planning of resource acquisition, for we can also estimate the probability that any given total amount of x can be acquired in a particular sale in terms of the total amount bid.

This example is typical of the problem of competition when the reaction of the competitor is sluggish. (It is interesting that in this particular study the solution had to be abandoned when some of the company personnel who had worked on it went to work for the rival companies.)

8. Company-Wide Models

Mathematics can be used to derive models which forecast the way in which a company can be expected to perform against changes in the environment. Some of these models are high-speed methods of doing complex and laborious arithmetic. For example, the cash position of a bank and its profitability can be derived by producing hypothetical balance sheets in terms of a wide range of external factors, such as state of the economy, rate of inflation, interest rates and so on. At a higher level of sophistication more complex deterministic models can be devised. For example, the operations of an oil company can be divided between prospecting, drilling and exploitation of reserves, transportation of crude oil, refining and marketing. Oil companies have evolved a series of models for all these separate operations. An example in this chapter showed how, in the acquisition of oil reserves, mathematical methods can be utilized. In the operation of a refinery the whole of the processes can be regarded as being linear in nature. Given the characteristics of the crude oils being refined, the output of the refinery in terms of grades of petroleum and the wide range of chemical by-products are linear in nature. Hence, in terms of the contribution to profits of each of the products involved, a production schedule which maximizes profits while satisfying all the marketing constraints can be derived.

Ocean transportation of crude oil can also be modelled in linear form. In this case relationships are not continuous but are of a step function form, because of the incremental nature of the oil carried in a particular tanker. By this means models can be constructed which link together the ocean transportation of crude oil, refinery operations and delivery to customers (this latter being a typical transportation problem of Chapter 3). A typical model for an individual oil company may consist of some 8000 variables linked by some 6000 equalities and constraints. In the development of these models the major limiting factor so far has been the capacity of high-speed computing. Such models enable management to play out alternative changes in the environment including sudden events, such as the closure of the Suez Canal, in order not only to understand the strategies they should adopt if such disasters occur, but also to understand in so complex a situation the consequences, for example, of a decision to build new oil refineries in particular locations.

At a less complex level mathematical models can be created to link together raw-material buying, production scheduling and marketing. In a food processing company, for example, the problem was faced of linking together the buying of seasonal raw materials, which could be substituted for each other, with production scheduling, with the allocation of sales offered in marketing finished goods in a highly competitive environment. Classically, the approach to such complex problems has been to divide them within the span of control and comprehension of single managers. This meant, in this instance, that one manager was responsible for purchasing, one for production scheduling, and one for marketing. Such a system can only operate if variabilities in separate parts have their effects uncoupled. Consequently, such a system will work at the expense of large stocks of raw materials and large stocks of finished goods. A mathematical model can extend the span of control of a man.

In this particular case, studies of mathematical and statistical relationships linked together, in an integrated plan, raw-material buying, production scheduling and marketing. In its turn this led to a reduction in raw material and finished goods stocks of 80%, and made the company significantly more profitable.

9. Industry-Wide and World Modelling

So far in the development of this chapter, and indeed of the book as a whole, the argument has been *deductive*. By this we mean that total systems have been considered in their separate parts, these parts have been analysed quantitatively and relationships have been deduced from these analyses. These separate tactical results have then been brought together in larger groupings to establish strategic relationships.

There is however a complementary approach to the problem of building models for large-scale problems. This is what can be called an inductive approach, in which the analyst imposes on the system models which stem from the observation that if life has any reason then certain things should be happening. We shall apply this first to company-wide systems (industrial dynamics) and then to world systems.

These approaches are stimulated by the pioneering work of Jay Forrester at the Massachusetts Institute of Technology. Industrial dynamics attempts to provide a single framework for integrating different areas of management – marketing, production, accounting, research, development and capital investment. The object is to provide a basis for the design of more effective industrial and economic systems.

The lines of approach in building systems dynamic models will be familiar to those who have read the earlier chapters of this book. Forrester lists them as:

(*a*) identify the problem;

(*b*) list factors which influence the problem;

(*c*) experiment where possible to see how the parts of the system influence the whole;

(*d*) trace the cause effect information feedback loops that link decisions to actions;

(*e*) formulate decision policies that describe how discussions result from available information;

(*f*) construct a mathematical model linking policies, information and interactions of components of the system;

(*g*) generate the behaviour of the system through time;

(*h*) compare the results and revise the model until it is accepted as an adequate representation of the system;

(*i*) redesign the organizational relationships and policies to find those changes which improve the behaviour of the system;

(*j*) use the systems model to alter the real system in the direction of improved performance.

It can be seen that this construction, which is condensed from Forrester's own writings, has two important differences from all the methods so far evolved in this book.

(*i*) So far we have concentrated on problems of data collection and analysis. We have used these analyses to deduce what large-scale relationships might be deduced. The Forrester approach starts with hypothetical relationships and then scales down to data and analysis.

(*ii*) The concept of feedback is introduced for the first time.

Feedback lies at the basis of any learning system. For example, if I am playing darts my throw is constantly affected by my observation of the results of my last throw(s). Hence I am much more likely to hit the inner bull if I can observe how my darts land than if I receive no information. This is a controlled feedback of me as decision-maker. But there are natural feedbacks. The behaviour of a volume of warm wet air as it rises, as described in Chapter 1, is conditioned by a feedback of nature. Were this feedback not to occur our atmosphere would never return and none of us would be worried by long-term forecasting! There are economic feedbacks; for example the reaction of consumers to rises in price is to withhold purchases, which can pull prices down again. In a similar situation the feedback can lead to instability: for a failure of customers to buy will reduce volume sold, which increases the average cost of an item, and this leads to a price increase and a spiral with which we are all too familiar in, say, public transport costing.

Consequently feedbacks are of two forms: there is a positive feedback which confirms and increases a trend and a negative feedback which reverses a trend.

In some cases a delay in information can cause an attempted negative feedback to act like a positive one and actually accentuate a trend. For example, suppose we are trying to keep a variable y, which varies over time, as near to some constant value as possible. Then at any time t the history and present value of y might be as shown in Fig. 6.12. At time t_1 we will want to influence the variable by giving it a downward push. If we have up-to-date information and can apply correction forces immediately we can reverse the upward trend of y to bring it towards the required constant level. Similarly, if we had had the same situation at t_0, we would at that time have imposed an upward force on y. But suppose all we know at time t is the situation as it was at time t_0. Our correction will then be in precisely the wrong direction and we are making the situation much worse via a positive feedback.

The problem of delay in response, even when we know the up-to-date situation, can be observed from watching a convoy of military vehicles. The front vehicle will be driving at (more or

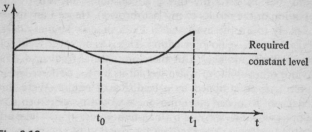

Fig. 6.12

less) constant speed while the others try to keep constant distance. But very slight variations in speed of the lead vehicle are, because of delays in reaction by successive vehicles, transformed into larger variations down the column until the last vehicle is either stationary or driving at high speed to catch up.

A similar problem arises, for example, in the holding of stocks of product in industry. A retailer may have almost constant demand for a product. If he sees (or thinks he sees) an increase in sales of, say, 5%, his routine monthly order may be increased by, say, 10%, as he knows there is always a delay before the wholesaler supplies him. The wholesaler observes the 10% rise and in similar way orders 15% more than usual and so on. A consequence of these increases in themselves is to increase the delays in supplying, since out-of-stock situations may develop and by the time the manufacturer is reached there may be extreme escalation of orders and an apparent insatiable demand for a product whose sales have in fact only increased by 5%. The converse applies when demand declines and the feast or famine situation at the manufacturer shows how the stop/go policies, so berated politically, may arise as a result of natural causes rather than deliberate policy.

We can illustrate this by an example (see Fig. 6.13). The diagram shows the way in which information flows through the system inwards and the resulting physical flow of goods outwards. In any such system there are, of course, delays and it is the nature and extent of these delays which determine how the system behaves. For example, in the diagram it is assumed that there is a one-week delay in translating orders from customers into deliveries to them. Since one would not continuously be informing the distributor of the rate at which the retailer is receiving orders, it is assumed that every three weeks customers' orders are added up and transmitted to the distributor in the form of a request for more goods to be sent to the retailer. It takes half a week for the order to come into the distributor and a further week for the distributor to get the stocks to the retailer. The distributor himself gathers together retailers' orders every two weeks and transmits these to the factory warehouse, the order taking half a week to reach the factory.

If orders remain constant all the characteristics of the system will remain fixed and, hence, weekly factory orders from the distributors, distributors stocks, stocks at distributors, retailers and manufacturer, delay in filling orders at the factory and so on are all fixed.

This is so for any level of sales (except that these fixed levels would of course be different). But if sales are suddenly (on 1 January, say) increased by 10% the system will behave in an oscillatory fashion until the new equilibrium state is reached, which may take nearly 18 months, as shown in Fig. 6.14.

As might be imagined, the behaviour of the system is even more ephemeral if, instead of a rise in fixed sales from one level instantly to another, we have an oscillation between these two levels with a 1-year cycle.

In a similar fashion the effect on all the system parameters of the single input variable, retail sales, can be forecast. It can be seen, for example, that the manufacturer's orders to factory and the factory output are both responding very sensitively to modest changes in retail sales.

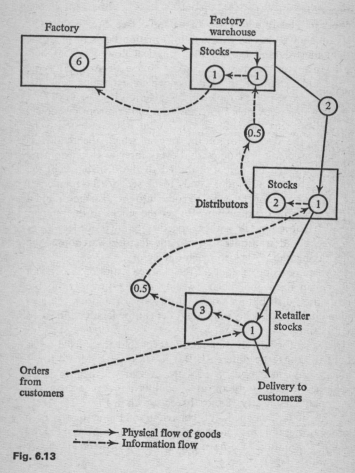

Fig. 6.13

Clearly such an approach makes it relatively easy to forecast the effect of changing the decision rules – such as the imposed delays – on the output parameters shown graphically in Fig. 6.15. Also, as can be seen, a similar method can yield forecasts on an industry-wide scale by taking, for example, the retailer–wholesaler–manufacturer stereotype above and using it as a building block in a problem with wider bounds.

10. World Models

One of the major growth industries at present is that of world modelling. These models are natural consequences of the methodology developed for industrial modelling and were also originally stimulated by Forrester. There are at least thirty such world models now in use; unfortunately many of them work on different assumptions and the computer languages and data bases used are often not compatible. Hence it becomes difficult to compare the results of these different models.

There are five major inputs to such models: population, industry, health, resources and environment. These five factors are interconnected. These connections produce effects which are measured not just in months but sometimes over centuries. Consequently all these models will be concerned with the world over the next century. The explosive ingredient in all this is the effect of exponential change. If something (say population) is growing at 7% a year, it will double in about 10 years. A 10% annual growth means a doubling every 7 years. (As a rough rule, divide the annual percentage growth rate into 70 to obtain the number of years to double up.) Hence all these models are very sensitive to assumptions made about growth rate. The other important concept is that of feedback which we noted earlier.

The feedback loop for population is represented in Fig. 6.16. Over the last 300 years the average life expectancy has increased from 30 years to 53 years, hence the negative feedback has decreased substantially. There has been only a slight decrease in the birth rate and so we have had the dominant effect of the negative feedback decrease and a consequential exponential

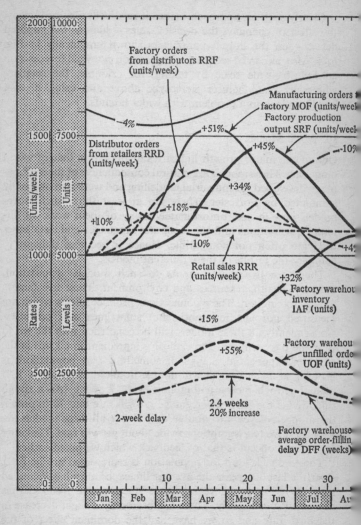

Fig. 6.14. Response of production–distribution system to a sudden 10% increase in retail sales

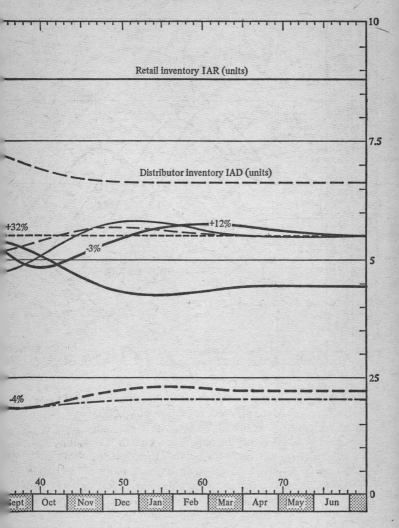

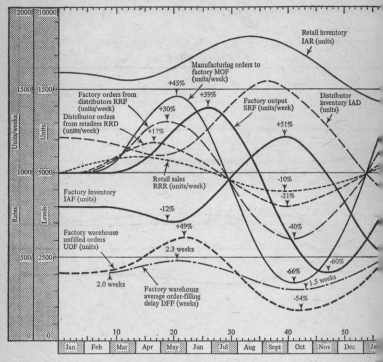

Fig. 6.15. Response of production-distribution system to a 10% unexpected

growth in population. It is here that the lag effects referred to now occur for most of the prospective parents of AD 2000 have already been born and unless there is a sharp rise in mortality, there is no prospect of levelling off the population growth before then.

Industrial output can also be structured by the same combination of positive and negative feedback loops as shown in Fig. 6.17. These two factors, population and capital, can be taken as basic variables. A link between them can be established via the amount of capital which is put into services (health and education) (Fig. 6.18). As can be seen the important feature of this logical connection is the existence of two pairs of feedback

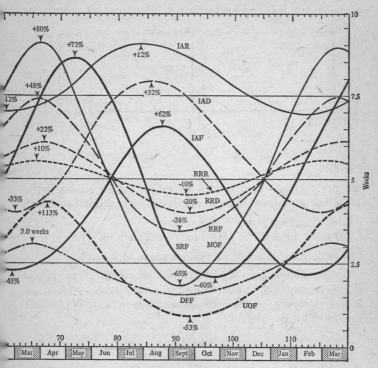

e and fall in retail sales over a 1-year period

loops (Fig. 6.19). Forrester's total world model consists of a number of submodels, linked, like Siamese twins, via certain aspects as can be seen from a study of Fig. 6.20.

The problem of dealing quantitatively with the underlying units and relationships of the above kinds of model is not easy. Formal mathematics by continuous functions breaks down in the face of the complexities involved. However, all is not lost. The methods of simulation (of Chapter 3) based on a series of snapshots, similar to those of Chapter 1, can be adapted. Starting from a base point description of the world of 43 variables, time is divided into a series of intervals and at each interval the state of

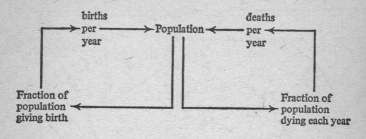

Positive feedback   Negative feedback

Fig. 6.16

the world system is calculated by evaluating 22 relationships. These relationships give parameters which are used in two ways:

(*i*) to describe the world at that time;

(*ii*) as an input to the next stage of calculations.

The basic variables and assumptions are as follows.

(*a*) Population growth is a function of consumption, population density and pollution. At poverty levels, population decreases rapidly while, with affluence, growth of population tends to 5·8% per annum.

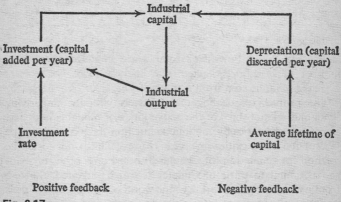

Positive feedback   Negative feedback

Fig. 6.17

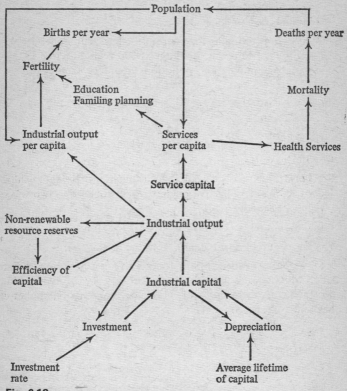

Fig. 6.18

(*b*) Pollution is generated by economic activity and absorbed by natural forces. The creation of pollution is proportional to population and increases with an increased capital–labour ratio.

(*c*) Output *per capita* is a function of (*i*) capital–labour ratio and (*ii*) remaining resources.

(Note this implies no technological progress.)

(*d*) *Per capita* food production is an inverse function of pollution and population. There is a slight effect of increasing capital investment in agriculture on increased food production.

(e) Capital investment.
(f) Resources.

Simulation runs, bringing together all the basic submodels, have produced sets of forecasts for the next 300 years in terms of the basic variables, population, pollution, resources, capital invest-

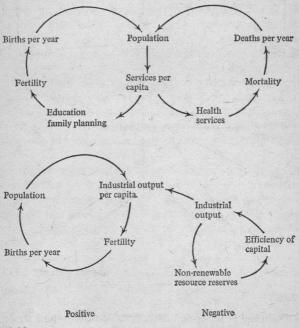

Positive   Negative

Fig. 6.19

ment, quality of life. The objective is to try to forecast the ultimate equilibrium or steady state. In Forrester's words:

'1. Industrialization may be a more fundamental disturbing force in world ecology than is population. In fact, the population explosion is perhaps best viewed as a result of technology and industrialization. (Medicine and public health are included here as a part of industrialization.)

'2. Within the next century, man may face choices from a four-pronged dilemma-suppression of modern industrial society by a natural-resource shortage; decline of world population from changes wrought by pollution; population limitation by food shortage; or population collapse from war, disease, and social stresses caused by physical and psychological crowding.

'3. We may now be living in a "golden age" when, in spite of a widely acknowledged feeling of malaise, the quality of life is, on the average, higher than ever before in history and higher now than the future offers.

'4. Exhortations and programs directed at population control may be inherently self-defeating. If population control begins to result, as hoped, in higher *per capita* food supply and material standard of living, these very improvements may relax the pressures and generate forces to trigger a resurgence of population growth.

'5. The high standard of living of modern industrial societies seems to result from a production of food and material goods that has been able to outrun the rising population. But, as agriculture reaches a space limit, as industrialization reaches a natural-resource limit, and as both reach a pollution limit, population tends to catch up. Population then grows until the "quality of life" falls far enough to stabilize population.

'6. There may be no realistic hope of the present under-developed countries reaching the standard of living demonstrated by the present industrialized nations. The pollution and natural-resource load placed on the world environmental system by each person in an advanced country is probably 20 to 50 times greater than the load now generated by a person in an under-developed country. With 4 times as many people in under-developed countries as in the present developed countries, their rising to the economic level that has been set as a standard by the industrialized nations could mean an increase of 10 times in the natural-resource and pollution load on the world environment. Noting the destruction that has already occurred on land, in the air, and

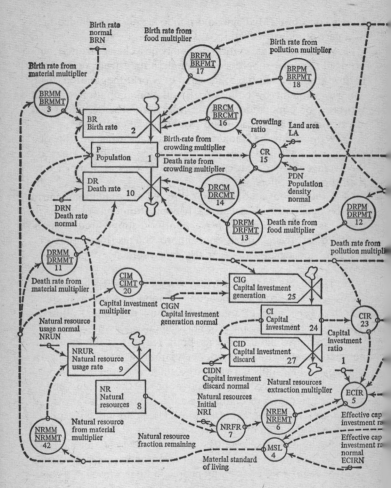

Fig. 6.20. Complete diagram of the world model inter-relating the five-level variables — population, natural resources, capital investment, capital investment in agriculture fraction, and pollution

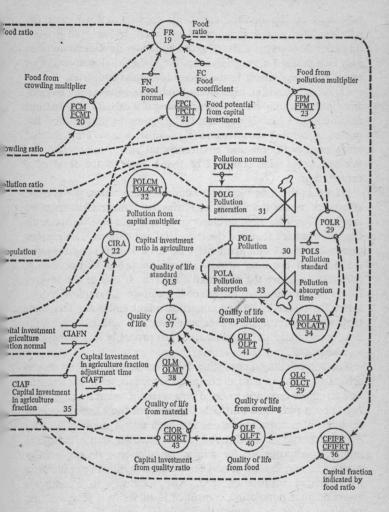

especially in the oceans, capability appears not to exist for handling such a rise in standard of living. In fact, the present disparity between the developed and under-developed nations may be equalized as much by a decline in the developed countries as by an improvement in the under-developed countries.

'7. A society with a high level of industrialization may be non-sustainable. It may be self-extinguishing if it exhausts the natural resources on which it depends. Or, if unending substitution for declining natural resources were possible, a new international strife over pollution and environmental rights might pull the average world-wide standard of living back to the level of a century ago.

'8. From the long view of a hundred years hence, the present efforts of under-developed countries to industrialize may be unwise. They may now be closer to an ultimate equilibrium with the environment than are the industrialized nations. The present under-developed countries may be in a better condition for surviving forthcoming world-wide environmental and economic pressures than are the advanced countries. If one of the several forces strong enough to cause a collapse in world population does arise, the under-developed countries might suffer far less than their share of the decline because economies with less organization, integration and specialization are probably less vulnerable to disruption'.

These modelling approaches have been subjected to severe criticisms. First, as with all computable models, there is a temptation to ignore data and to build on a series of 'what if' questions. It has been remarked that although most of the 22 relationships are plausible, not a single one has been tested empirically. The population model is criticized on the grounds that although it might be appropriate for animal populations, it is generally rejected by demographers and economists for human populations living above subsistence level. Also, contrary to Forrester's assertion that population control of itself will not solve all our problems there are population policies, other than those he studied, which will. A slow decline in population by itself will relieve all the growth pains in world dynamics. There are several similar criticisms of other aspects of the model, and the reader is referred to Nordham's trenchant critique.

However, the most elegant criticism is of a mathematical nature. In such simulations and in the models the flow of time is arbitrary. Suppose therefore that we start from the ultimate steady states adduced by Forrester for the year 2000 and reverse the runs. Will we now return to anything like the conditions of today's starting point? It has been shown by Curnow and Cole that backtracking as a technique can determine four sources of error in dynamic controlling, viz. errors introduced by integration procedures, errors introduced by the imprecision of finite digital computing, transient disturbances and inconsistent starting conditions. But the main criticism of these global models does not rest on back-tracking alone. If the flow diagrams are examined in detail it can be seen that the feedback loops all concentrate on the assumption of the finiteness of the availability of natural resources. The models neglect the possibility of the renewability of natural resources. Even a minor change of the basic assumptions of the global model will defer for centuries a predicted collapse of the economic and material system.

FURTHER READING

It is possible that this chapter will stimulate the reader to browse around in some of the relevant literature. In some cases this will be in order to put flesh on the skeletons which have been presented and in other cases it will be to develop further the arguments and the mathematics.

On the processes of planning generally there is an excellent (non-mathematical) perspective by Russell Ackoff (*A Concept of Corporate Planning*, Wiley, New York, 1970). This book gives a logic for planning, making plain that planning is not a case of writing out a set of decisions but is rather a reactive learning process.

The literature on manpower planning is growing rapidly. A good, non-technical overview is given by John Lawrence ('Manpower and Personnel Models in Britain', *Personnel Review*,

Gower Press, 1970), while more mathematical approaches are outlined by R. E. Laslett (*A Survey of Mathematical Methods of Estimating to Supply of and Demand for Manpower*, EITB Occasional Paper No. 1, Engineering Industry Training Board, 1972) and David Bartholomew ('Note on the Measurement and Prediction of Labour Turnover', *JRSSA*, 122, No. 2, 1959). The simulation approach is given in a paper by C. W. Walmsley ('A Simulation Model for Manpower Management', in D. Bartholomew and A. Smith (editors), *Manpower and Management Science*, English Universities Press, 1971).

Competitive tendering is presented in a case study by Fred Hanssman and Patrick Rivett in the *Operational Research Quarterly* ('Competitive Bidding', Vol. 10, No. 1, March 1959).

Finally there remains the important work in systems dynamics. The original work by Jay Forrester is presented in *Industrial Dynamics* (MIT Press, 1961) and *World Dynamics* (Wright-Allen Press, Cambridge, Mass., 1971), of which there is an excellent (but non-mathematical) summary in *Limits to Growth* by D. H. Meadows *et al.*, Potomac Associates. Some of the criticisms to which this work has been subjected are given in *A Critique of the Limits to Growth* by R. Curnow, Sussex University Press, May 1973, and 'Backcasting with World Dynamics Models', by R. Curnow and S. Cole, *Nature*, 243, No. 5402, May 1973.

These references will lead the reader through the first stages of any follow up. They also, of course, contain many suggestions for second-stage study.

*

Figs. 6.14 and 6.15 are reprinted from *Industrial Dynamics* by Jay Forrester (MIT Press, Cambridge, Massachusetts, 1961) with kind permission of the publishers.

Fig. 6.20 and the quotation on pp. 422, 423 and 426 are reprinted from *World Dynamics* by Jay Forrester, second edition, copyright © 1973 (Wright-Allen Press Inc., 238 Main Street, Cambridge, Massachusetts, USA) with kind permission of the publishers.

Epilogue

As was stated in the preface, this book has been concerned with newer uses of mathematics. This has, of necessity, meant excluding coverage of fascinating developments in engineering, chemistry, physics, which are well-established areas in which mathematics can be of service.

But we have been concerned with *new* uses. Hence although it is difficult at first sight to think of areas as different as weather, industry and commerce, biology, finance and world modelling, we hope we have established the way in which a language, such as ours, can be the servant of so many and varied concerns.

Mathematics is essentially a language which discusses, powerfully, quantity and logical relationships. For if I state 'I am concerned with two quantities. Three times one and twice the other is seven. Their sum is three' then it is not obvious what they are. But if I write

$$3x + 2y = 7$$
$$x + y = 3,$$

then the problem is trivial.

Extending this, although I might 'solve' (as the mathematician would say) the above in my head (if I was smart enough), it would be almost impossible to solve in my head

$$3x + 2y + z = 9$$
$$2x + y + 2z = 8$$
$$x - 2y + z = -1,$$

but mathematically it would take about a minute for a school boy. Mathematics then translates certain descriptions into a language. We can see the power of this language if we try to translate every line of an algebraic argument into English. For example the development:

$$x^3 - 2x^2 - 5x + 6 = (x + 2)(x^2 - 4x + 3)$$
$$= (x + 2)(x - 1)(x - 3)$$

is obvious in algebra, but in English it is impossible to follow.

In some of this book the translation from some reality into mathematics has been clear. For example, in Chapter 3 the problems as outlined have clearly been able to be put into a mathematical language. This language has been 'continuous', that is, we have been able to describe an external world by continuous functions. But it is of equal importance that, when we have described the world in this way, we shall be able to evaluate the functions in a *useful* way, that is, in a way related to our purposes. In particular in the linear programming example of a textile mill, in Chapter 3, all the verbal descriptions of the mill, the constraints under which management operates with reference to machine capacity and the markets and, most importantly, the goal of maximizing contribution to profits could be expressed in mathematical form. The mathematical form was very special (linear equalities and inequalities) and hence a special kind of routine for its solution (called an algorithm) could be deployed.

In some cases it is easier to replace the continuous functions by a set of separate and distinct instants at which the function is evaluated. This was the case in Chapter 1, where although weather (as measured by temperature, pressure, humidity and air velocity) is clearly continuous, it is more convenient to deal with the inter-relationships of these variables by working on a series of 'snapshots'. For exactly the same reason the global modelling of the final chapter is calculated in the same way, the difference being that the former is based on well-established laws of physics and a wealth of data, whereas the latter is based on hypotheses selected by the researcher and (often) very sketchy data.

This method of breaking a continuous function into separate pieces is most natural when time is involved. It is of interest that in many chapters this was the case. In finance a year is a natural epoch as it is also in education planning. In education planning and in biological models we noted the change from the determinism of weather, world modelling and finance, which stemmed from the introduction of probability and chance. By their nature probabilities have to be applied to specific separate situations and so the methods of 'continuous' mathematics are excluded. But the manipulation of probabilities demands an appeal to mathematics.

In all problems of that external world in which man lives we are conscious of varying degrees of chaos. The scientist is sometimes fortunate enough to be able to separate the phenomena he is studying from environmental effects (a laboratory is a 'machine' for isolating certain things from their environment). In this book, however, all the problems we have dealt with are affected by the external world and at first sight it is curious that a subject as pure and passionless as mathematics can have anything useful to say about that messy, ill-structured, chancy world in which we live. It certainly would not have anything relevant to say if the world was, of its essence, chaotic. Fortunately, in all these areas of new uses of mathematics, we will find that, whenever we comprehend what was previously mysterious, there is at the centre of everything order, pattern and common sense. It has been order, pattern and common sense that the botanist has found beneath his microscope and the astronomer in the reaches of the universe. Even processes dominated by chance are dominated by *patterns* of chance. But this common sense is not what we put into the problems we study, it is what we find there. These are problems with structure, with relations of causes and effects, involving quantities of different things. The natural language of structure and quantity is that of mathematics.

B. H. P. RIVETT